अत्त दीपो भवथ,
अत्तसरणा अनञ्ञसरणामति ।

***To lighten oneself
To take one's own refuge
Don't take other's refuge.***

Useful for undergraduate and post-graduate students
&
Central/State Civil Services, Forest Services, Banking Services, B.Sc. (Ag.)/M.Sc. (Ag.) Entrance Test and Other Competitive Exams.

Fundamentals of Agriculture

Agronomy
Soil Science
Extention Education
Agriculture Economics
Farm Management

By :
Arun Katyayan

Vol -1

(Fully Revised & Enlarge edition)

KUSHAL PUBLICATIONS AND DISTRIBUTORS
VARANASI

Publisher :
Kushal Publication's and Distributor's
(Higher Educational Publisher)
Gyanmandal Plaza
IInd Floor, Shop No. 28
Maidagin, **Varanasi**-221001 U.P. (India)
Ph. : 0542-3291887
Mob. : 09839040484
e-mail : kushalpublication@yahoo.co.in

I.S.B.N. : 81-86099-70-0

First Edition : 2001
Fifth Edition : 2008
Reprint : 2009

Selling Agents :

Jain Brothers
16/873, East Park Road
Karol Bagh, **New Delhi**

Rawat Book Depot
Shopping Complex
Pant Nagar (U.S. Nagar)

International Book Distributing Co.
Khusnuma Complex, Basement,
7-Meerabai Marg, **Lucknow**

Sasta Pustak Bhandar
Opp. Rawatpur Rly. St.
Rawatpur, **Kanpur**

Printed at : **Modern Deepak Press,** Nadesar, Varanasi. Phone : 2502048, 2500894

FOREWORD

Agriculture is the backbone of our economy.

Students of Agricultural science play the important role in the development of agriculture. To fulfil their needs the book **Fundamentals of Agriculture** by **Mr. Arun Katyayan** is a welcome effort.

To understand the fundamentals clearly and easily the author presented the book in a scientific and systematic manner which is the beauty of this book. The author has consiiderable experiences in teaching and designing the book for competitive purposes also.

I personally congratulate Mr. Arun Katyayan for his attempt at designing such a valuable textbook. I am confident that this book will be widely accepted among the students. I extend my best wishes to Mr. Arun Katyayan for the success of his book.

9.12.2000

(A.C. Ranjan, I.A.S.**)**
Secretary
Deptt. of Agriculture
Cooperation, Animal
Husbandry and Fisheries
Govt. of Jharkhand, Ranchi

PREFACE TO THE FIRST EDITION

There are few books in Agriculture which satisfies the requirements of students particularly for those who are preparing for competitive examinations. This book is designed to satisfy them and is presented in a systematic manner to understand fundamental aspects clearly and easily. This book cover all the important chapters of Agronomy, Soil Science and Ext. Education. A set of objective questions with answer is also included which shows the pattern of questions to be asked in various competitive exams.

Before the compilation of this book, author has consulted various textbook, periodicals and journals from which matters have been heavily drawn in this book and author heartly acknowledges his indebtedness for the same. In some cases, a few tables and figures have been reproduced, acknowledgement is duly credited beneath each reproduction. In very few cases, where it has not been possible to contact and obtain permission for such reproduction, the author and publisher offer their sincere appologies.

My special and sincere regards are due to my parents, family members and wife who share each and every achievement made by me and to whom I am very much indebted.

Dhamma Vijayadashmi, 2k.

(Arun Katyayan)

PREFACE TO THE FIFTH EDITION

Since its first edition in the year 2001, the book **Fundamentals of Agriculture** is gone through various inclusions and enlargements especially on the request of the readers and the students. It is reported to me that the readers have welcomed the book zealously due to the systematic presentations of the fundamentals and its very simple as well as easy to understand language.

Suggestions given by the readers and the earnest request made by the instrumental publisher Mr. Madan Mehrotra compelled me to manage the time to revise the book word to word. In the revision I feel that this volume requires some new additions in almost each and every chapter and the same have been incorporated.

Up-to date informations and findings and the required changes are made sinceraly although your suggestions for the improvement of the book will be acknowledged.

Buddha Purnima, 2008 **(Arun Katyayan)**

CONTENTS

❑❑❑

1 History of Agriculture

It is supposed that man was evolved on earth about 15 lakh years ago. This man was the evolved form of monkey who started to move by standing erect on his feet. Such man has been called ***Homo erectus*** or Java man (Homo → continuous, erectus → erect). Later on Java man was transformed into cro-magnan and cromagnan into modern man. The modern man is zoologically known as ***Homo sapiens*** (sapiens means learning habit) due to his continuous learning habit. And according to trinominial nomenclature modern man got the name ***Homo sapiens sapiens***. Such modern man was evolved first in Africa about 35000 Years ago what generally accepted today. In the beginning such man had been spending his life wildly but during the period of 8700 -7700 B.C., they started to pet sheep and goat, although the first pet animal was Dog which was used in hunting.

The cultivation of crop started during 7500 B.C. and thus agricultural Science came into existence. To produce the crop human beings tilled the soil. Tilling of soil or in another word cultivation of soil is called Agriculture.

Agriculture consists of two **Latin** words i.e.,

Agric/Ager + Cultura = Agriculture

Agric/Ager ↓ Soil

Cultura ↓ Cultivation

Thus agriculture is an art and science both in which we study all the human activities related to use of soil.

Discovery and use of Iron revolutionised the Agriculture to fulfill the needs of rising population and earnest desires. Later on it was realized that due to lack of field management , the productivity of soil was started to be diminished. Therefore, Agriculturists had to study the management of field as well as principles and practices of crop production .

Thus the term ' agronomy' came into existence Agronomy consist of two **Greek** words viz.

Agronomy = Agros + nomos = Greek words

Agros ↓ Field; nomos ↓ to manage

Agronomy is that branch of agricultural sciences which deals with the management of field to provide favourable environment to the crop for higher productivity in terms of quantity and quality both.

Peter Decresenzi (1230 -1307 A.D.) collected many literatures related to Agronomy in his book 'Opus Ruralium Kamo Daram' for the first time. That is why he is called the **Father of Agronomy.**

Jethrotull (1674-1741 A.D.) wrote a book 'Horse Hoeing Husbandry' on which basis 'seed drill' and ' Horse Drawn cultivator' were developed. The words 'weed' and 'zero tillage'were also used firstly by him.

In India , scientific cultivation has been started with the commercialisation of sugarcane.cotton and tobacco . On 27 th April 1871, a joint department of agriculture,revenue and commerce was established by Lord Mayo on the request of *A.O. Hume*. Due to direction issued by Famine commission 1880 , a separate central department of agriculture was established in 1881.

IARI : Imperial Agricultural Research Institute was established in **1905** at Pusa (at that time Bengal but presently in Bihar) under the viceroyalty of Lord curzon. In **1934**, its buildings were damaged due to earthquake. Therfore in **1936**. IARI was transfered to Pusa Road , New Delhi. After the Independence the word 'Imperial' was substituted by 'Indian' and now it is called **Indian Agricultural Research Institute**. In **1958**, IARI was given the status of deemed university by university Grant Commission.

ICAR : The Royal Commission on Agriculture was constituted in **1926** under the viceroyalty of Lord Linlithgow. On the recommendations of the Royal Commission Report of **1928,** The Imperial Council of Agricultural Research Was established on 16th July **1929**. It was registered as a society under the Societies Registration act - 1860. Its first president was Md. Habibullah and secretary was S. A. Hydari. In March, 1946, it was decided to change the 'imperial' into 'Indian under the presidentship of Jogendra Singh and now it is called **Indian Council of Agricultural Research**. In 1966, ICAR was reconstituted into full autonomous body and its first *Director General* (Chief executive) was Dr. B.P. Pal. The mandate of ICAR is

a) To plan undertake, aid promote, and co-ordinate education, research and its application in agriculture, agro forestry, animal husbandry, fisheries, home science and allied sciences.

b) To act as a clearing-house of research and general information relating to agriculture, animal husbandry, home science and allied sciences, and fisheries matters through its publications and information system, and instituting and promoting transfer of technology programmes.

c) To provide, undertake and promote consultancy services in the fields of education, research, training and disssemination of information in agriculture, agroforestry, animal husbandry, fisheries, home sciences and allied sciences.

d) To look into the problems relating to broader areas of rural development concerning agriculture, including post-harvest technology by developing cooperative programmes with other organizations such as the Indian Council of Social Science. Research, Council of Scientific & Industrial Research (CSIR), Bhabha Atomic Research Centre (BARC) and the Universities.

e) To do other things considered necessary to attain the objectives of the society.

Agricultural Colleges & Universities :

During 1901-1905, six agricultural colleges were established in India at Pune , Kanpur, Sabour, Nagpur, Llyalpur (now in Pakistan) and Coimbtore.The course study of Agriculture Extension was firstly started at Sabour (Bhagalpur, Bihar).

Based On the recommendations of First Education Commission (1949), headed by late. Dr. S. Radhakrishnan and the recommendation of the First & Second Indo American Teams, **First** Agriculture University was established at **Pantnagar,** Uttarakhand, (erstwhile part of Uttar Pradesh) in 1960, (At that time the Chief minister of U.P. was Pt. Govind Ballabh Pant.). A joint Indo - American Team was appointed by the Govt. of India in 1954 and The other in1959. In India there are 38 State Agricultural Universities and one Central Agricultural University (Imphal) with around 210 constituent Colleges. Four Institutes IARI, NDRI, IVRI, CIFE of ICAR and Allahabad Agricultural Institute, are deemed to be universities and four Central Universities have distinct agricultural faculties.

IRRI: International Rice Research Institute at Manila (Philipines) in 1960.

CIMMYT: CIMMYT is in mexican word which denotes international centre for **Maize** & **Wheat** improvement at Mexico in 1966.

In India due to modern agriculture, Green Revolution occured in **1965-66**. Later on a series of revolutions i.e., **White** revolution (Milk production/ Operation Flood); **Blue** revolution (Fisheries); **Round** revolution (Potato); **Silver** revolution (Eggs and poultry); **Brown** revolution (Biomass or Fertiliser Production); revolutionised our country. Presently a varieties of crops and horticultural plants are grown whole the year in glasshouse by using the principle of green – house effect.This development in agriculture is the resultant of the progress of man. Agriculture shares 22% in Gross Domestic Product (GDP) of India in the year 2005-2006. It was 24% during the year 2000-2001 and 25.2% in 1999-2000. Here Agriculture includes grains, livestocks, forest resources and fisheries.

Agricultural Universities

- Acharya N G Ranga Agricultural University Rajendranagar, Hyderabad (Andhra Pradesh) 500 030
- Anand Agricultural University, Anand (Gujarat) 388 110
- Assam Agricultural University, Jorhat (Assam) 785 013
- Bidhan Chandra Krishi Viswavidyalaya, Mohanpur, Nadia (West Bengal) 741 252
- Birsa Agricultural University, Ranchi (Jharkhand) 834 006
- Chandra Shekhar Azad University, of Agriculture and Technology Kanpur (Uttar Pradesh) 208 002
- Chaudhary Charan Singh Haryana Agricultural University, Hisar (Haryana) 125 004

- Dr Panjabrao Deshmukh Krishi Vidyapeeth, Akola (Maharashtra) 444 104
- Dr Yashwant Singh Parmar University of Horticulture and Forestry Nauni, Distt Solan (Himachal Pradesh) 173 230
- Govind Ballabh Pant University of Agriculture and Technology Pantnagar (Uttarakhand) 263 145
- Karnataka Veterinary, Animal and Fisheries Sciences University Bidar (Karnataka) 585 401
- Ch. Sharwan Kumar Krishi Vishwavidyalaya, Palampur (Himachal Pradesh) 176062
- Indira Gandhi Krishi Vishwavidyalaya, Raipur (Chhattisgarh) 492 012
- Jawaharlal Nehru Krishi VishwaVidyalaya, Jabalpur (Madhya Pradesh) 482 004
- Junagadh Agricultural University, Junagarh (Gujarat) 362 001
- Kerala Agricultural University, Vellanikkara, Distt Trichur (Kerala) 680656
- Dr Balaesahib Sawant Konkan Krishi Vidyapeeth, Dapoli (Maha - rashtra) 415712
- Mahatma Phule Krishi Vidyapeeth, Rahuri (Maharashtra) 413722
- Marathwada Agricultural University, Parbhani (Maharashtra) 431402
- Maharana Pratap University of Agriculture and Technology, Udaipur (Rajasthan) 313001
- Narendra Dev University of Agriculture and Technology, Faizabad (Uttar Pradesh) 224229
- Navsari Agricultural University, Navsari (Gujarat) 396450
- Orissa University of Agriculture and Technology, Bhubaneshwar (Orissa)751003
- Punjab Agricultural University, Ludhiana (Punjab) 141004
- Rajasthan Agriculture University, Bikaner (Rajasthan) 334 002
- Rajendra Agricultural University Samastipur, Pusa (Bihar) 848125
- Sardar Krushi Nagar Dantiwada Agriculture University Dantiwada (Gujarat) 385506
- Sardar Ballabh Bhai Patel University of Agriculture and Technology Modipuram, Meerut (Uttar Pradesh) 250 110

- Sher-e-Kashmir University of Agricultural sciences and Technology Srinagar (Jammu and Kashmir) 191121
- Sher-e-Kashmir University of Agricultural Sciences and Technology 45-B, Gandhinagar Jammu (Jammu and Kashmir) 180012
- Tamil Nadu Agricultural University Coimbatore (Tamil Nadu) 641 003
- Tamil Nadu Veterinary and Animal Sciences University, Chennai (Tamil Nadu) 600051
- University of Agricultural Sciences, GKVK Campus, Bangalore (Karnataka) 560065
- University of Agricultural Sciences, Dharwad (Karnataka) 580 005
- West Bengal University of Animal and Fishery Sciences, 68KB Sarani Kolkata (West Bengal) 700 037
- Maharashtra Animal Sciences and Fisheries University, Nagpur (Maharashtra) 440 006
- UP Pandit Deen Dayal Upadhyaya Pashu Chikitsa Vigyan Vishwa Vidyalaya evam Go Anusandhan Sansthan, Mathura (Uttar Pradesh) 281001
- Uttar Banga Krishi Vishwavidyalaya, Pundibari, Cooch Behar (West Bengal) 736165
- Sri Venkateswara Veterinary University, Tirupati, Chittoor (Andhra Pradesh) 517 502 (at ANGRAU, Hyderabad Andhra Pradesh 500 030 till April 2006)

Central Agricultural University

Central Agricultural University, Imphal (Manipur) 795 004

Central Universities

- Aligarh Muslim University, Aligarh (Uttar Pradesh) 202 002
- Banaras Hindu University, Varanasi (Uttar Pradesh) 221 005
- Visva Bharati Sriniketan, (West Bengal) 731 236

Deemed-to-be Universities

1- **IARI** - Indian Agricultural Research Institute, Pusa, New Delhi 110012

2- **IVRI** - Indian Veterinary Research Institute, Izatnagar (Uttar Pradesh) 243122

3- **NDRI** - National Dairy Research Institute, Karnal (Haryana) 132001

4- **CIFE** - Central Institute of Fisheries Education Jaiprakah Road Seven Bungalows, Versova Mumbai (Maharashtra) 400061

5- **AAI** - Allahabad Agricultural Institute, Allahabad (Uttar Pradesh) 211007

Agricultural Research Institutes

Agricultural Sciences

1. CARI : Central Agricultural Research Institute, Port Blair Andman and Nicobar Group of Islands.(Andaman & Nicobar.Island.744101)
2. CAZRI : Central Arid Zone Research Institute Jodhpur (Rajasthan) 342003.
3. CIAE : Central Institute of Agricultural Engineering, Berasia Road, Nabi Bagh Bhopal, (Madhya Pradesh) 462038
4. CIAH : Central Institute for Arid Horticulture, Bikaner (Rajasthan) 334006
5. CICR : Central Institute for Cotton Research,Nagpur (Maharashtra 440010).
6. CIPHET : Central Institute of post- Harvest Engineering and Tech nology, Ludhiana (Punjab 141004)
7. CIRCT : Central Institute for Research on Cotton Technology, Adenwala Road, Matunga Mumbai (Maharashtra) 400019.
8. CISTH : Central Institute of Sub-tropical Horticulture, Rehma - nkhera, PO Kokori Lucknow (U.P.) 227107
9. CITH : Central Institute of Temperate Horticulture Old Air Field, Post Sanat Nagar Srinagar (Jammu and Kashmir) 190 007
10. CPCRI : Central Plantation Crops Research Institute, Kasaragod (Kerala) 671 124
11. CPRI : Central Potato Research Institute Shimla (Himachal Pradesh) 171 001
12. CRIDA : Central Research Institute for Dryland Agriculture Santoshnagar, P O Saidabad Hyderabad (Andhra Pradesh) 500 059

13. CRIJAF : Central Research Institute for Jute and Allied Fibres Barrackpore, Distt 24 Paraganas (West Bengal) 700 120

14. CRRI : Central Rice Research Institute, Cuttack (Orissa) 753 006

15. CSSRI : Central Soil Salinity Research Institute, Zarifa Farm, Kachwa Road, Karnal (Haryana) 132 001

16. CSWCRTI : Central Soil and Water Conservation Research and Training Institute, 218 Kaulagarh Road, Dehra Dun (Uttarakhand) 248 195

17. CTRI : Central Tobacco Research Institute, Rajahmundry (Andhra Pradesh) 533 105

18. CTCRI : Central Tuber Crops Research Institute, Sreekariyam, Thiruvananthapuram (Kerala) 695 017

19. IASRI : Indian Agricultural Statistics Research Institute, Library Avenue, Pusa Campus, New Delhi 110 012

20. IGFRI : Indian Grassland and Fodder Research Institute, Pahuj Dam Gwalior -Jhansi Road Jhansi (Uttar Pradesh) 284 003

21. IIHR : Indian Institute of Horticultural Research, P.O. Hassaraghatta Lake Bangalore (Karnataka) 560 089

22. IIPR : Indian Institute of Pulses Research, Kanpur (Uttar Pradesh) 208 024

23. IISS : Indian Institute of Soil Science, Nabi Bagh, Bhopa (Madhya Pradesh) 462 038

24. IISR : Indian Institute of Spices Research, P O Marikunnu Kozhikode (Kerala) 673 012

25. IISR : Indian Institute of Sugarcane Research, P O Dilkusha, Lucknow (Uttar Pradesh) 226 002

26. ILRI : Indian Lac Research Institute, Namkum, Ranchi (Jharkhand) 834 010

27. IIVR : Indian Institute of Vegetable Research, Varanasi (Uttar Pradesh) 221 005

28. NIRJAFT : National Institute of Research on Jute and Allied Fibre Technology, 12 Reagent Park,Calcutta(West Bengal) 700 040

29. SBI : Sugarcane Breeding Institute, Coimbatore (Tamil Nadu) 641 007

30. VPKAS : Vivekanand Parvatiya Krishi Anusandhan Sansthan, Almora (Uttara khand) 263 601

Animal Sciences and Fisheries

1. CARI : Central Avian Research Institute, Izatnagar (Uttar-Pradesh) 243 122

2. CIFRI : Central Inland Fisheries Research Institute, Barrackpore (West Bengal) 700 120

3. CIBA : Central Institute of Brackishwater Aquaculture, 75 Santhome High Road R A Puram, Chennai (Tamil Nadu) 600 028

4. CIFT : Central Institute of Fisheries Technology, Willingdon Island, PO Matsyapuri Cochin (Kerala) 682 029

5. CIFA : Central Institute of Freshwater Aquaculture, Kausalyaganga Bhubaneshwar (Orissa) 751 002

6. CIRB : Central Institute for Research on Buffaloes, Sirsa Road, Hisar (Haryana) 125 001

7. CIRG : Central Institute for Research on Goats, Makhdoom, Mathura (Uttar Pradesh) 281 122

8. CMFRI : Central Marine Fisheries Research Institute, Tatapuram, Kochi (Kerala) 682 014

9. CSWRI : Central Sheep and Wool Research Institute, Avikanagar, District Tonk Via Jaipur (Rajasthan) 304 501

10. NANP : National Institute of Animal Nutrition and Physiology, Adugodi, Bangalore (Karnataka) 560 030

Other

1. NAARM : National Academy of Agricultural Research and Management Rajendranagar, Hyderabad (Andhra Pradesh) 500 030
2. FRI : Forest Research Institute, Dehradoon (Uttarakhand)
3. ICARDA : International Centre for Agricultural Research in Dry land areas, Aleppo (Syria)
4. ICRISAT : International Crop Research Institute for Semi Arid Tropics, Patancheru, Hyderabad. (Andhra Pradesh Aug. 1972)
5. NCERT : National Council of Educational Research & Training. New Delhi.
6. PBCS : Potato Breeding & Certification Station. Kufri (Shimla) Himachal Pradesh.

NATIONAL BUREAU

Agricultural Sciences

1. NBPGR : National Bureau of Plant Genetic Resources, FCI Building, Pusa New Delhi 110 012
2. NBSSLUP : National Bureau of Soil Survey and Land Use Planning. Shankar Nagar Amravati Road, Nagpur (Maharashtra) 440 010

Animal Sciences

1. NBAGR : National Bureau of Animal Genetic Resources, Karnal (Haryana) 132 001
2. NBFGR : National Bureau of Fish Genetic Resources, Canal Ring Road PO Dilkusha, Lucknow (Uttar Pradesh) 226 002
3. NIBAIM : National Bureau of Agriculturally Important Micro - organisms, Kusmaur, Mau Nath Bhanjan (Uttar Pradesh) 275 101

Courtesy : Handbook of Agriculture (ICAR)

❑❑❑

2 Cropping System & Pattern

Shifting Cultivation/ Land Rotation/Jhumming :

1. Forest land is cleared and cultivated. Due to cultivation of the same crop generally rice on the same cleared forest land year after year; soil productivity is lost. And the crop is shifted to other slashed and burnt land.
2. Here same crop is grown year after year. In this case land is rotated but crop is fixed. Therefore it may also be called land rotation.
3. Shifting of land hence called shifting cultivation. Also called Jhum cultivation.
4. It causes soil erosion.
5. Practised in northeastern states of India, Chhotanagpur pleatau of Jharkhand, M.P. and in Hilly areas.

Crop Rotation :

1. Crop rotation is about reverse of land rotation . Here land is fixed but crop is rotated year after year.
2. On a certain land, repeated cultivation of crops or crop and fallow land in a certain sequence is called crop rotation.

   Rice- Wheat – Moong : 2006 – 07 (Agl. Year)

   Rice – Mustard – Chilli : 2007 - 08

   Here Wheat is rotated by mustard and Moong by chilli in the next Agriculture year.
3. Maintains and even improves soil fertility and stabilise income .
4. It checks the soil erosion and conserves moisture.

5.. Ensures a balanced programme of work throughout the year.

6. Suitable crop rotation is the key of modern scientific agriculture which aims to produce max yield by maintaining soil productivity.

Sustainable Agriculture / Eco- Farming :

The concept of sustainable agriculture has come up because yields from modern farming technique (modern Commercial Agriculture) reaching a plateau and the environmental problems due to excessive use of chemicals and fertilizers and pesticides residue in food chain.

	Types of Agriculture	Objective	Quantity of Input	Output
1.	Subsistence Agriculture i.e. low level equilibrium	To Sustain life and family need	Low	Low
2.	Commercial Farming i.e. High leve equilibrium	To obtain high Income	High	High
3.	Sustainable Agriculture i.e. Natural equilibrium or ecological equilibrium	Ecological Balance	Low	High

Sustainable agriculture is that form of farming which produces sufficient food to meet the needs of the present generation without eroding the ecological assets and productivity of life supporting systems of future generations. **Natural farming** is an excellent illustration of sustainable Agriculture. It is also known as ecological farming/ecofarming or Organic farming or **Permaculture**. It is called eco farming because ecological balance is given importance and organic farming because organic matter is the main source for nutrient management. It is a system of cultivation with use of manures, crop rotation and minimal tillage. Sustainable agriculture also involves agroforestry and multi- level cultivation and integrated animal husbandry. The term sustainability denotes the characteristic of a process that can be maintained indefinitely and sustainable use of the eco system refers to making use of the system without impairing its capacity for renewal or regeneration.

Cropping Systems :

System means a set of elements/components that are inter–related and interacting among themselves. The objective of any cropping system is efficient utilization of all resources viz. land, water and solar radiation, maintaining stability in production and obtaining higher net returns . The efficiency is measured by the quantity of produce obtained per unit resource in a unit time.

The objective of traditional agriculture was to increase the production by two means-

(a) by increasing the area of cultivation.

(b) by increasing the productivity per unit area of the crop.

But two more dimensions are added to modern agriculture-

(c) to increase the production per unit time.

(d) to increase the production per unit space.

In the traditionl cropping systems, mixtures and rotations were developed by the farmers over years of experience by trial and error to suit specific ecological and socio economic conditions to attain yield stability . Whereas modern scientific cropping system has three pillars -

(i) Genotype, (ii) Geometry of planting and (iii) Management Practices

(i) Genotype means genetic make up of seed.

(ii) Geometry of planting means.

(a) **Shape** of planting pattern on the land surface.

(b) Space of the area available for the individual plant. Geometry of planting may be circular, rectangular, square type or cuboidal. It is indirectly related to plant population. Cuboidal pattern of planting has maximum plant population. Plant population may be defined as -

(a) Size of area available to the individual plant.

(b) Number of plants per unit area.

(iii) Management practices : include all the practices of crop production. For the cropping systems, management includes.

(a) type and arrangement of crops in time and space i.e. cropping pattern.

(b) Choice of variety.

(c) Method of stand establishment.

(d) Pest management and harvest

Cropping Pattern : means

(i) Crop rotation practiced by a majority of the farmers in a given area or locality.

(ii) Type and arrangement of crops in time and space.

(iii) Yearly sequence and spatial arrangement of crops or of crops and fallow on a given area.

(iv) The proportion of area under various crops at a point of time in a unit area.

Cropping System : means

(i) Cropping patterns and its management to derive benefits from a given resource base under specific environmental condition.

Cropping system = Cropping Pattern + Management

It changes when place and environment are changed . It means cropping system is location specific.

(ii) The cropping patterns used on a farm and their interaction with farm resources, other farm enterprises and available technology which determine their make up.

(iii) Pattern of crops taken up for a given piece of land, or order in which the crops are cultivated on a piece of land over a fixed period, associated with soil management practices such as tillage, manuring and irrigation.

Farming Systems :

It represent an appropriate combination of farm enterprises viz. cropping system, livestock, poultry, fisheries, forestry and the means available to the farmer to raise them for increasing profitability. The term 'farming systems' and 'mixed farming' are used interchangeably but have some subtle differences.

Mixed Farming :

Defined as a system of farming on a particular farm which includes crop production, raising livestock, poultry, fisheries, bee keeping etc. to sustain and satisfy as many needs of the farmer as possible. The objective of mixed farming is subsistence while higher profitability without altering ecological balance is important in farming systems. Cropping system is an important component of a farming system.

Sole cropping/solid Planting :

It is opposite of inter cropping . It is defined as the cultivation of one crop variety alone in pure stands at normal density in a certain time and place.

Cropping Scheme :

The plan according to which crops are raised on individual plots of a farm with an object of getting the maximum returns from each crop without impairing the soil fertility. Soil fertility and crop yields are important factors in determining which cropping programme to follow, but at the same time, the amount of labour and capital required and prices in relation to the resources influence the adoption of a particular cropping scheme.

Monoculture :

Monoculture also called single cropping or Monocropping. A system of growing the same crop on the same land year after year.

Cropping Intensity/Cropping Index –

C1 = No. of Crops grown in a year x 100

Cropping Intensity in India is about 136 %

Types of Cropping Systems :

→Mono Cropping – C I is always 100%

→Multiple Cropping – Growing of two or more crops on same field in a year . It is the intensification of cropping in **time** and **space** dimensions i.e. more number of crops within a year and more no. of crops on the same piece of land at any given period. The Principal limiting factors in adoption of multiple cropping are low intensity of irrigation and preponderance of long duration varieties.

Recently one more dimension i.e. to maintain the soil fertility, is added in the definition of multiple cropping. Thus multiple cropping is defined as the cultivation of two or more crops on the same field in a year without deteorating the soil fertility.

Multiple Cropping :

- Inter-Cropping
- Mixed Cropping
- Sequential or sequence cropping or non – overlapping cropping
- Relay cropping or overlapping cropping.

Intercropping :

Growing two or more crops simultaneously on the same piece of land with a definite row arrangement or in a fixed ratio is called intercropping e.g.

Wheat + Mustard = 9:1

Setaria + Red gram = 5:1

Here croppying intensity in space dimension is achieved. Such crops are usually grown along with those crops which have larger row interval . Intercropping includes alley cropping , strip cropping, contour cropping , paired row cropping, skip cropping , parallel cropping . But the major intercropping are-

(i) Parallel Cropping : Cultivation of such crops which have dif ferent natural habit and zero competition e.g.
Black gram/ Green gram + Maize
The peak nutrient demand period for Urd or Mung is around 30-35 days after sowing (DAS) while it is 50 days after sowing for Maize.

(ii) Companion Cropping : Such intercropping where the produc tion of both intercrops is epual to that of its solid planting . e.g.
Mustard/ Potato/ Onion + Sugarcane.
The growing of the above three intercrops individually with sugar cane, the production of none is hardly affected.

(iii) Multistoreyed/Multitiered/Multilevel Cropping : Cultivation of two or more than two crops of different heights simultaneously on a certain piece of land in any certain period. e.g.
Sugarcane + Mustard + Onion/Potato

(iv) Synergetic Cropping : The yields of both crops are higher than of their pure crops on unit area basis e.g. Sugarcane + Potato . On the basis of percent of plant population used for each crop in intercropping system, intercropping is of two types viz- additive series and replacement series –

(a) Additive series intercropping : In such type of inter cropping, one crop is main crop or base crop and another crop is intercrop. Intercrop is introduced into the base crop by adjusting or changing crop geometry i.e. addition of intercrop to the base crop. Here plant population of base crop is same to what recommended population in pure stand whereas that of intercrop is less. Cropping husbandry is according to the base crop. This type of intercropping is prevalent in India and its main objective is to get additional income and to cover risk. e.g. sowing of potato in the field of sugarcane in between the rows of cane at a 90 cm.

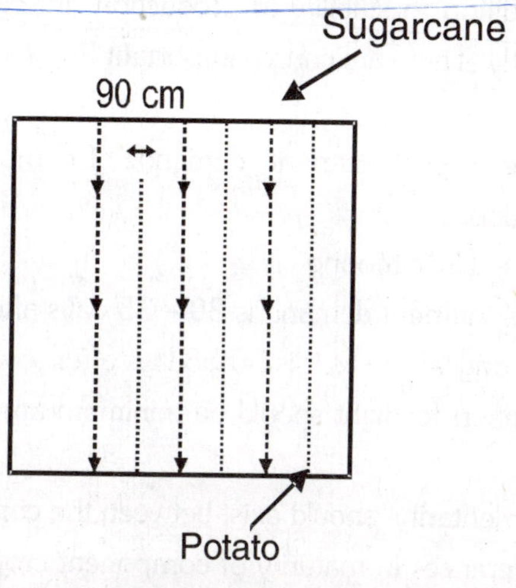

Fig.2.1

(b) Replacement Series Intercropping : Both the crops are component crops. Neither is the base crop nor is the intercrop. It means the plant population of both component Crops is less than their recommended population in pure stand. Such type is widely practised in western countries.

Here 10 th row of wheat crop is replaced by mustard and row to row distance of wheat crop is somewhat reduced to adjust the mustard crop.

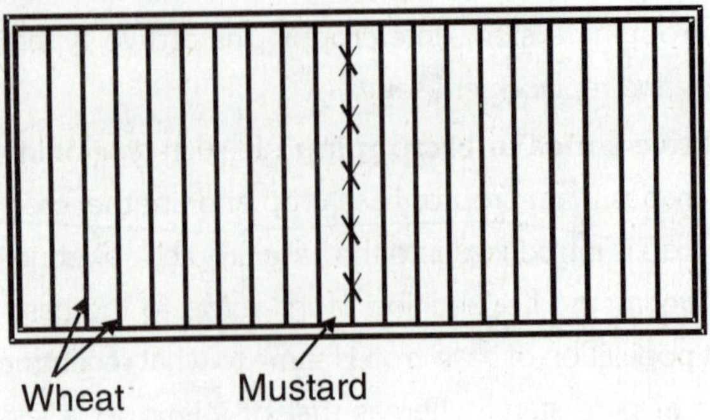

Fig. 2.2

Pre- requisites of Successful Intercropping :

The main objective of intercropping is to get higher productivity per unit area in addition to stability in production. Intercropping utilizes resources efficiently. There are certain important Pre- requisites of intercropping :-

(i) The time of peak nutrient demands of component crops should not overlap.

Maize + Urd/ Moong.

The peak nutrient demand is 30 – 35 days after sowing in case of urd/ moong where as it is 50-55days after sowing for maize.

(ii) Competition for light should be minimum among the component crops.

(iii) Complimentarity should exist between the component crops.

(iv) The differences in maturity of component crops should be at least 30 days.

(v) Competition for CO_2 and water should also be minimum among component crops.

Mixed Cropping :

Cultivation of two or more than two crops simultaneously on the same land without definite row pattern or fixed ratio. Sowing of seeds is generally by broadcasting method and commonly practiced in dryland area of India . It is subsistence farming. Its main objective is to lessen the risk of total crop failures, and to satisfy the farmers in food and fodder. Intercropping is the new concept of mixed cropping . Scientific study of mixed cropping was firstly done by La – Flitze (1928).

Sequence/Sequential Cropping :

Growing of two or more crops in quick succession on the same piece of land in a farming year. The sowing of the succeeding crop and harvesting of the preceeding crop may be done simultaneously or in a quick succession e.g. just after the harvesting of Maize,Potato is sown, and just after digging of potato, Chilli is sown. Sequential cropping is also called non – overlapping cropping because there is no overlap between the two or more cropping .

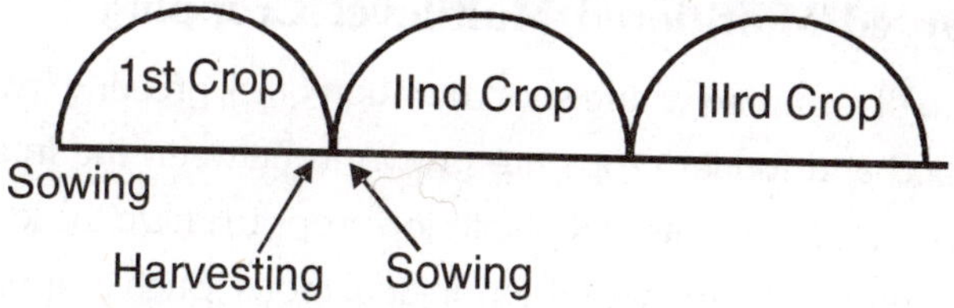

Fig.2.3

Ratooning or Ratoon cropping may also be classified under sequential cropping.

Relay Cropping :

It's concept has been derived from Relay race in which four runners run in the field having flag in their hand. The first runner passes on its flag to the succeeding partner and 2_{nd} to 3_{rd} and 3_{rd} to the 4_{th} runner. Relay cropping can be defined as growing two or more crops simultaneously during the part of the life cycle of each. Succeeding crops planted before harvesting the preceeding crop. Generally 2_{nd} crop is planted after the first crop has reached its reproductive stage of growth but before it is ready for harvest e.g.Potato is planted before the harvesting of Maize and Radish is shown before harvesting of potato. Paira (Bihar & W.B.) or Utera (M.P.)

cropping may be example of relay cropping. Paira or utera cropping means sowing of Lathyrus or lentil (generally pulses) before the harvesting of rice in lowland area. Planting of nursery crop may also be an example of relay cropping. Relay cropping is also called overlapping cropping.

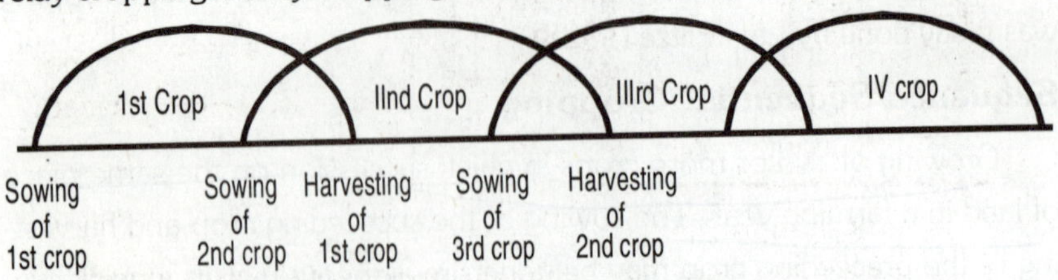

Fig.2.4

Dr. Sher Singh Bains has suggested the following relay cropping pattern for north west part of India—

Moong - Maize - Potato - Wheat

Multistoryed/Multitiered/Multilevel Cropping :

System of cropping where short duration quick growing fruits and vegetables, food and fodder crops are grown in between the interspaces of principal horticultural fruits and plantation crops to utilize the leisure period of growth and development of the principal crops. Such cropping is useful in both plains and hilly areas It utilizes solar energy and soil layers

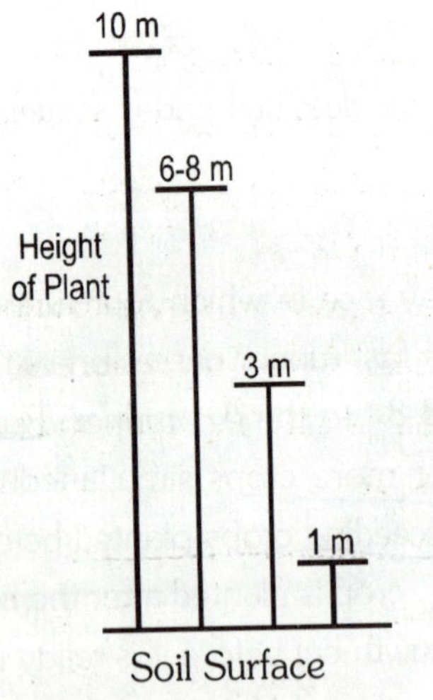

Hieght of different Plant

Fig. 2.5

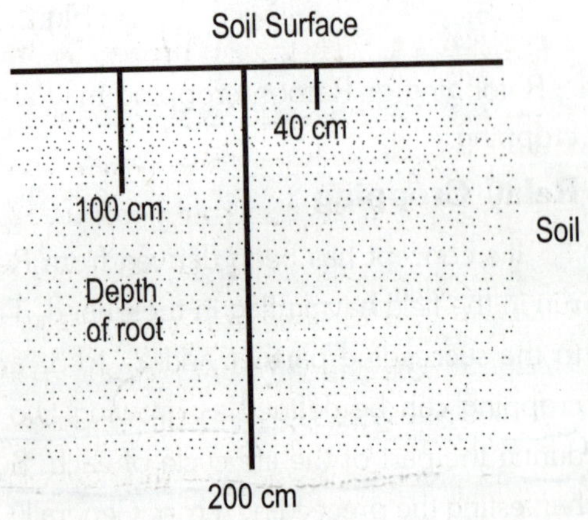

Depth of root of different Plant

Fig.2.6

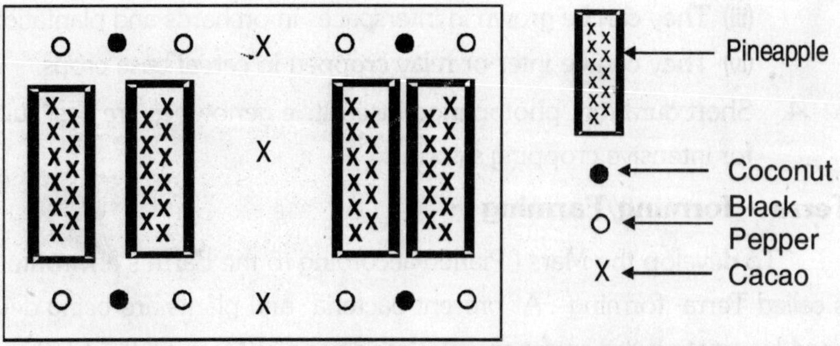

Fig. 2.7 Planting of Different crops in multistoryed Cropping in coconut garden

effectively. The canopy orientation of component crops and localization of root system at distict zones are the tools of this system. The practice of growing different crops of varying, rooting pattern and duration is called multi- storeyed cropping. The objective is to utilize the vertical space more effectively. It is based on the principle of annidation in space e.g.

	Coconut	+ Black pepper	+ Cocao	+ Pineapple
	10 m ht. of plant	6-8 m ht.	3 m	1 m
	175 palms/ha	-	600 plants/ha	3500/ha
	100 cm	-	200 cm	40 cm.

Choice of Crops for Multiple Intensive Cropping :

1. The crops should be arranged in such a way that Allelopathic effect, temporary immobilization of nutrients and depletion of nutrients from the same layer of soil do not occur.
2. Legumes have a definite place in any cropping system because of their limited demand on resources like water, nutrients and light, their adaptability to varying environmental conditions and capacity to fix atmospheric N in root nodules.
3. Vegetables deserve their due place in intensive cropping because
   (i) of high cash and nutritional values,
   (ii) most vegetables can be raised as seedlings and planted, thus reducing main field duration and competition with the associated crop.

(iii) They can be grown in interspaces in orchards and plantations.

(iv) They can be inter or relay cropped in cereal base crops.

4. Short duration, photoperiod insensitive genotypes are best suited for intensive cropping systems.

Terra – forming/Farming :

To develop the Mars (Planet) according to the Earth's environment is called Terra- forming . At present bacteria and plants are being developed by creating the environmental conditions of Mars on the Earth.

Aeroponics -

In such technique plants are grown without soil in the aerated moist atmosphere with the help of very small quantity of water. Through this technique any plant can be grown at any time. Since plants are grown in clean and controlled atmosphere, there is no chance of spread of insect-pests and diseases. It saves 98% water, 60% chemical fertilizer and 100% pesticides. This technique was come into existence when American Space Agency NASA in 1997 had tried to grow pods of bean in Mir Space Station of Russia. Here seeds or propagules are hanged in the air and its roots are grown in controlled aerated atmosphere where roots are kept always moist with the nutrient rich spray.

❏❏❏

3 Annidation and Allelopathy

Annidation

The complimentary interaction between intercrops in the intercropping system is known as Annidation .When plants are grown as components of intercropping system, interactions between the component species occur. Such interactions are essentially the response of one crop species to the environment as modified by the presence of another species. One component species helps the other. Annidation occurs both in space and time .

(a) Spatial Annidation (Annidation in Space) :

Here the complimentary interactions between the intercrops occur with regard to spatial position (space/place). This principle is used in Multistorey cropping. The component species occupy different vertical layers by spreading out their canopies or roots either in air or in soil. As far as aerial annidation is concerned the taller intercrops occupy higher vertical layers and shorter intercrops the lower layers. Such taller component species are comparatively tolerant to strong light and high evaporative demand than shorter one. On the other hand, shorter component species are relatively shade- loving and acclimatised to high relative humidity. The principle of spatial annidation may also occur in soil, where rooting patterns and its systems of component species exploit nutrients from different soil layers. Therefore for efficient utilization of soil resources, intercrops of different root system viz. deep rooted species + shallow rooted species are grown.

(b) Temporal Annidation (Annidation in Time) :

The complimentary interaction between the intercrops in the time aspect is called Temporal Annidation. Such intercrops have different natural habit and zero competition. Both the component species have widely varying duration and different peak demands for light and nutrients, Thus competition between the intercrops is reduced. In the Maize + Urd/Moong intercropping, the peak nutrient demand for Maize is 50 days after sowing (DAS) whereas around 30-35 DAS for Urd or Moong. Other intercropping utilising this principle are Groundnut + Redgram, Shorghum + Redgram, etc.

Allelopathy :

The interactions between the crops may also occur through other means. i.e. competitive or non –competitive The competitive interaction between the crop species in intercropping and sequential cropping through the release of chemical substances or toxins is called Allelopathy. Such chemical substances or toxins generally exuded by the roots or produced by decomposition of the residues of one crop species have direct or indirect harmful effect on the other crop species. It affects germination, establishment and growth of the associated crops. Such crops are unsuitable for intercropping or sequential cropping. Sunflower, Sorghum, Walnut, Cucumber,Peach and Eucalyptus are the examples of such crops which are known for allelopathic effect. Sunflower affects the associated crops through the release of allelopathic chemicals in the soil by the roots. It inhibits the germination process of subsequent crops. Sunflower residues also produce allelo chemicals during decomposition. Therefore a sufficient time period of 15-20 days is allowed to lapse between the harvest of sunflower and sowing of the subsequent crop. It results into benefiting the subsequent crop in two ways-

(i) Subsequent crop escapes from the allelopathy effect.

(ii) The rapid decomposition of sunflower residues causes mineralisation of soil nitrogen.

Types of Allelopathy :

(a) True allelopathy.

(b) Functional allelopathy.

(a) True allelopathy :

The direct or indirect harmful effect on the other crops through the release of toxic substance as such from the plant.

(b) Functional allelopathy :

When precursor is released which is converted into active substances by some microorganisms, is categorized under functional allelopathy.

Allelopathy is observed in two ways –

(i) Allo – inhibition : The chemical substances released by one species may inhibit species of plants other than one releasing it.

(ii) Auto- inhibition : The toxins may inhibit more strongly plants of the producer species itself.

Allelo-chemical :

Such types of chemicals, released by plants which show allelopathy are known as allelo-chemicals.The genetic make-up of the plant and the surrounding environment may affect the type and quantity of allelochemicals. Some of the organic substances exudated by the roots may inhibit the growth of the neighbouring species. The known examples are walnut, cucumber and peach whose living roots exudates allelochemicals and inhibits the growth of the plants growing near them. Some of the allelochemicals may be produced by the aerial portions of the plant. These allelochemicals may reach the ground through raindrops, falling leaves or insects or animal activities. The leaves of ***Eucalyptus globules*** exudate some allelochemicals which drastically reduce the germination of mustard seeds when sown underneath . The crop residues of sunflower may produce allelochemicals after its decompostition which adversely affects the germination, establishment and growth of the subsequent crop . When sunflower stalks are applied as such, it inhibits the establishment and growth of sorghum in a untilled plot.

On the contrary, the non-competitive interaction between the crop species through the release of organic substances may also be possible. The release of hormone- like substances by one crop species may stimulate the growth of the other crop species . It may be called Negative Allelopaathy. But the release of Nitrogen from the root nodules of legumes is not considered as negative allelopathy.

Legume Effect :

The beneficial effects of the legumes in any crop rotation and intensive cropping system is termed as Legume effect. Inclusion of legumes in the cropping system is beneficial in many ways-

(i) Legumes fix atmospheric nitrogen in root nodules and thus improves the nitrogen status of the soil.

(ii) It saves upto 25%of recommended level of Nitrogen application to the associated cereals when grown as intercrop.

(iii) The crop residues and root nodules of legumes release nitrogen during decomposition for the use of the succeeding crop.

(iv) Legumes absorb soil phosphorus more efficiently and part of this mobilized phosphorus in organic form is available to the succeed ing crop. It means legumes convert **Inorganic Phosphorus** into **Organic form of phosphorus** and thus is able to extract insoluble forms of soil phosphorus.

(v) Legumes have greater root cation exchange capacity (CEC) than cereals. The plants with greater root CEC are capable of absorbing more of divalent cations such as Ca^{2+} and Mg^{2+} But they can not compete effectively with the cereals in the absorpt ion of monovalent cation like K^{+}.

Shorghum Effect :

The nutrient status of soil is exhausted by any fast growing cereal . The crop residue of such cereal having wide C:N ratio takes a long time to decompose. During the process of decomposition, part of soil nitrogen is temporarily immobilised affecting the succeeding crop. Such effect is more pronounced in sorghum in low fertile soil causing to temporary deficiency of nitrogen in the soil for the succeeding crop. To reduce shorghum effect, 25% more nitrogen is applied at the time of first fertiliser dose of the succeeding crop. It hastens the process of decomposition and overcomes the immobilised nitrogen. Ragi is the exception because its residues decompose rapidly resulting in mineralization of soil nitrogen.

Cotton Effect :

Cotton feeds in the deeper layers of the soil , and removes comparatively smaller quantities of nutrients. The succeeding crop having the shallow root system is able to tap on the unused pool of nutrients in the surface layers of the soil . This effect may be termed as cotton effect.

❑❑❑

4 Crop Ecology

The term 'Ecology' is derived from ' Oikos' meaning home. Crop ecology means the relationship of a crop or crop communities with its surroundings say environment . A local climatic condition that differs from surrounding areas because of differences in relief, exposure or cover is called Microclimate . From the agricultural point of view the microclimate is studied at two levels i.e. at plant level or at soil level.

Micro – climate at plant level :

The ecoclimate of the plant i.e. the climate just above and beneath the plant's canopy, is different from the climate of the surrounding environment . The microclimate at the plant level i.e. the ecoclimate varies with the type of plants, morphological features of plant i.e. height, condensed canopy, shape and size of leaf and stem and soil colour . Soil colour modifies the microclimate by affecting the sunlight, temperature, humidity, CO_2: O_2 level, air velocity and organisms. The plants having large morphological features reduce the air movement and the penetration of sun rays to the ground surface. Even emerging seedling alters the climate near the soil surface by reducing air movement and by shading the ground. In the vegetation covered areas, the extreme variations in the temperatures of the soil surface are reduced and the eco- climate of the plant is more humid and cooler than the surroundings which favours the spread of insectpest and diseases. The plant canopy changes the intensity and quality of light when the light rays pass through it. In the day time when sunrays directly fall on the leaves, the leaf temperature may be higher than the air temperature and at night may be lower than air temperature.

Micro – climate at soil level :

Soil colour is one of the factors which affect microclimate. Due to soil colour, part of incoming radiations are reflected. The percentage of the incoming radiation that is reflected, is called **Albedo.** Albedo varies according to soil type, plant cover, moisture content of soil, soil organic matters, etc. Due to differences in albedo of the soil surfaces, different thermal regimes are set up even when incoming radiations are indenticals. For example, a mulch on the soil surface becomes very hot by absorbing most of the incoming radiations and keeps the soil cool during day time and vice – versa during night hours.

Wet soil absorbs more incoming radiations than dry soil. Most of the absorbed energy is used to evaporate the soil water due to which moist soil takes longer time to heat up or to cool down. The heat gain or heat loss by the soil is five times quicker than by the water because the specific gravity of the soil is 0.2 and the sp.gr.of the water is 1.0.High soil temperature restricts the root growth. In the arid region warm (20^0 c-30^0 c) moist soil is the favourable medium for most of the cultivated crops.

Harvest Index :

The effectiveness of photosynthesis depends on-

(i) a large and efficient assimilating area.

(ii) An adequate supply of solar energy and CO_2

(iii) Favourable environmental conditions.

The total product of photosynthesis throughout the lifetime of the crop growing in the given circumstances depends upon the following factors-

(a) the size of the assimilating area.

(b) The efficiency of assimilation and CO_2 fixation.

(c) The period of efficient assimilation and CO_2 fixation.

In the Agronomy , the efficient utilisation of assimilation and CO_2 fixation i.e. photosynthesis is expressed in terms of Harvest Index .H.I. is also called 'Coefficient of Effectiveness.'

$$\text{H.I.} = \frac{\text{Economic yield (e.g. grain)}}{\text{Biological yield (e.g.grain + straw)}} \times 100$$

$$= \frac{\text{Sink}}{\text{Source}} \times 100$$

Grain yield can be calculated by the formula-

$Y = a \times b \times c \times d$

Where. y → grain yield per unit area.
a → no. of plants per unit area.
b → no. of fertile tillers per plant
c → no. of grains per ear
d → weight of the individual grain.

In wheat, barley and similar cereals, the photosynthetic activity of the ear, which is situated at the top of the stalk, makes a considerable contribution to grain formation. Practically all the drymatter of the grain is produced by the part of the shoot above the flag leaf node .Of this dry matter, the ear contribute about 50% in wheat and upto 70% in barley.

Sink – Source Ratio :

Source : Such part of the plant where photosynthesis occurs; generally leaves are the source.

Sink : Where photosynthates accumulate , is called sink. Generally grain is the sink but stem (e.g. potato),root (e.g.Sugarbeet) or leaf (e.g.Palak) may be the sink.

For higher HI, development of sink is the most concerned. But the development of sink depends on three factors—

(a) Good Source
(b) Partitioning of Photosynthates , and
(c) Translocation of Assimilates

In spite of good source if partitioning of photosynthates is not well developed, sink will be poorly formed and thus is the lower H.I. Therefore for higher sink-source ratio, the followings are necessary—

(i) Source should be good.
(ii) Partitioning of source i.e. photosynthates should be well developed.
(iii) Translocation of photosynthates must be good.
(iv) And lastly photosynthates do accumulate as sink.

Formation of Photosynthates as Source → Partitioning of Photosynthates → Translocation of Assimilates → Storage of Assimilates as Sink

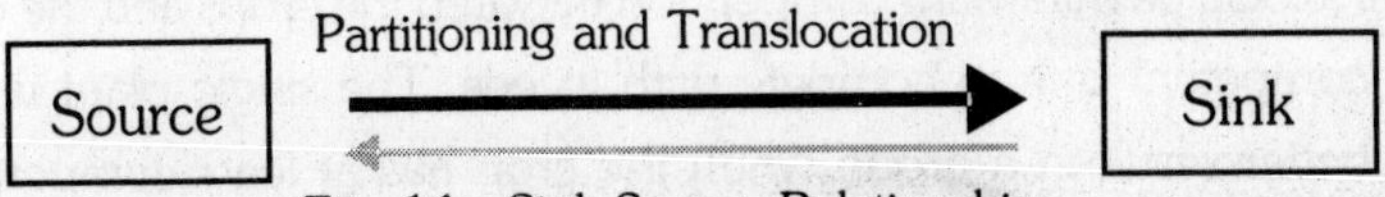

Fig. 14 : Sink-Source Relationship

The formation of sink from source is a reversible process but source to sink is more pronounced.

With the appearance of inflorencence, competition for assimilates between the leaves and the inflorescence takes place . When the seed formation starts, most of the assimilates move to the grain. This internal competition is more severe as competition between plants increases i.e. plant population pressure, which is illustrated by the figure.

Ideotypes (Model Plants) :

The production of the every plant is affected by the plant's type, climatic condition, soil type and management factors. According to Army and Greiver (1967), the production of the plant might be increased by changing plant's type and increasing the period of grain – filling in a certain climatic condition where management and soil factors are not limiting. There is a direct relationship between the plant's type and the crop production because the orientation and number of leaves play the important role in the CO_2 assimilation (Photosynthesis).

The concept of Ideotype was given by **Donald** in 1968. He illusrated

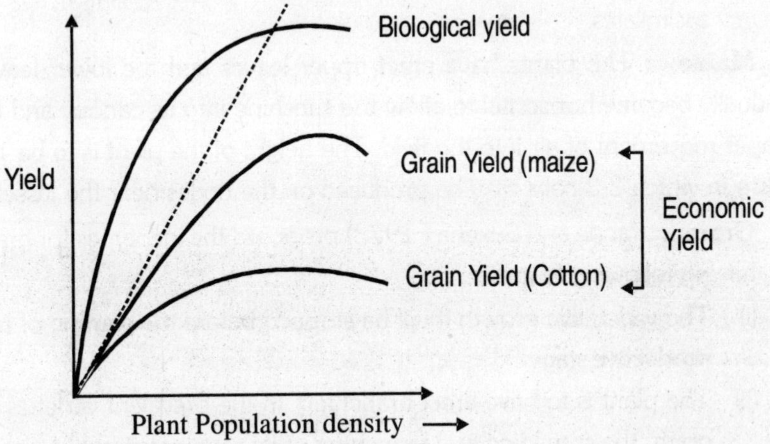

Fig. 4.2 : Biological and Economic Yield

that there should be minimum competition between the crops and the crop must be competent one to compete with weeds. The single plant would give the better result in a group when the crop has at least competition with the same type of the crop. Ideotype is the model plant which may be defined as " a biological model which is expected to perform or behave in a predictable manner within a defined environment". On the basis of environment Donald & Hamblin (1976) identified two forms of ideotypes i.e. isolation ideotypes and competition ideotypes. Competition ideotypes are suitable for mixed cultivation.

Ideotypes for Some Crops :

Wheat : According to Donald (1968), the ideotype for wheat or the similar crop has following features-

(i) To avoid lodging, there should be short strong stem.

(ii) Few small erect leaves to allow the sunshine into its canopy .

(iii) A large erect ear.

(iv) The no. of fertile florets per unit area should be more so that HI is more.

(v) There should be awns to increase the photosynthetic area.

(vi) A single culm to avoid wasteful vegetative growth .

(vii) Resistant to insect pests & diseases.

(viii) There must be proper partitioning and translocation of assimilates.

Maize : The plants have erect upper leaves and the lower leaves gradually become horizontal to allow the sunshine into its canopy and for proper movement of air into the field. The height of the plant is to be 1.5 metre in which 2-3 cobs may be produced on the nodes near the tassel.

Gram : Pande and Suxena (1973) proposed the ideotypes for gram having following features.

(i) The vegetative growth must be stopped before the starting of reproductive stage.

(ii) The plant is to have erect branching .In the prevalent varieties of gram, the spreading and branching of its canopy is just like umbrella which interferes to penetrate the sunshine into its canopy causing humid conditions favourable for insect pests & diseases.

(iii) To harness the long photoperiod and favourable temperature at the time of flowering , there should be 2-3 longer pods in the leaf axis and 2-3 seeds in each pod.

Arhar : The growth of Arhar varieties in the beginning is too less i.e. in the first two months only one or two branches are come out. Therefore Arhar is unable to harness the solar energy properly in the first two months. The flower's drop is also a major problem.The filling of pods according to Hydro-dynamic model sets up the competition between vegetative and reproductive phases. Considering all the views pande & Suxena suggested ideotypes having following features-

(i) The fast growth of plant's canopy at least in the beginning.

(ii) The reproductive phase starts after the closure of vegetative growth.

(iii) Long floral axis having 2-3 flowers in each trifoliate axis.

(iv) Synchronised flowering i.e. all the flowers come out at the same time.

(v) Active root nodules for the long time .

(vi) Resistant to insect pests & diseases.

Rainfed upland Rice :

(i) Short growth duration (85-100 days)

(ii) Effective deep root system.

(iii) Dwarf (less than 100cm) plant having erect leaves and thick stem.

(iv) Early strong fertile tillering .

(v) Synchronised flowering .

(vi) Good number of panicles at higher density about 400panicles per square metre.

(vii) Highest number of grains per panicle.

(viii) Moderate seed dormancy.

(ix) Resistant to insect pests & diseases e.g. resistant to *Helminthosporium sp.*

Rainfed Wheat :

An ideotype of rainfed wheat was conceived and proposed by Dr.

Asana-

(i) Large no. of spikelets,

(ii) Large peduncle,

(iii) Strong and deep root system,

(iv) Flat leaves parallel to soil and

(v) A grain development period coinciding with mean max. temp. of 25^0C.

Ideotypes for Dryland Farming :

(i) Short growth duration.

(ii) Effective root system.

(iii) Drought tolerance.

(iv) High yield potentiality with altered morphology viz.

(a) Plant with few leaves just sufficient to maintain photosynthetic output and growth (to minimise the use of water).

(b) Leaves horizontally disposed for better light interception con trary to vertically disposed most effective under irrigated conditions.

Agro-Climatic Regional Planning :

Agro – climatic Regional Planning in India was initiated in 1988 by the planning commission . It means under the seventh five year plan; Regional planning was come into force on the basis of agro- climatic factors for balanced regional growth .The agroclimatic factors are Rainfall, temperature, soil, topography , cropping and farming system and water resources.

Agro- Climatic Zones :

Agro- climatic zone is a land unit is terms of major climates suitable for a certain range of crops and cultivars (FAO, 1983).

On the basis of criteria of homogenity in agro- characteristics such as **water surplus and water deficit**, India was divided into 11 agro- climatic zones.

Later on, India was divided into 15 agro- climatic zones on the basis of **water surpulus, water deficit** and **cropping system.**

An ecological region is an area of the earth's surface characterised by distinct ecological responses to macro – climates expressed by soils, vegetation fauna and aquatic systems.

Agro- ecological region is the land unit cut out of agro climatic region when superimposed on land form and the kind of soils and soil conditions that act as modifiers to climate and length of growing period.

The classification is on the basis of :-

(i) Growth period (ii) Rainfall/potential evapotranspiration

(iii) Soil (iv) Districts.

There are two components of ecological zones-

(i) Biotic factors viz Fauna and Flora.

(ii) Abiotic factors viz water, minerals, light etc.

At present India has been divided into 21 agro ecological zones. The detailed descriptions of 15 agro- climatic zones are as follows-

1.Western Himalyan Region : Includes Jammu and Kashmir, Himachal Pradesh & Hilly areas of U.P. climatic is cool and humid. Annual rainfall- 1650-2000 mm. Important crops are Wheat, Maize, Paddy and Potato . It is divided into 3 subregions. Cropping Intesity is lowest in J & K and highest in H.P. the productivity level of all the crops is below the national average. To develop this region-

(i) Intensify the soil and water conservation planning .

(ii) Land utility planning should be implemented viz. Agriculture for 30% sloppy land, Horticulture for 30-50% slope, paddy for above 50% slope.

(iii) Good storage for transportation, marketing & processing.

(iv) Irrigation management like water harvesting technique

(v) Livestock management etc.

2. Eastern Himalayan Region : Sikkim , Hills of Darziling, Arunachal Pradesh ,Nagaland, Meghalaya ,Manipur, Tripura, Mizoram, Assam, Jalpaiguri area of W.B. and some parts of Bihar,Climate humid & subhumid, 1840-3528mm.annual rainfall, 5 sub-regions,main crops-Maize , Paddy, Jute, Rapeseed. Shifting cultivation on $1/3_{rd}$ of cultivated area, soil erosion more.

For the development of this region-

(i) Soil and water conservation for each watershed .

(ii) Educate the people about the harmful effect of land erosion at panchayat level.

(iii) Facilities of good supply of inputs, marketing transportation, processing & storage.

(iv) Adoption of Integrated Farming system because of different Dietary pattern of local people.

(v) Good communication & Transport facility.

3. Lower Gangetic Plain : West Bengal; Productivity of rice is lower than the national average 15 q/ha but this region contributes 12% of the total rice production in India. Crops :Rice,Jute, Rapeseed and Wheat. Rapeseed , Maize & Potato are newly introduced; moist humid and dry humid climate , 4 subregions. 1302-1607 mm. annual rainfall. Followings are negative points- (i) lack of good water management;drainage problem. (ii) Use of local variety (iii) Less use of HYV (iv) poor extension work (v) Inadequate supply of inputs (vi) Poor marketing & processing (vii) Lack of transportation and storage (viii) Dependent on Monsoon (ix) Soil & water erosion (x) Uneducated & poor Farmers (xi) Poor electric supply.

4. Middle Gangetic Plain : Eastern parts of U.P. and Bihar plains. 40% of total cropped area is irrigated cropping intensity 142% (India's national CI 136%). Low productivity of rice, 3subregions of earstern U.P. and 9subregions of Bihar ,moist sub humid to dry sub humid climate. 1211-1470 mm. annual rainfall, main crops - Sugarcane, Paddy, Maize ,Wheat.

5. Upper Gangetic Plain : Remaining parts of UP and 3 sub regions, **dry sub humid** to **sub dry climate,** 721-979 mm. annual rainfall. Main Crops- Rice, Wheat, Maize and Sugarcane. Irrigation intensity is 131% and CI 145% . Irrigation by canal and tube well. Special planning to grow vegetables & fruits scientifically. Milk production and processing units are being established.

- Maximum dry areas: Western dryland region.
- Gujarat Plains and Hills are called oilseed region.
- Plantations & spices are grown in western coast and ghats.
- Coconut cultivation is Island region.

6. Transgangetic Plains : Punjab, Haryana, Shiwalik region, Sriganganagar, Delhi semi –dry areas of plains, dry areas of Thar (Rajasthan), Maximum Net shown area and irrigated area. Cropping Intensity highest, low poverty, maximum use of ground water.Followings points for the development of this region-

(i) good water management .

(ii) management and reclaimation of user soil.

(iii) Proper use of ground water.

(iv) Diversification of crops.

(v) Introduction of Leguminous crops in cropping system.

(vi) Proper livestock & pasture management.

(vii) Increasing of area under vegetables, fruits and pastures

7. Eastern Plateau & Hills : Wainganga, Eastern Hilly areas of M.P; inland Orissa, Chhotanagpur plateau, Platean and Hilly regions of West Bengal, Chhattisgarh.5 subregions, moist sub humid to dry subhumid. 1271-1436mm annual rainfall, Main crops: Rice , Wheat,Maize , Ragi, 67% small and marginal farmers have 25% of agricultural land only. Shallow and undulating land . Horticulture development planning is being intensified. Irrigation through tanks & tubewells.

8. Central Plateau and Hills : about 46 districts of M.P. Chhattisgarh, U.P. and Rajasthan. Undulating soil topography, Ravines present, 14 sub regions, 400- 1550 mm annual rainfall, semi-arid to dry sub-humid climate, Main crops : Wheat, Gram, Jowar,Bajra,Paddy. 75% area is rain affected where following programmes are being run-

(i) Watershed management .

(ii) Crop diversification.

(iii) Reclamation of Ravine area.

(iv) Other programmes.

9. Western Plateau & Hills : Main regions of Maharashtra , major parts of M.P. and some parts of Rajasthan. 4 subregions,semi-arid climate, 602-1040 mm annual rainfall. It covers maximum parts of peninsular area. Major crops- Jowar,Bajra, Cotton and Wheat. Net sown area 65%, 11% forest area, irrigated area 12.4% . Irrigation mainly by canal.

50% Jowar of our country from this region .Best quality of orange, grape and banana. Following suggestion for this zone-

(i) Management of minor irrigation .

(ii) Improvement in the conditions of cattle and buffaloes.

(iii) Cultivation of costly crops.

10. Southern Plateau & Hills : about 35 districts of Andhra, Karnataka, Tamil Nadu, typically semi arid zones.81% dryland farming and 111% cropping intensity. 6 subregions, 677-1000mm annual rainfall, cultivation of low value crops. Suggestions-

(i). Proper utilization of dryland technology in watershed area.

(ii). Increase fertilizer use efficiency.

(iii). Minor irrigation programme.

11. East Coast Plains : East coast of T.N. , Andhra & Orissa . Soils are mainly alluvial and coastal sands. Irrigation through canals & tanks. 6 sub regions semiarid and dry sub humid climate, 780-1287 mm annual rainfall. Main crops : Rice, Groundnut, Ragi, Jowar & Bajra. 20.3% of total rice production in India is from this zone & 17.5% groundnut prevalence of alkaline and saline soils. Development of fisheries due to waterlogging .75% area rainfed. In place of shifting cultivation, integrated Horticulture Agriculture programme should be adopted by tribal people.

12. West Coast Plains and Ghats : Westcoast of T.N. Keral, Karnataka, Maharashtra and Goa. Main occupation is cultivation of spices and plantation crops; 4 subregions, dry sub- humid to humid , 2226-3640mm annual rainfall, main crops – Rice, Ragi, Groundnut and Tapioca, suggestions –

(i) Storage of rain water.

(ii) Proper use of ground water,

(III) Increase in area under fruits and vegetable.

(iv) Fisheries management

13. Gujarat Plains & Hills : 19 districts of Gujarat, 7 sub regions, arid and low rainfall, 340-1793 mm annual rainfall, 32.5% irrigated land through wells and tubewells. Main crops- Maize, Wheat,Rice, Groundnut, Tobacco , Cotton ,Jowar, Bajra,This zone is famous for oilseed crops hence called **oilseed region**. CI : 114% . About 60% area is drought prone. 78% rainfed; suggestions –

(i) Rain water harvesting and its management.

(ii) Dryland farming & management of canal & ground water.

(iii) Development of Agro- forestry and arid- Horticulture.

14. Western Dry Region : 9 districts of Rajasthan, hot sandy desert, erratic rainfall, high evaporation, scanty vegetation, no perrenial rivers, general occurence of drought, 95 mm annual rainfall, 1.2% forest area, 4.5% pasture area 6.3% irrigated area, CI : 105%. Main crops- Bajra, Gram, Wheat, Rapeseed.Two main plannings-

(i) Indira Gandhi Cannal and

(ii) DDP (Drought Development programme).

15. Island Regions : Island territories of Andaman & Nicobar and Lakshadeep. 3000mm annual rainfall spread over 8-9 months. Largely forest zone with undulating land. Coconut cultivation on 50% area of total cropped area,humid climate, main crops- Rice & Coconut . Main emphasis on –

(i) Crop Improvement .

(ii) Water Management.

(III) Fisheries.

Agro- Ecological Zones :

India has been divided into 21 agro-ecological zones-

1. Western Himalaya I.: Cold arid region with shallow sedimentary soils, covering Laddakh and Gilgit. Temp less than8^0C, Rainfall less than 150 mm which is 15% less than annual PET , Growing period less than 90 days. Cover 15.6 million hectare (4.7% of total geographical area).

2. Western Himalaya II : Warm subhumid with brown forest & podzolic soils, J & K, H.P; northwest Hilly areas of U.P. (Uttarakhand); 17.7 mha (5.4%), Rainfall 1600-2000mm, exceeds PET, water surplus, Growing period 150-210 days .

3. Western Plains : Hot arid region, with desert and alkaline soil. Covers west Rajashtan, Southwest Haryana, Punjab, Kutchha Peninsula, north of Kathiawada. 29.6 mha (9% of total geographical area). Rainfall

300mm, PET: 1500-1900 mm, Deficit 1200-1600mm. Growing period less than 90 days.

4. Duccan Plateau I : Hot arid region , with mixed red & black soil, part of Duccan plateau including district of Raichur, Bellary of Karnataka & Anantpur of Andhra. 4.7 mha (1.4%) Rainfall 400-500 mm, PET : 1800-1900mm, Growing Period less than 90 days.

5. North Plain and Central Highland : Hot semiarid region with alluvial derived soil. Parts of North plain, Central High land and Gujarat plains. 32.9 on ha (10% of total geog. Area.), Rainfall 400-800mm, PET : 1400-1800mm, Growing period 90-150 days.

6. Central (Malwa) Highland and Kathiawada Peninsula : Hot, semiarid region with medium deep black soil , west M.P., East Rajasthan, and Gujarat, Kathiawada, 1.84 mha (5.6%) Rainfall 600-900 mm,PET : 1600-2000mm Growing period 90-150 days.

7. Central Highland (Malwa & Bundelkhand). : Hot subhumid region with medium & deep black soil, Central highland including distt of Raipur, Sagar, Bhopal, Sehore, Shajapur, Hoshangabad . 8.2 mha (2.5%) Rainfall 1000-1500mm,PET: 1300-1500mmGrowing period 150-180 days.

8. Deccan Plateau II : Hot semiarid with shallow & medium black soil, West Maharashtra, North Karnatka, A.P. 33.0 Mha (10%), Rainfall 600-1000mm, PET : 1600-1800mm, Growing period : 90-150 days.

9. Deccan Plateau and Eastern Ghat : Hot semi arid with red and black soil .Major part of Andhra Pradesh, 20.8 mha (6.3%), Rainfall 600 -1000 mm, PET : reduction 40% growing period : 90-150 days.

10. Eastern Ghat (TN upland) and Deccan Plateau : Hot semi arid region with red loamy soil, South of Deccan plateau, T.N. upland, West Karnataka, 22.7 mha (6.9%) Rainfall 600-1000mm, PET: Deficit 400-700 mm, growing period 120-150 days.

11. Deccan Plateau and Central Highland (Bundelkhand) : Hot and subhumid region with red and black soils. Bundelkhand part of M.P.,North Vidarbha, 13.7 mha (4.2%)Rainfall 1000-1500mm, PET : 1500-1600mm. Deficit : 500-700mm, growing period : 150-180 days.

12. Eastern Plateau (Chhatisgarh Region) : Hot subhumid with red and yellow soils; parts of eastern plateau covering Chhatisgash state, Palamu of Jharkhand and Kaimur of Bihar . 13.2 mha (4%) Rainfall 1200-1600mm,PET : 1400-1500mm,(Deficit 500-700m) growing period : 150-180 days.

13. Eastern (Chhotanagpur) Plateau & Eastern Ghat : Hot sub humid region with red loamy soil, Chhotanagpur plateau of Jharkhand, West of W.B., Eastern Ghat of Orissa, Baster region of M.P., 27.8mha (8.5%) Rainfall 1000-1600mm, PET : 1400-1700mm, growing period 150-180 days.

14. Northern Plains : Hot subhumid region with alluvial soil, part of Indogangetic plain including plain of west Himalaya, Area 12.2 mha (3.7%) Rainfall 1000-1200 mm, PET : 1400-1800mm,growing period 150-180 days.

15. Eastern Plains : Hot subhumid region with alluvial soil , North east UP, North Bihar, Foothill of eastem Himalaya, '9.3 mha (2.8%) Rainfall 1400-1600mm,PET : 1300-1500mm,growing period : 180-210days.

16. Assam & Bengal Plains : Hot humid with alluvial soil. Plains of the Bramhaputra and Ganges rivers, parts of Assam & W.B. , 11.7 mha (3.6%) rainfall increases from 1400/ 1600mm in the Ganges plain to 1600/ 2000mm in the Bramhaputra plain .PET : 1000-1400 mm, growing period more than 270 days.

17. North Eastern Hills (Purwanchal) : Warm humid region,Red & laterite soil, north east states of Nagaland, Meghalay, Manipur, Mizoram, South Tripura, 1'0.7mha (3.3%) . Rainfall 1600-2600mm exceeds PET : Dryspell 100-150mm in post – monsoon period, growing period more than 270 days.

18. Eastern Himalaya : Warm humid , brown – hill soils, Northern part of W.B., North Assam, Most part of Arunachal, Sikkim,8.0 mha(2.4%) Rainfall exceeds 2000mm occurs throughout year, PET : less than precipitation , growing period more than 270 days.

19. Eastern Coastal Plain : Hot and sub - humid alluvial derived soil, East Coast plain, The kavery delta to Gangetic delta, 8.1 mha (2.5%) Rainfall 1200-1600mm, PET: 400-500 mm, deficit 20%less, growing period: 150-210 days.

20. Western Ghat & Coastal Plain : Hot humid red laterite, alluvial sahyadris ,west Coast of Maharashtra, Karnataka, Kerala, 10.1 mha (3.0%) , Rainfall exceeds 2000mm, PET : 1400- 1600mm, Deficit 300 -400 mm dryspell in Feb- April,growing period more than 270 days.

21. Islands of Andaman, Nicobar & Lakshdweep : Hot humid, red loam- sandy loam, Andman Nicobar & Lakshdweep groups, 0.8 mha (0.3%) rainfall: 1600-3000mm, exceed PET : except 300-400mm deficit during December to March, suitable for plantation crops, growing period more than 270 days.

❑❑❑

5 Dryland Agriculture

Dryland Agriculture means cultivation of crops entirely under rainfed condition . According to the 'Principles of Agronomy' written by Reddy and Reddi, Dryland Agriculture may be classified into three groups on the basis of annual rainfall .

(a). Dry Farming: Cultivation of crops in areas where annual rainfall is less than 750 mm and crop failures due to prolonged dry spells during crop period are most common. Dry farming is practiced in arid regions with the help of moisture conservation practices. Alternate land use system is suggested in this region.

(b). Dryland Farming : Cultivation of crops in areas where annual rainfall is more than 750 mm but less than 1150 mm is called Dryland Farming. Dry spells may occur, but crop failures are less frequent . Higher Evapotranspiration (ET) than the total precipitation is the main reason for moisture deficit in these areas. The soil and moisture conservation measures is the key for dryland farming practice in semi – arid regions.Drainage facility may be required especially in black soils.

(c) Rainfed Farming : Means cultivation of crops in regions where annual rainfall is more than 1150 mm.There is less chances of crop failures due to dry spells. There is adequate rainfall and drainage becomes the important problem in rainfed farming. This farming is practiced in humid regions.

United Nations Economic and Social Commission for Asia and the Pacific differentiated between dryland farming and rainfed farming with the following constituents in the given table-

Table 5.1 Dryland vs Rainfed Farming

S.No	Constituent	Dryland Farming	Rainfed Farming
1.	Rainfall (mm)	< 800 (less than)	> 800
2.	Moisture availability to the Crop	Shortage	Enough
3.	Growing Season	< 200 days	> 200 days
4.	Growing Regions	Arid and semiarid as well as uplands of sub-humid and humid regions	Humid & Subhumid regions
5.	Cropping system	Single Crop or Intercropping	Intercropping or double cropping
6.	Constraints	Wind & Water erosion	Water erosion

Problems & Prospects of Dryland Agriculture :

In our country, irrigated area constitutes 33% and 67% is dryland and rainfed , out of 143.8 million hectares of cultivated land. Total cropped area is 187.94 mha . Such dryland agriculture contributes about 44% of national food grain production . The maximum irrigated area will be achieved 50% after the full utilization of all sources of irrigation . In this way 50% cultivated area will remain unirrigated. After full exploitationof dryland ,it may contribute upto 75% of total foodgrain production. Pulses and oil-seeds are mainly grown in such areas. Important commercial crops viz Cotton, Castor, Groundnut and all coarse grains viz Jowar, Bajra, Maize crops are rainfed. The major part of milk , meat, wool hides, bonemeal etc are also from this area.

Dry farming areas are characterised by very low and highly variable and uncertain yields. Problems of dry land agriculture are-

(a) Inadequacy and uncertainty of rainfall and its erratic distribution. Expected rainfall in a year is 650 mm but 80% of it is received in 75-90 days of the monsoon season.

(b) Late onset and early cessation of rains.

(c) Prolonged dryspells during the crop period .

(d) Low moisture retention capacity.

(e) Poor soil fertility condition.

(f) Socio – economic constraints particularly because of the predomi-nance of small and marginal farmers.54% of the holdings is less than one hectare.

(g) Technological and developmental constraints.

(h) Limited infrastructure development and unproper and untimely availability of credits and agricultural inputs. There are three types of agriculture possible in Dryland areas- viz .

  (i) Crop production

  (ii) Animal husbandry with pastures management and

  (iii) Agro forestry.

Drought and Agricultural Drought :

Indian Meteorological Department, Pune defined: Drought as a situation occuring in any areas in a year when annual rainfall is less than 75% of the normal . When deficiency of rainfall is above 50% of the normal, called severe drought .

However, agricultural drought is the moisture deficit which results when amount of water available in the soil is not sufficient to meet the demands of potential evapotranspiration (PET).

Areas where drought has occurred in 20% of the years during the period are considered Drought areas and where it has occurred in more than 40% of the years, called chronic drought areas.

ICRISAT : International Crop Research Institute for the Semi-Arid Tropics : established on Oct 11,1972 near Hyderabad (Patancheru).

ICARDA : International Centre for Agl.Research in Dry Areas in 1977 at Aleppo (Syria).

Moisture Availability Index (MAI) is the ratio between rainfall (weekly/monthly) at 50% probability level to potential evapotranspiration of the corresponding period.

$$MAI = \frac{\text{Rainfall at 50 \% probability}}{PET}$$

Aridity Index (A.I.)

$$Ia = \frac{WD}{PET} \times 100$$

Where WD ⟶ Water Deficit

PET ⟶ Potential evapotranspiration

Moisture Deficit Index (MDI)

Where P : Precipitation

$$Im = \frac{P-PET}{PET} \times 100$$

Thornthwaite Moisture Index (TMI)

$$Im\ a = \frac{AE}{PET}$$ Where AE : Percent probability of Actual evaporation.

Crop Adaptation for Dry Growing Conditions

Escaping Drought

(i) Ephemerals Drought

(ii) Early Maturing Variety

Drought Resistance

Avoiding Stress
(Maintaining Favourable Water balance)

Conserving Water (Water Savers)

(a) Early closure of stomata
(b) Increased Photosynthtic efficiency
(c) Low rates of cuticular transpiration
(d) Lipid desposites on leaves
e.g. Soybean; Sorghum
(e) Reduced leaf Area
(f) Morphology of Leaf surface
(Thick cuticle; Waxy surface; Spines etc)
(g) Effects of Awn : Awned varieties yield more in dryland. Awns contribute 12% of photosynthates to grain
(h) Water Storage in plants: The pineapple leaves contain substantial amounts of water in special nonchlorophyllous tissue which is utilised during drought.

Improving Water Uptake (Water Spenders)

(1) Efficient root system
(2) High root/top ratio
(3) Increased Plant Osmotic potential

Tolerating Stress

Mitigating Stress by maintaining high water potential

(a) Resistance to dehydration
(b) Prevention of cell collapse (leaf with thick cuticle)
(c) by maintaining higher osmotic pressure by accumulating higher amounts of solutes

High degree tolerance to

(a) Metabolic strain
(b) Plastic strain

Ima is much superior to Im in explaining the variability in crop yields from year to year.

Watershed Management :

Watershed means an area of land whose runoff drains to a common point i.e. in the river or waterpool. Watershed may be defined as a natural unit of land whose runoff collects & flows out of the area through a single common outlet into a river or other waterbody. It is a drainage basin which is demarcated by ridges or gullies. Watershed, drainage area or catchment area are generally interchangeably used meaning for the same. Ordinarily the catchment area (water collecting area) is larger but the area of watershed is comparatively smaller.Watershed area is the unit from the both angles i.e. land unit and water body unit,but there are separate units for both land and water in the catchment area.Therefore catchment area comprises of all the areas from which water flows out into a common river or waterpool. The different river's tributaries having different water catchment areas flow out into a common larger river. The larger drainage area is called river basin. The water catchment area of a tributary is called sub-catchment area and the different catchment areas may be identified in it for its certain parts. The catchment area of this certain part is called watershed area. There may be variations in the watershed area. When these variations are more in number it means the area is hilly, undulating or rolling and we get smaller watershed area whereas in the plain we get larger.The watershed for the agricultural purpose may be several fields of farmers that contribute runoff water to the flow at the lower point. The runoff move-

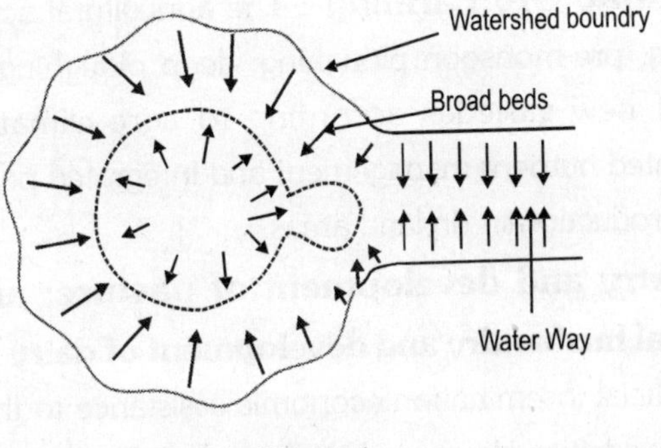

Fig. 5.1 : Watershed

ment depends on the land topography.Runoff from the upper fields enters into the lower fields in a watershed causes soil and water erosion.

Soil conservation measures in an isolated manner are not sufficient to check the erosion. Therefore, it is necessary that soil conservation measures should be adopted on the watershed basis instead of personal individual field. In the dryland areas, soil and water erosion, flood etc. are the major problems.Considering these points, the Govt. of India launched a watershed Management Programme in the seventh five year plan to develop dryland areas on the basis of watershed .For soil and water conservation measures,watershed is demarcated into sub watersheds & micro-watersheds.Micro-watersheds are the basis for planning and execution.In the beginning,there was a plan to develop 4000 watersheds in different agro-climatic regions.The Department of Agriculture provided the funds for this programme and the responsibility of evaluating 47 watersheds management programmes initiated in different states was given to ICAR.This programme was accelerated in 1983 when world bank gave the assistance in the development of four watersheds. Watershed management programme has following features—

(a) Soil & Water Conservation Measures : Soil and water conservation measures on the watershed basis include all those measures which are effective in preventing or delaying the movement of soil and rock particles.In it the emphasis is given on soil surveying,contour and graded bunding,repairing of old bunds and dams, land reclamation,digging of farm ponds etc.

(b) Scientific Dry Farming : The agricultural activities like contour ploughing, pre-monsoon ploughing, deep ploughing, mulching, development of new varieties according to agro-climatic zones,weed control,integrated nutrient management and integrated pest management improve the production in dryland areas.

(c) Forestry and development of pasture; and

(d) Animal husbandry and development of dairy : Proper animal husbandry,artificial insemination,economic assistance to the milk producing cooperative socities etc.are to be adopted.

Aims and Objectives of Watershed Management

1. To protect.conserve& improve the land resources for efficient and sustained production;

2. To protect and enhance water resource, moderate floods &reduce silting up of tanks,increase irrigation and conserving rainwater for crops and thus mitigate droughts.

3. To utilise the natural local resources for improving agriculture and allied occupation or industries so as to improve socio-economic conditions of the local residents.

Thus, watershed management is the planned use of watershed lands in accordance with pre-determined, objectives, such as the control of erosion, stream flow, sedimentation, and the improvement of vegetative cover and other related resources. Such watershed management integrate plans for soil conservation, soil improvement. irrigation drainage, flood prevention and development of water supplies for all purposes. In case of agricultural micro-watersheds, important aspects are soil conservation measures, soil improvement, storage and utilisation of runoff water. In dry farming areas, watershed approach provides an ideal means for integrated development. There is a need for multipronged approach to maximise crop production & also to ensure stability in rural income for integrated watershed management having following steps—

(i) Water harvesting

(ii) Location specific technology for crop production

(iii) Adoption of intercropping and crop substitution according to soil suitability

(iv) Afforestation on cultivable wastes and marginal lands

(v) Practice of dryland horticulture to minimise risk. Mango, guava, ber, pumegranate, sapota and tamarind have good potential on marginal land.

(vi) More efficient exploitation &use of ground water and irrigation

(vii) Development of dairying and pasture management

(viii) Stall fed goat and sheep farming needs to be popularized

(ix) Popularisation of sericulture and mulberry cultivation

(x) Processing of farm & horticultural produce

(xi) Sound extension and training programmes for the farmers

Water Harvesting/Runoff Concentration/Rainfall Precipitation: In humid climates, runoff usually occurs only when rain falls on a saturated soil. Under semi-arid conditions, the total annual precipitation may or may not be sufficient, but most of the rain falls in a short period causing humid climate conditions and resulting into runoff. In both conditions, the excess rain water should be safely guided and collected in nearby **ponds** for **recycling** in drier period for life saving irrigation.

Water harvesting is defined as the collecting and storing water on the surface of soil for subsequent use.It is a method to induce,collect,store and conserve local surface runoff for agriculture in arid and semiarid regions.Water harvesting was developed to a fine art about 2000 years by the Nabateans.The term 'water harvesting' was first time used by Myors.

Normally water harvesting is more effective in areas situated near hill side or in the areas where large portion is bare soil and cultivation is done on small portions. The technology of water harvesting differs according to soil types,topography,annual rainfall and the crops to be grown. In India three types of water harvesting techniques are generally practiced—

(i). Inter-row water harvesting : Practiced in arid areas with light soil where annual rainfall does not exceed 400-450 mm. The crop is sown in narrow strips between wide intervals that are ridged as artificial miniature watersheds. Later on these are compacted to increase runoff to the crop rows . It is more practical since no land is sacrificed for harvesting water.

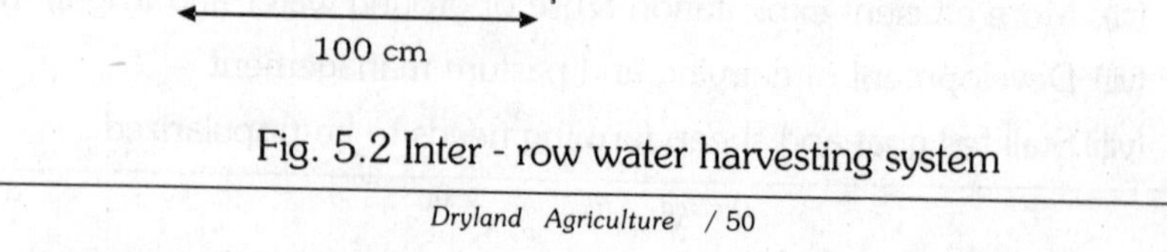

Fig. 5.2 Inter - row water harvesting system

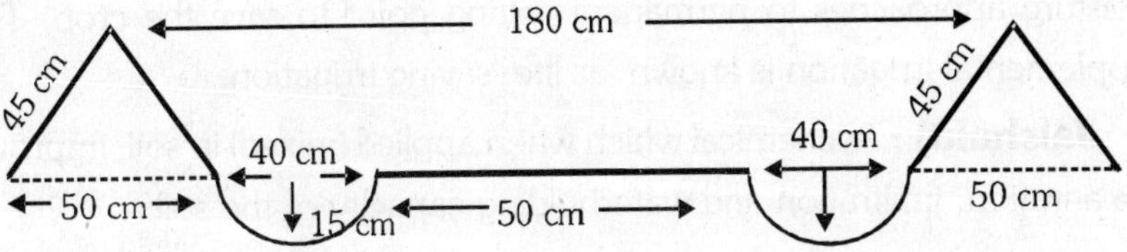

Fig. 5.3

Fig. 5.3 Modified inter - row water system

ii). Inter- plot or Microplot Water Harvesting : In this case water is harvested in the passages or furrows between the plots when rainfall is comparatively more. Runoff from the sloping area supplements rainfall for raising crop on level land.

(iii). Water harvesting in farm ponds & Reservoirs : Surface runoff from small watersheds is stored in farm ponds & reservoirs for utilization as supplemental or life saving irrigation. This harvested water can either be used as life saving irrigation to Kharif crops at the time of drought due to irregular monsoon or as supplemental irrigation to winter crops. Since the stored water is limited, steps should be taken to reduce losses from seepage and evaporation. Suitable lining materials for pond beds and anti- evaporants should be found out and used.

Runoff can be increased by either by land alteration or by chemical treatment . Impermeable catchments are prepared by spraying asphalt compounds. Thin plastic films or metal sheets can also be used to cover ground surface for water collection. Waterborne asphalt emulsions are much effective in controlling seepage loss. Bentonite has excellent sealing properties if kept continuously wet.

In the case of Broad Bed & Furrow system, excess water during rainy season is collected through the furrows and passed through grassed waterways into farm pond or reservoirs.

The stored water should be recycled most judiciously so as to get maximum return out of it. Only one or two irrigations are possible with the limited water stored. Therefore supplemental irrigation can be scheduled by two approaches viz. soil moisture depletion approach and critical stage approach. Irrespective of the crop stage, irrigation is scheduled when soil moisture approaches to permanent wilting point to save the crop. This supplemental irrigation is known as life- saving irrigation.

Jalshakti : A chemical which when applied (mixed) in soil, improves the aeration, infiltration and waterholding capacity of the soil.

Evapo- transpiration (ET) :

Soil moisture is the most limiting factor in dryland agriculture. About 60-75% of the rainfall is lost through evaporation. The evapo-transpiration is the evaporation from the soil surface and transpiration from the plant surface. Such losses can be reduced by (i) mulches (ii) antitranspirants (iii) wind breaks and (iv) weed control.

Antitranspirants Such materials are applied to transpiring plant surfaces to reduce water loss from the plant. Antitranspirants are of four types-

(a) Stomata Closing Type : Such antitranspirants reduce water loss through stomatal closing . e.g. phenyl mercuric acetate (PMA), herbicides like **Atrazine** in low concentrations, **ABA** and **CO_2**

(b) Film Forming Type : Retard moisture loss due to formation of thin film as physical barrier.e.g. Mobileaf, hexadecanol, silicone, oils, waxes.

(c) Reflecting Type : Such materials reflect the radiation and thus reduce leaf temp and vapour pressure gradient from leaf to atmosphere and ultimately transpiration is reduced e.g. celite (a diatomaceous earth product) and Kaolin. Application of 5% Kaolin spray reduces transpiration loss.

(d) Growth retardants : Such chemicals reduce shoot growth and increase root growth and thus enable the plants to resist drought . It also induces stomatal closure e.g. **cycocel (ccc).**

Antitranspirants generally reduce photosynthesis . Therefore, their use is limited to save the crop from death under severe moisture stress. They have some practical use in nurseries and horticultural crops.

Windbreaks and Shelter belts

Windbreaks are such structures which break the wind – flow and reduce wind speed while shelterbelts are rows of trees or shrubs planted for protection of crop against wind . They provide a protective shelter against

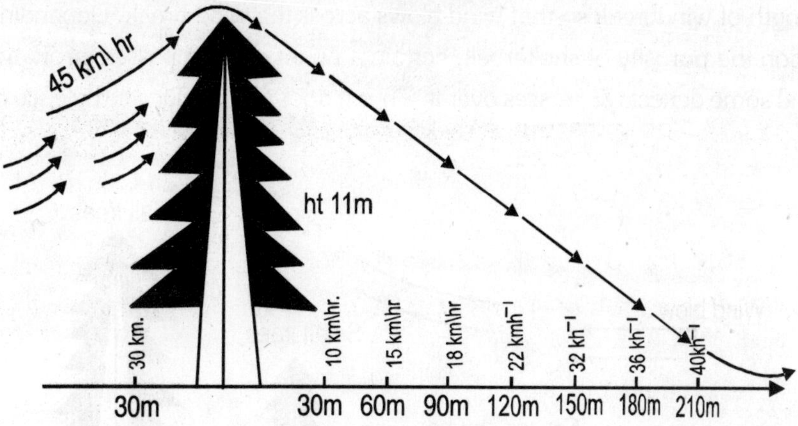

Fig. 5 : Planting of trees & shurbs as shelter belt

Wind and suitable habitat for birds and honeybees as well as produce cattle feed and fuel wood. In the wind erosion areas, windbreaks & shelterbelts are to be planted. They prevent soil erosion, arrest the march of desert and protect the agricultural and residential lands from dust- storms. A rows of trees & shrubs planted across the wind – direction is the most effective. It reduces the wind speed upto 60-80% on leewide side . The height of tall tree and length of windbreak determine the extent of protection provided to soil. Windbreak & shelterbelt provides the protective shelter against dessicating winds to the extent of 5-10 times the height of tall tree on windward side and upto 30 times on leeward side. For example, a 10-11m tall windbreak when encountered by 45-50 km/hr. wind, it reduces on windward side to 20-30 km/hr and to 10 km per hour on just leeward side which is illustrated by the figure 5.4.

The wind–break reduces the windspeed on the leeward side at 200m away upto 20% only and in the area as at 300m away, there is no effect at all. Therefore at a interval of 300m, such windbreaks should be repeated.

To control weed erosion, the capacity of shelterbelt depends upon the speed and direction of wind. In case of high windspeed, the protective area is reduced and in such areas, the interval between two shelterbelts is to be reduced. To counter wind's direction, it is necessary to be long length of windbreak so that wind blows across the wind break. Depending upon the porosity of shelterbelt, certain amount of wind passes through it and some deflects & crosses over it. Thus it does not produce turbulence of air.

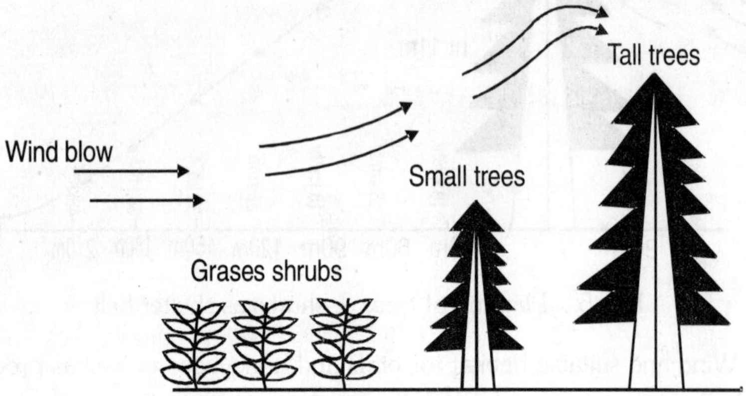

Fig. 5 : Planting of trees & shurbs as shelter belt

Reduction in wind speed, reduces the evaporation losses and makes available more water to crops. According to planting pattern of trees & shrubs as shelterbelt, the grasses & shrubs are planted on the outer rows which train the wind to rise much above the ground surface. The inner rows are of small trees and the outer rows are planted by tall trees which further raise the wind level.

The beneficial effects of shelterbelts are more clearly seen in drought areas. In such areas windbreaks of 3-7 rows and 15-30m wide are more effective . It modifies the micro- climate favourable for crop production and shelters for birds, honey bees & pet animals.

❑❑❑

6 Erosion and Soil Conservation

Soil erosion is the process of the removal of soil particles from the parent body and transportation of such particles by wind and/or water.

Runoff :

That portion of the precipitation which is not absorbed by the soil but finds its way into the streams after meeting the persistent demands of evapo-transpiration including interception and other losses. In another words, runoff is the excess water from precipitation that moves out of field and finds its way to river, lakes and oceans etc.

Runoff = Rainfall - [**infiltration** + Interception + Surface Storage + Surface detention]

Factors Affecting Runoff :

(A) Climatic Factors

(i) Rainfall characteristics : Amount, Intensity, Duration and distribution of rainfall.

(ii) Snowfall

(iii) Evapo-transpiration.

(B) Physiographic Factors :

(i) Watershed Characteristics : Topography of soil , Soil types, Orientation, Geology, Vegetal Cover, shape and size of watershed.

(ii) Channel Characteristics : size, slope , cross- section and roughness of channel bed.

(C) Geological Factors :

Lithological, structural and hydrological characteristics of the aquifers, permeability, porosity, transmissivity etc.

Rainfall, type of vegetation and soils are the important factors that influence erosion. High intensity rainfall of long duration causes severe erosion. Rainfall intensity of more than 5cm/hr is considered as severe. Due to presence of vegetation on the soil, there is no break down of soil aggregates.

Soil Loss Equation due to Water Erosion :

Based on the mechanism and factors influencing soil erosion, an universal soil loss equation was developed .It is useful for predicting soil loss due to water erosion.

A = RKLSCP

Where A → Predicted soil loss (t/acre/year)

R → Rainfall and runoff factor

K → Soil erodibility

L → Slope Length

S → Slope gradient or steepness

C → Soil cover and management

P → Erosion Control Practice.

Active soil erosion by water and wind is prevalent over 140 million hectares resulting in the loss of 6000 million tonnes of fertile soil containing about 5.53 million tonnes of N.P.K.

Soil Loss Equation Due to Wind Erosion :

E = IRKFCWDB

Where E → Soil loss by wind erosion

I → Soil cloddines factor

R → Surface cover factor

K → Surface roughness factor

F → Soil textural class factor

C → Local wind factor

W → Field width factor

D → Wind Direction Factor

B → Wind Barrier factor

Wind erosion control measures should aim at reducing wind velocity or altering soil characteristics.

Soil Conservation Measures :

Soil conservation is the preservation of soil against deterioration and loss by using it within its capabilities, and applying the conservation practices needed for its protection and improvement.More specially,soil conservation consists of using the land within the limits of economic practicability while safeguarding it against impoverishment or depletion by erosion, deposition,exhaustion of plant nutrients(through leaching excessive cropping or overgrazing).accumulation of toxic salts, burning,water logging (inadequate drainage), improper cultivation or any type of improper use or failure to protect the land from soil or impairment of productiveness.The fundamental principles of soil conservation are—

(i) Land use based on its capability

(ii) Conservation of soil and moisture to avoid damage to the soil, and

(iii) Use of best soil crop management practices, correction of acidity, alkalinity and drainage etc.

Land Capability Classification :

The first requisite to conservation of land is to fit the crop to the capabilities of the soil and the water availability.Lands are mainly used for agriculture,pastures and forestry. The capability of land to grow crops depends on the nature and properties of soils. On the basis of capability or limitations,the lands are grouped into two major groups viz—

(A) Lands suitable for cultivation : Such lands are used for agriculture or cultivation of crops.Such land has four classes.These four classes are differentiated on the basis of limitations.These limitations are (i) Erosion :may be water erosion or wind erosion denoted by 'e'.(ii) Climate by (c) (iii) Water characteristics by (w) due to excess water,drainage problem (iv) Soil by (s) : low water holding capacity or low plant nutrient content of soil.

(B) Lands not suitable for cultivation : Such lands has also four classes which all are not capable of supporting cultivation of crops.These lands are used for growing grasses,forestry and supporting wild life.

Thus lands are together grouped into eight classes by U.S. soil conservation service.The first four classes viz class I,II,III, and IV are used for cultivation and the classes from V to VIII are not used for cultivation.But for the regular cultivation,only first three classes viz I,II,&III are used

Class I : No limitations hence well suited for intensive crop cultivation, have permanent irrigation system, well drained level lands with high water holding capacity. Such lands need only crop management practices to maintain their productivity like use of fertilizers,manures, crop rotation etc. Example of class I are alluvial soils of Indo-Gangetic plains.

Class II : Moderate limitations which reduce choice of crops,require moderate conservation practices. Limiting factors are : (a) gentle slope (b) moderate erosion hazard (c) inadequate soil depth (d) less than ideal soil structure and workability (e) slight to moderate alkali or saline conditions and (f) somewhat restricted drainage. Example of class II soils are deep red soils and black soils.Management practices are strip cropping,contour tillage,rotation involving grasses and legumes and grassed waterways.

Class III : Severe limitations requiring special conservation measures, limitations are moderately steep slope, high erosion hazard, very slow water permeability, shallow depth and resticted root zone, low water holding capacity, low fertility, moderate alkali and salinity and/or unstable soil structure. Example : Shallow red soils,slightly saline black soils. Conservation measures require management practices mentioned for Class II having higher proportion of grasses or legumes in crop rotation.Tile and other drainage may also be needed.

Class IV : Very severe limitations on choice of crops;suitable for occasional cultivation,best use for pasture/hay; extensive use of close growing crops; such land requires very careful management e.g.shallow soils, saline soils, alkaline soils.

Class V : Limitations are (i) interference from stream flow (ii) short growing season (iii) stony or rocky soils (iv) Ponded areas where drainage is not possible. Pastures can be developed on these soils, used for grazing and forestry e.g. Arid soils, rocky soils, uneven or rolling soils which are not suited for cultivation.

Jatinder Garthi 0161-5090413 99153-12527

ClassVI : Moderate limitations on use for grazing/forestry.

Class VII : Severe limitations which restrict their use for grazing, wood land or wildlife.

Class VIII : Extremely rough land, not suitable for any kind of crop production, its use is restricted to recreation, wildlife, aesthetic purpose and watershed protection. e.g. sandy beaches, river wash etc.

Conservastion Measures for agl. Lands are grouped under two heads viz.-

(a) Crop management according to the capability of land and moisture.

(b) Agronomic and Engineering Measures.

Factors influencing measures are (i) soil (ii) land slope and (iii) rainfall characteristics of the area.

Measures :

(A) Agronomic Measures : are adopted where slope is less than 2% and erosion problems are not severe. These measures help to –

(i) intercept raindrops and reduce the splash effect.

(ii) to obtain a better intake of water rate by improving the content of organic matter and soil structures.

(iii) to retard and reduce the over land runoff through the use of-

(a) Contour Cultivation : Cultivation of crops along the contours of a slope. It has following advantages -

(i) conserve soil and water (ii) conserve soil fertility

(iii) increase crop yield (iv) much less power required

(v) less wear and tear of implements and less time is required.

Disadvantage: The establishment of contour farming on undulating land is tedious.

(b) Mulching : Mulch is any material applied on the soil surface to check evaporation and improve soil water. It is defined as a natural or artificially applied layer of plant residues or other materials on the surface of the soil. Mulches are used for various reasons but water conservation and erosion control are the most important for agriculture in dry regions. e.g. crop residues leaves manures, straw, plastic films etc.

Effects of Mulching :

(i) Mulching affects soil water through (a) runoff control (b) Increased infiltration (c) Decreased evaporation (d) weed control.

(ii) affects soil temp . through–
(a) radiation shielding (b) Heat conduction and trapping
(c) evaporative cooling.

(iii) improves soil nutrients status through. (a) organic matter addition (b) differential nitrification (c) mineral solubility.

(iv) improves soil structure.

(v) affects soil biological regime through (a) O.M. additions (b) microbial and soil fauna populations (c) plant root distributions.

(vi) affects soil erodibility.

(vii) affects soil salinity through–
(a) Leaching and (b) Evaporation control. Mulches by reducing evaporation, reduce subsequent return of the salts to the leached zone.

Types of Mulches :

(i) Soil mulch or dust mulch : If the soil surface is loosened, it acts as a mulch for reducing evaporation. It is called soil mulch or dust mulch. Intercultivation creates soil mulch in growing crops. Its use fullness is doubtful in alfisols but helps in closing deep cracks in vertisols.

(ii) Stubble mulch : Crop residues like wheat straw or cotton stalks etc. are left on the soil surface as a stubble mulch.

(iii) Straw mulch : Straw is used as a mulching material.

(iv) Plastic mulch : Plastic materials like polyethylene, polyvinyl chlo ride are used as mulching materials.

(v)Vertical mulching : Subsoiling is probably the most effective method of breaking hardpans to improve root penetration, aeration and water percolation.To prolong the beneficial effect of subsoiling,a method called vertical mulching has been developed.The object of vertical mulching is to fill slots with O.M. and keeping them open and functional for a longer period.In black soils,to improve infiltration and storage of rain water,vertical mulches are formed. It consits of digging narrow trenches across the slope at intervals and placing the straw or crop residues in these trenches. This is mostly practiced in coffee gardens. The bepruned plant material is placed in contour trenches formed between rows of coffee or in trenches around the plants in concentric each year in one circle.

(c) **Dense growing crops :** provide maximum cover to soil e.g. cowpea & moong. Growing of Bidi tobacco causes maximum loss of soil and water. Therefore a cover- cum – manure crop is grown during the early monsoon before tobacco transplanting. Intercropping of Maize + Arhar/ Urd, does not reduce water and soil loss .

(d) **Strip Cropping** : In strip cropping , two or more than two crops are grown simultaneously in strips wide enough to permit independent cultivation but narrow enough for the crops to interact agronomically . Strip cropping is essentially another form of rotation. It may be defined as the process of growing the series of alternate strips of various types of crops laid out so that all tillage and management practices are performed across the slope or on the contour. It controls the runoff erosion and maintains the soil fertility. It is of four types-.

(i) **Contour Strip Cropping :** Growing of soil- exposing and erosion – permitting crops in strips of suitable widths across the slopes on contour, alternating with strip of soil – protecting and erosion – resisting crops. Such cropping shortens the length of slope, checks the runoff, helps to desilt it and increases water absorption. Here care of contour is of utmost importance.

(ii) **Field- strip Cropping :** Planting of farm crops in more or less parallel strips across fairly uniform slopes but not on exact contours. Here care of land slope is taken . It is useful for soils of regular slope and of high infiltration rates.

(iii) **Wind Srtip Cropping :** Planting of tall growing crops such as jowar, maize etc and low growing crops in alternately arranged straight and long, but relatively narrow, parallel strips laid out right across the direction of the prevailing wind, regardless of contour. Here wind direction is cared. Objective is to control wind erosion rather than water erosion.

(iv) **Permanent or Temporary Buffer Strip Cropping :** To take care of critical i.e. steep or highly eroded, slopes in fields under contour strip – cropping.It does not from part of crop rotation unlike others. Strips of perennial legumes, grasses or shrubs on a permanent or temporary basis .Buffer strips are counter strips of grass or other erosion resisting vegetation between or below cultivated strips or fields. The width of the strips of erosion resisting crops and erosion permitting crops depends upon slope, soil texture, type of crops and rainfall characteristics. Strip cropping is not adopted in India on a large scale because of the small size holdings.

(e) Cropping systems like crop rotations, strip cropping, inter-cropping and crop mixtures.

(B) Mechanical Measures are adopted to supplement the agronomical practices and when land slope is more than 2%.

Objectives :

(i) to increase the time of concentration by intercepting the runoff and thereby providing an opportunity for the infiltration of water.

(ii) to divide a long slope into several short ones so as to reduce the velocity of runoff and thus prevent erosion.

(a) Basin – Lisitng : Making of small interrupted basin along the contour; effective on retentive soils having mild slopes.

(b) Sub-soiling : Breaking of hardpan by the sub soiler at 30-60 cm depth and 90-180 cm interval.

(c) Contour bunding :making a comparatively narrow based embankment at intervals across the land slope on a level that is along the contour,Contour bunding is adopted in arid and semi arid areas with high infiltration and permeability and slope of about 6%. Spacing between bunds should not exceed 150 cm vertical drop or 67.5 cm horizontal spacing.

(d) **Graded bunding/channel terraces :** recommended where rain water is not readily absorbed either due to high rainfall or low intake of the soil e.g. in areas receiving rainfall of more than 800 mm per year irrespective of soil texture but in clay soil even less than 800 mm/year rainfall. There are two types of Graded bunding viz. Narrow based and Broad based. Graded bunds are spaced at the same intervals as contour bunds. Broad based bunding is recommended where farming is practiced with tractors.

In general bunding is suitable for lands having slopes from 2-10%. Contour and Graded bunding are extensively used in India.

(e) **Broad bed and furrow (BBF) system :** suitable for managing rainwater in black soils where surface drainage during the monsoon period is a problem.

(f) **Bench terracing :** Usually practiced on slope ranging from 16 to 33% on steep sloping and undulating land,intensive farming is possible only with bench –terracing.The vertical drop may vary from 60 to 180 cm depending on the slope and soil conditions.

(g) **Zing terracing :** adopted in lands with 3 to 10% slopes.Zing terraces are constructed in medium to deep soils in moderate to high rainfall areas.It is adopted to cut cown from upper area for the benefit of crops grown in lower side and to ensure– adequate drainage during periods of heavy rainfall.

Conservation Measures for Hilly Slopes :

(a) **Contour Trenching and Afforestation :** Trenches are made along the contour or along a uniform level, used both on hill slopes and on degraded and bare waste lands for soil and moisture conservation and afforestation purposes. These trenches break the slope lengths, reduce the velocity of surface runoff and consequently retard its scouring action and carrying capacity. The size of trenches varies with slope, rainfall and depth of soil available. The trenches are usually of 60 cm to 48 cm in size. The spacing varies from 10 to 30 m. The trenches are half refilled diagonally with excavated material and remaining half of the soil forms the spoil bank. The water retained in the trenches help in conserving the moisture and provide advantageous sites for sowing and planting. Such trenches are not

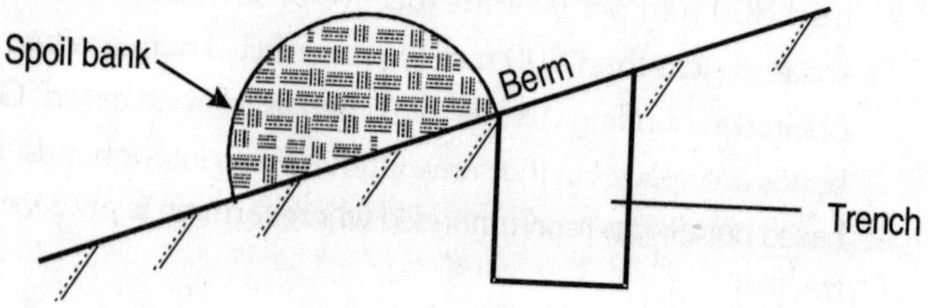

Fig. 6.1 : Cross- Section of Contour trench

advisable when slope is more than 20% either technically or economically. It is used in both high rainfall and low rainfall conditions,varying soil types and depths,for soil and water conservation and afforestation.

Protection of the trenched area, till it is fully covered by vegetation,is essential to achieve the desired results.

(b) Bench Terracing :Consists of construction of series of plateforms along contours cut into hill slope in a step like formation. These plateforms are separated at regular intervals by vertical drops and protected by vegetation and sometimes by packed stone.Bench terraces convert the long intercepted slope into several small strips and make protected plateforms available for farming.It is adopted for converting sloping lands to irrigated fields or for orchard plantations.

Bench terracing is the suitable and common measure in hiily areas for the conservation of soil and water. The crop is cultivated in hilly areas like Himachal Pradesh,Jammu & Kashmir, Uttarakhand, Assam and other eastern states of India by adopting bench terracing measure. Bench terrace reduces the slope length and gradient and is easily adopted upto 15% slope but may be used upto 33% slope. Bench terrace is possible only where depth of soil is more so that cutting and filling of land work can be performed.

Types of Bench Terraces-

1. According to purpose :

(i) Hill-type bench terraces :used for hilly areas with a reverse grade towards the hill.

(ii) Irrigated bench terraces : Level benches are adopted under irrigated conditions and where slope is upto 8%.

(iii) Orchard bench terraces :Narrow width terraces of about 1 m for individual trees are adopted.

The conservation of land into bench terraces over a period of time is reefed to as gradual bench terracing.

2. Acording to slope of benches and rainfall conditions

(i) **Bench terraces sloping outwards —** Used in low rainfall areas (<750 mm) with permeable soil of medium depth.

Bench terraces with narrow width (1m) are constructed for orchard plantations called orchard bench terracing.

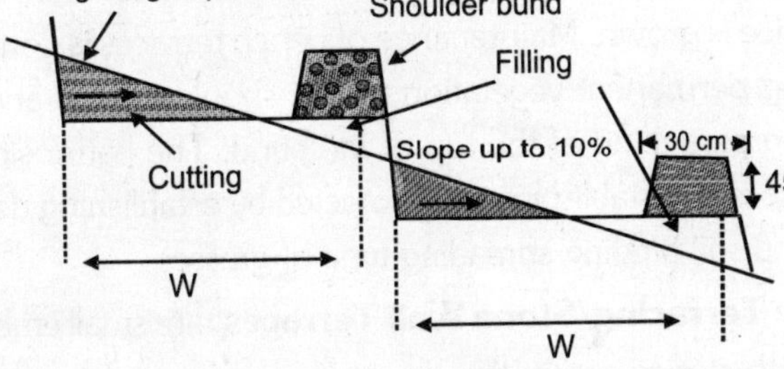

Fig. 6.2 : Slope outward

(ii) **Bench terraces sloping Inwards —** It is also called hill type bench terraces; adopted in high rainfall areas (>750 mm annual) where a major portion of the rainfall is drained as surface

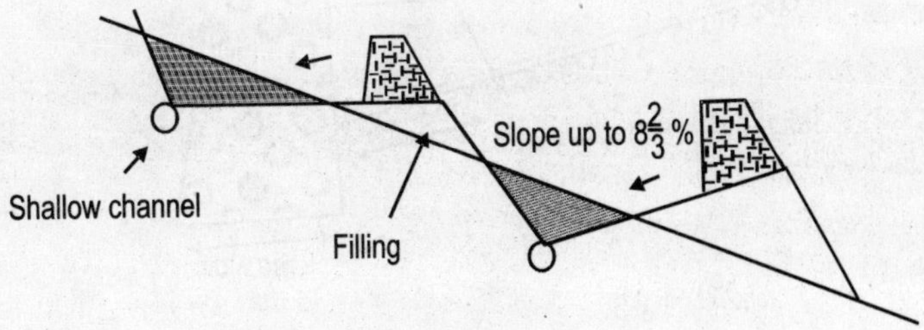

Fig. 6.3 : Slope inward

run off. Such soils have poor infiltration rate. A suitable drain at the inward end of each of these terraces is provided to drain the runoff. It is also called Hill type Bench terrace. Such terracing is followed in H.P; Uttrarakhand, J &K and Nilgiri Hills where cultivation of Maize, Tomato and Potato is generally practiced.

(iii) Bench terraces with level and table top - For medium rainfall areas, evenly distributed and having deep and

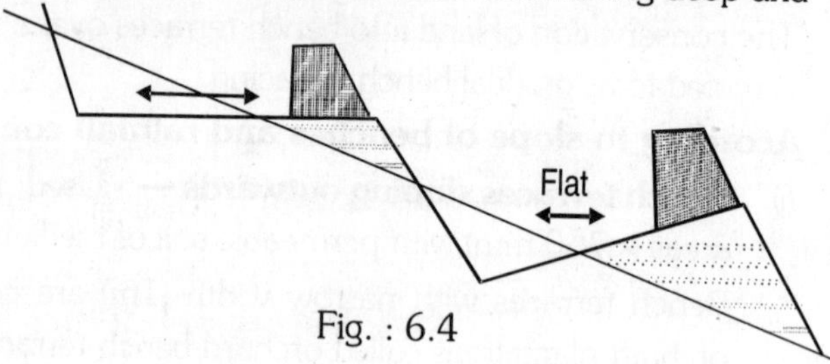

Fig. : 6.4

highly permeable soils. It is also used where irrigation facilities are available called irrigation bench terraces. Here generally Rice is grown. Maintenance of bench terraces is done by planting permanent vegetation on the shoulder bund and by avoid ing ploughing of the top of the bund. The batter slope of ter races be established and protected by establishing deep rooted and soil binding spreading type of grasses.

(C) Stone Terracing/Stone Wall Terraces : are small embankments constructed with stones across the slopes. It is adopted on any slope with spacing from 10 to 30 m depending upon slope.

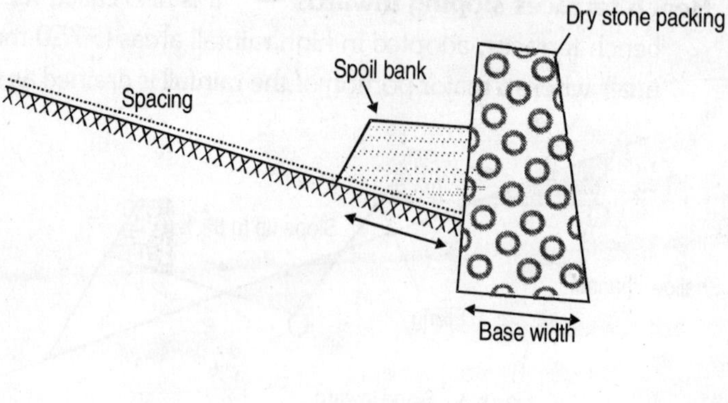

Fig. 6.5

❐❐❐

7 Classification of Crops

(A)Taxonomic Classification of Crops :

1. Poaceae (Graminae) or Grass family : Cereals, **Sugarcane, Napier, Paragrass.**
2. Papilionaceae (Leguminosae) or Pea family : Pulses, Legumes, **Groundnut, Sunnhemp.**
3. Cruciferae/Mustard Family :Mustard, Radish, Cabbage, Cauliflower, Knolkhol.
4. Cucurbitaceae/Gourd family : All gourds (e.g. bottle gourd, bitter-gourd), Cucumber, Pumpkin.
5. Malvaceae/Cotton Family **: Cotton, Bhindi, Roselle.**
6. Solanaceae/Brinjal family : Brinjal, Potato, Tomato, tobacco, Chillies.
7. Tiliaceac **: Jute, Phalsa.**
8. Asteraceae (Compositeae) : Sunflower, Safflower, Niger.
9. Chenopodiaceae : Spinach, beet Sugarbeet.
10. Pedaliaceae : **Seasame**(Sesamum)
11. Euphorbiaceae :Castor,tapioca.
12. Convolvulaceae :Sweetpotato
13. Umbelliferae : Coriander, Cumin, Carrot
14. Aliaceae **: Onion,** Garlic
15. Zingiberaceae : Ginger, turmeric.

(B) Special Classification :

1. **Arable Crops :** Which require preparatory tillage e.g. potato, tobacco, rice, sugarcane, maize etc.

2. **Alley Crops :** Such arable crops which are grown in alleys/passages formed by trees or shrubs,established mainly to hasten soil fertility restoration, enhance soil productivity and reduce soil erosion Slight shade tolerance and non-trailing habit are the pre requisites of Alley crops e.g. Arable crops like sweet potato, urd, turmeric & ginger are grown in the passages formed by the rows of Eucalyptus, subabool and cassia.
3. **Augmenting Crops :**Such crops are sown to supplement the yield of the main crops e.g. Japanese mustard with Berseem, Chineese Cabbage with mustard. Here Japanese mustard and Chinese cabbage help in getting higher yield in the first cutting.
4. **Avenue crops :**Such crops are grown along farm road and fences e.g.Arhar, Glyricidia, Sisal etc.
5. **Border crops/Barrier/Guard Crops :**Such crops help to protect another crops from trespassing of animals or restrict the speed of wind and are mainly grown as border e.g. safflower (thorny oilseed crops) is planted around the field of chickpea.
6. **Brake crops :**
   - to break the continuity of agro-ecological situation of the field under multiple cropping systems.
   - to reduce the inoculum of soil-borne harmful biotic agents such as weeds,pest and improves soil condition for crop growth e.g. Legume in rice-wheat system.
   - are also used to designate guard crops — particularly those which help to break(retard)the wind speed and protect crops from wind hazards.
7. **Cash crops :** Such crops are grown for sale to earn hard cash e.g. Jute, Cotton, Tobacco, Sugarcane.
8. **Catch/contingent/Emergency crops :** Such crops are cultivated to catch the forthcoming season when main crop is failed.They are of very short duration,quick growing,fast bulking,harvestable or usable at any time e.g. greengram,urd,cowpea,onion radish etc.
9. **Cleaning crop :** Whose agronomical practices makes the field clean e.g. potato,maize etc.

10. **Cole crops :** 'Cole' is derived from colewart.Colewart is the ancestor of wild cabbage.Therefore,cole crops are essentially cold weather crops belonging to the cruciferae capable for withstanding considerable frost e.g. cabbage,cauliflower and Brussels sprouts.

11. **Contour Crops :**Grown on or along the contour lines to protect the land from erosion e.g. marvel grass etc.

12. **Commercial crops :**Such crops are grown to earn money e.g. Jute, Cotton, tobacco, Sugarcane etc.Cash crops are certainly com-mercial crops.

13. **Cover crops :** Able to protect the soil surface from erosion through their ground covering foliage and/or root mats e.g. lobia, groundnut, urd, paragrass, sweet potato.

14. **Complementary crops :** Each other crop is benefited in intercropping e.g.Jowar+Lobia.Jowar receives nitrogen from lobia and lobia requires support from jowar.

15. **Competitive crops :** Such crops compete to each other and are not suitable for intercropping e.g.two cereals.

16. **Supplementary crops :** Such crops are neither competitive nor complementary e.g.maize+cucurbits.

17. **Exhaustive crops :** Such crops leave the field exhaustive after growing e.g. cereals.

18. **Energy crops :** To obtain liquid energy such as ethanol and alcohol e.g.sugarcane, potato, maize, topioca.

19. **Fouling crops :** Whose cultural practices allow the infestation of weeds intensively e.g. direct seeded upland rice.

20. **Ley crops :** Any crop or combination of crops is grown for graz-ing or harvesting for immediate or future feeding to livestock e.g. Berseem+mustard. Such cropping is called ley cropping.

21. **Mulch crops :** to conserve soil moisture,such crops are grown e.g. cowpea.

22. **Nurse crops :**Such crops help in the nourishment of other crops by providing shade and acting as climbing sticks e.g. rai in peas, jowar in cowpea.

23. **Paira/Utera Crops :** Growing of such crops shown a few days or weeks before harvesting of standing mature crops is called paira/utera cropping and the sown crop is called paira/utera crop e.g. Lathyrus in rice, paira cropping in succession may constitute relay cropping.

24. **Paired row cropping :** Each third row is removed or growing of crops in paired row is called paired row cropping.It is suitable for dryland and objective is to conserve soil moisture.

25. **Restorative crops :** Such crops provide a good harvest along with enrichment or restoration or amelioration of soil e.g. Legumes.

26. **Riparian crops :** Grown along irrigation and drainage channels or waterbodies e.g. waterbind weed (kalmi sak),para grass. They help to protect the soil from erosion.

27. **Skip Cropping** : A line is left unsown in the regular row series of sowing is called skip cropping.

28. **Silage crops :** Such crops are grown to preserve in pits in a succulent condition by a process of natural fermentation or accidi fication for feeding livestock during lean months or offseason e.g. cowpea, jowar etc.

29. **Smother crops :** Able to smother (suppress)the population and growth of weeds by providing dense foliage and quick growing ability e.g.cowpea, mustard.

30. **Soiling crops :** Grown to harvest while they are still green and fed fresh to livestock in stalls e.g. burseem, napier etc.

31. **Trap/Decoy Crops :** Grown to trap insect-pests and soil-borne harmful biotic agents such as parasitic weeds e.g. cotton red bug is trapped by growing Bhindi around the cotton and **orobanche (weed)** is trapped by **solanaceous plants and striga by sorghum.**

32. **Truck crops :** Grown to market fresh e.g. Bhindi,spinach.

33. **Ware Crops :** Such crops are grown for temporary storing as intact in warehouse for future use or sale e.g. potato.

❑❑❑

8 Crop production

Kharif Crops : Paddy, **maize**, jowar, bajra, **Groundnut, Til, Cotton**, Sunhemps, Moong, Urd, **Soyabean.**

Rabi Crops : Wheat, Barley, Gram. Mustard, **Sugarcane, Tobacco,** Berseem.

Crops and Botanical Name :

Cereals		**Graminae**
Rice	:	***Oryza sativa***
Wheat	:	***Triticum aestivum***
Maize	:	***Zea mays***
Sorghum/Jowar	:	***Sorghum bicolor***
Bajra	:	***Pennisetum typhoides*** (Peart millet)
Barley	:	***Hordeum vulgare***
Madua/Ragi/Finger Millet	:	***Eleusine coracana*** (good for diabetes Patient)
Cheena/Proso millet	:	***Panicum miliaceaum***
Sawan/Barnyard millet	:	***Echinochloa frumentacea***
Kodo	:	***Paspulum scrobiculatum*** (Coarsest of all foodgrain)
Kakun/Italian/German Millet/Foxtail	:	***Setaria italica***
Pulse Crops		**Leguminosae**
Gram/Chickpea/Bengal Gram	:	***Cicer arietinum***
Lentil	:	***Lens esculenta***
Pea	:	***Pisum*** *spp.*
Arhar/Redgram/Pigeon		

Pea	:	***Cajanus cajan***
Greengram/Moong	:	***Vigna radiata***
Blackgram/Urd	:	***Vigna mungo***
Cowpea/Lobia	:	***Vigna sinensis***
Soybean	:	***Glycine max***

Oilseed Crops :

1.Groundnut/Peanut/Earth Nut/Monkeynut/ Manillanut	:	***Arachis hypogea*** (Leguminosae)
2.Sesamum/Til	:	***Sesamum indicum*** (Pedaliaceae)
3.Castor	:	***Ricinus communis*** (Euphorbiaceae)
4.Rapeseed &Mustard	:	***Brassica spp.*** (crucifereae)
5. Linseed/Flax	:	***Linum usitatissimum*** (Linaceae)
6.Safflower	:	***Carthamus tinctorious*** (Compositae)
7.Sunflower	:	***Helianthus annus***

Fibre Crops :

1.Cotton	:	***Gossypium spp.*(malvaceae)**
2.Jute	:	***Corchorus capsularis*** **(White Jute)** } Tiliaceae
		Corchorus olitorius **(tossa)** }
3.Sunhemp/Banaras Hemp/ Sann-Hemp/Bombay Hemp/IndianHemp/ Warangal Hemp	:	***Crotalaria juncea*** (Leguminosae)
4.Mesta (Bimli in Andhra)	:	Andhra Pradesh is the leading state in both area & production.
		Hibiscus spp. — *sabdariffa* / *cannabinus*

Forage :

1.Oat	:	***Avena sativa*** (graminae)
2.Napier/Elephant grass	:	***Pennisetum purpureum*** (graminae)
3.Berseem/Egyptian Clover	:	**Trifolium alexandrium** (Legumi nosae)

4. Lucerne/Alfalfa	: ***Medicago sativa***	Legu-minosae
5. Guar	: ***Cyamopsis tetragonoloba***	

Sugar Crops :

1. Sugarcane : ***Saccharum officinarum*** (graminae)
2. Sugarbeet : ***Beta vulgaris*** (Chenopodiaceae)

Miscellaneous :

1. Potato	: ***Solanum tuberosum***	Solanaceae
2. Tobacco	: ***Nicotiana tabacum***	

Root and Tuber Crops (Starchy) :

Five crops on global basis —

1. Potato : ***Solanum tuberosum***
2. Sweet Potato : ***Ipomea batatas***
3. Cassava/Tapioca : ***Manihot utilissima***
4. Yams : ***Dioscorea spp.***
5. Cocoyam :
   - Taro **: Colocasia esculenta**
   - Tannia **: Xanthosoma sagittifolia**

In addition to above five crops, three more starchy crops are grown in India—

6. Elephant foot yam/ (Zimikand in Hindi) : ***Amorphophallus campanulatus***
7. Chinese Potato(Coorka): ***Coleus parviflorus***
8. Arrow root/Koova/ ararut : ***Marantha arundinaceae***

Origin :

Cereals :

1. Rice : Indo-Burma(Indo-Myanmar)
2. Wheat : South West Asia
3. **Maize** : Mexico
4. Jowar : Ethiopia&Sudan(Africa)
5. Bajra : Africa
6. Barley : Abyssinia(Ethiopia)

Pulses :

1. Gram	: South West Asia
2. Field Pea	: Mediterranean region
3. Arhar	: Africa(Nile river&Angola)
4. Mung/Urd	: India(Central Asia)
5. Cowpea	: Central Africa
6. **Soybean**	: China

Oilseeds :

1.Groundnut	: Brazil
2. Sesame	: South West Africa
3. Castor	: Ethiopia
4. Rai	: China
5. Rapeseed	: Afghanistan, Pakistan & India
6. **Sunflower**	: Mexico&South USA(Central America)

Others :

1. Cotton	: India
2. Corchorus olitorius	: Africa
Corchorus capsularis	: Indo-Burma(Indo-Myanmar)
3. Tropical Cane	: Oceania(New Guinea)
Indian Cane (Saccharum spontaneum)(Kans)	: North Eastern India
4. Sugarbeet	: Mediterranean region
5. **Potato**	: South America (Peru)
6. **Tobacco**	: Mexico &Central America

Protein Content in Cereals :

Carbohydrate	: Around 70% in all cereals
Rice	: 6-7%
Wheat	: 11-12%
Maize	: 10%
Sorghum	: 10-12%
Bajra	: 11-12%
Barley	: 11.5%(albuminoids)

Pulses (Carbohydrate : Around 60%) :

Gram	:	21.1%
Lentil	:	25%
Pea	:	22.5%
Arhar	:	21-25%
Mung	:	25%
Urd	:	24%
Cowpea(Vegetable meat)	:	23.4%

Oilseeds :

Soybean	:	42%
Groundnut	:	26%
Sesame	:	18-20%
Linseed	:	36%
Safflower	:	40-45%(oilcakes)

Area &Production :

1. Rice	:	West Bengal(both area &production)
2. Wheat	:	Punjab (both area & production)
3. Bajra	:	Gujarat (both area &production)
4. Maize	:	U.P.(both area &production)
Summer Maize	:	Bihar
5.Total Pulses	:	M.P. (both area & production)
6.Total oilseeds	:	M.P.(both area &production)
7. Gram	:	M.P.(both area &production)
8. Arhar	:	Maharashtra(both area & production)
9. Groundnut	:	Andhra Pradesh (both area &production)
10. Rape &Mustard	:	Rajasthan(both area &production)
11. Soybean	:	M.P. (Nearly 70% of production)
12.Sugarcane	:	U.P.
13. Cotton — Area	:	Maharashtra
Production	:	Punjab
14.Tobacco	:	Gujarat(70% of total area &80% of total production)

- Irrigated Wheat : Harayana (Max. area)
- Foodgrains = (Cereals +Pulses) only
- Production of Cereals in descending order in India.
  (1) Rice (2) Wheat (3) Maize (4) Bajra (5) Jowar
- Area of cereals in India.
  (1) Rice (2) Wheat (3) Bajra (4) Jowar (5) Maize
- Production of Cereals (world)
  (1) Wheat (2) Maize (3) Rice (dehusked) (4) Jowar
- Oilseeds in India Production wise
  (1)Groundnut (2) Rape & Mustard (3) Soyabean (4) Sunflower (5) Sesame (6) Niger (Edible) (7) Castor(non-edible) (8) Linseed (9) Safflower (Maharashtra)
- Kharif crops : Generally short day plant.
- Rabi crops : Generally Long day plant

Rice (2n=24)

3 types of season for rice cultivation in India—

1. **Aus/Autumn** :Word 'aus' derived from Sanskrit 'ashu'meaning quick or early.Rice varieties grown during pre-monsoon period is called aus rice and harvested in August-September hence called Autumn rice due to harvest time.
2. **Aman/Kharif/Winter :** 'Aman' derived from Arabic words means safety which indicates stability of crop (Aman-chain). Aman rice is also called winter rice due to harvesting time in winter.
3. **Boro/Summer** :Boro refers to rice grown in submerged land lower in elevation,grown during jan-feb to Apl-May,also called summer crop because Boro rice is harvested in summer season.

- Botany = Oryza has 24 spp : only two Oryza sativa & Oryza glaberrima (only in South Africa) are cultivated.

Oryza Sativa has three varietal types —

(a) Indica Rice : It is tropical rice,grown in India,awnless or short awn, late in maturity, long stem.

(b) Japonica Rice : Temperate or sub-tropcial rice, grown in Japan, early maturity, photosynthetically very active,short stem,very responsive to nitrogenous fertilizer, No lodging.

(c) Javanica Rice :Wild form of rice,grown in Indonesia.

- Rice Inflorescence is known as **Panicle**
- Hull = Lemma + Palea together
- Test weight (weight of 1000grains) = 25g
- Rice is self pollinated and short day plant.
- Plumule is covered by coleoptile and Radicle by coleorhiza in embryo
- Cardinal temperature :30-32^0 C
- Hulling% = 70-75%

$$= \frac{\text{Weight of rice grain}}{\text{Weight of paddy grain}} \text{ X } 100$$

$$= \frac{\text{Wt. of hulled rice}}{\text{Wt. of unhulled rice}} \text{ X } 100$$

$$= \frac{\text{Wt. of rice without husk}}{\text{Wt. of rice with husk}} \text{ X } 100$$

- TN-1 was developed after II nd world war in Taiwan.

T.N. - 1	= **Dee-geo-woo-gen** X	Tasai Yung Chung
(Taichung Native)	(dwarf & N-responsive)	(tall and drought resistance)

Dr. T.T. Chang brought TN-1, Dee-geo-woo-gen and I-geo-tze from Taiwan to IRRI, Manila.

- **I R-8** was evolved by breeder Henry M.**Beachell** at IRRI Manila One of the parent of IR-8 was Dee-geo-woo-gen.
  IR-8=Dee-geo-woo-gen X Peta (from Indonesia)
- First time in India,TN-1(2kg in 1964-65) was introduced by G.V.Chalam (G.M. of NSC) from IRRI.
- IR-8 was introduced in India in 1966 and IR-8 outyielded TN-1.
- Literally meaning of Dee-gee-woo-gen : **Brown tipped short legged.**
- Jaya = T.N.-1 (from Taiwan) X T-141 (Indian Variety); evolved by Dr. Shastry, first rice variety is developed under India's rice programme. It outyielded T.N.-1& IR-8 both;hence called 'Miracle Rice'in India.

- Padma is the reverse cross product of the parent of Jaya.
  Padma = T.141 X TN-1
- Jagannath :Mutant Variety of T.141.
- CR-1014 = T-90 X Urang Urangan
  (Indica) (Javanica)

  Released in 1988, popular in Orissa, Andhra Pradesh and West Bengal; It is super fine grain variety capable of yielding 3-4 tonnes/ha. under submerged condition in semi-deep water regime.
- Killer diseases of rice are Bacterial leaf blight and Tungro virus.
- **Pusa Basmati-1** : World's first high yielding dwarf variety under quality rice has been developed by IARI through convergent breeding.
- deep water rice varieties : Chakia -59, Madhukar, Jalamagan, Jaisuria, Jaladhi 1 & Jaladhi 2,Pantdhan II,Jalapriya.
- Aromatic (scented)Rice :Sabarmati, Basmati-370, Karnal local, Dehradun Basmati (T-3), Pusa Basmati-1,Kasturi, Taraori Basmati, Basmati-385, HaryanaBasmati-1.
- Non Basmati superfine rice -
- IR-64,PR-'106,Gaurav & Punjab No. -1
- Max.Rice exporter : (1) Thailand (4 million tonnes) (1990-91)
  (2) US (2.4 m tones)

India at the 5th position. Half of the world market is now captured by Thailand which is managing rice on the principles of the corporate sector.

- Highest Productivity **:** Japan (58q/ha);av.Average Productivity in India 18.5q/ha.
- Rice yields are generally lower in direct- seeded uplands due to high tiller mortality after maximum tillering phase.The tiller mortality is attributed to lag phase N-uptake and low level of leaf nitrogen (<2%).Timely top dressing to maintain leaf N-level above 2% in the post-maximum tillering stage has been found to minimise the percentage mortality and increase thereby yield level.

Yield constraints in cultivars of different maturity groups : In the early varieties, grain number per panicle;In medium duration varieties-spikelet sterility and in Late varieties-number of panicles per unit area.Under low light intensity late maturing varieties suffer due to tiller mortality.

- DAP has been found superior as basal dose to SSP. Ammonium polyphosphate, a new complex P-fertilizer is superior to both diammonium phosphate (DAP) and single super phosphate (SSP), especially in acid clay loams.
- Zinc deficiency is a major problem in intensively cultivated and saline-sodic-soils. In normal or neutral soils apply $ZnSO_4$ @ 40 kg/ha for every three crop seasons.In Saline-sodic soils initial application of $ZnSO_4$ @ 100kg/ha to make allowance for high Zn-fixation, followed by normal application of Zn after 3 successive crop seasons has given higher grain yields.
- Use of gypsum @ 6 tonnes/ha for sodic soils with p^H 9.5-9.7 rapidly brings exchangeable sodium from 70%to 50% in the very first season. A combination of gypsum (3 tonnes/ha)+ green-manure (6 tonnes/ha) is equally effective in reclaiming saline-alkali soils.
- **Iron toxicity** is the major problem to rice production in highly acid Ultisols,Oxisols and acid-sulphate soils. **Varietal tolerance** is only solution.'Phalguna' variety exhibits some tolerance to Fe-toxicity.

Weed Control-

1. **Propanil (Stam F-34) :** @ 3kg a.i (8 lit) in 400-600 lit. water/ha at 6-8 days after transplanting (DAT) when weeds are in 1 to 3 leaf stage.
2. Butachlor (Machete)@ 2 kg a.i./ha as pre-emergence.
3. Fluchloralin (Basalin) :Soil incorporated at the time of puddling or 1-3 DAT @ 1kg a.i./ha.
4. Nitrofen (TOK-E-25) :Pre emergence application @2 kg a.i./ha.

Critical Stages for Water :

(i) Booting Stage :most critical stage for water critical stages.

(ii) Tillering stage (0-20 days) and

(iii) Primordia growth to flowering (40-60 days) in dwarf rice. At these stages, submergence (5 cm) of water is must.

- **V'-shaped rice cultivation :**

Field experiment by **Matsushima** (1967)

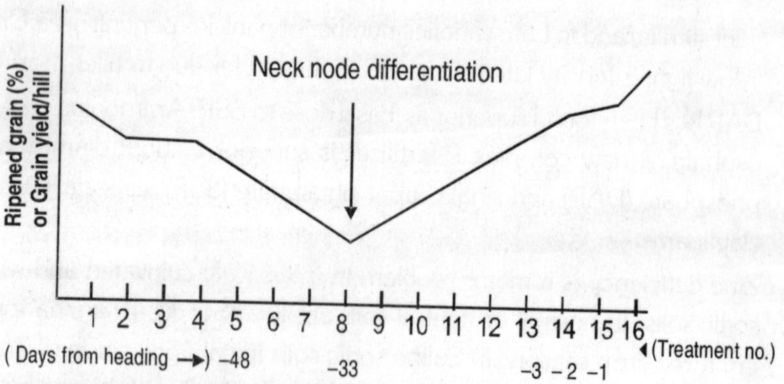

Fig. 11 Min. at the top dressing at neck node differentiation stage (about 33 days before heading)

After heavy application of Nitrogen at 5 days interval at successive growth stage;a curve between percentage of ripened grains or grain yield per hill i.e. 'V'-shaped line is obtained.

- **Dapog method** of rice cultivation : Prevalent in Philippines and Japan. In this method, **seedlings** become ready for transplanting on 12th day. Rate for nursery sprouted seeds @ 1.5 kg/m^2 or 50 times of test weight/m^2 on wooden planks, trays, concrete floor,seed bed covered with polythene sheet.Seedlings from 1m^2 are sufficient for 200m^2 field. (Seed rate : 2.5 times higher than other method).Crop flowers 4 days earlier when rice is transplanted by this method.

Transplanting :

- Ideal Age for transplanting :Seedling having 4 leaves,may be 3 leaves but in no case 5 leaves .Seedlings of 20-30 days in kharif (spacing 20 cm X 15 cm) and 30-35 days old in Rabi (Spacing 15 cm X 15 cm).

Hybrid Rice :

In the last decade of 1970,hybrid rice was first developed by China by using cytoplasmic male sterility-Fertility restore gene system. After China,in India,in 1994,First set of five hybrid rice varieties have been released for commercial cultivation.Out of these five hybrids,four varieties APRHI,APRH 2 (Andhra),KRH 1 (Karnataka),MGR-1 (Tamilnadu),have been released by A.P., Karnataka & T.N. After two years, Another three hybrids– CNRHI,KRH 3 and DRRH–1 have been released by West Bengal & Andhra

Pradesh respectively.PHB-71 is the only hybrid rice which is released by private organization.

Wheat :

Triticum spp and chromosome no-

Triticum aegiloploids = 2n = 14-***T. monococcum***

T.durum = 2n = 28 = ***T. dicoccum***

T.spelta = 2n = 42 = ***T. macha/T.vulgare/T.aestivum***

In India, following Triticum spp. are grown—

1. Emmer Wheat ***(T.dicoccum)*** :Grown in some parts of south states like Maharashtra,Tamilnadu and Karnataka and good for South Indian dish "Uppumav".This wheat is called Ravva, Godhumalu and Samba Wheat in South states.
2. Marconi Wheat ***(T. durum)***/duram Wheat :Very old, best wheat for drought condition or under restricted irrigation, used for semolina (suji) preparation, semya, sphagetti, vermicelli. Important Varieties : Malvaika, HD 4530, Jairaj, Meghdoot. It is 2nd most important wheat grown in country. Area : MP, parts of Gujarat, Rajasthan, Maharashtra & Karnataka.
3. Bread Wheat ***(T.Vulgare)*** :Typical tall wheat for rainfed condition in Indo-gangetic plains.Varieties : **C-306, K-65, K68, C13.**
4. Indian Dwarf wheat ***(T. Spherococcum)*** called club wheat in western countries, grown in limited areas of M.P.,U.P. very short and compact heads.
5. Mexican Dwarf Wheat ***(Triticum aestivum)***Presently grown in India everywhere and called common bread wheat; it was evolved by **Dr.N.E. Borloug** of Mexico at **CIMMYT.**

- Chapati making quality depends on **gluten** strength.Strong gluten varieties are most suitable for bread and for expression of high loaf volume, attractive brownish crust,silky and small granular structure of crumbs. Such varieties are HI 977, NI5439, HW657 and DWR-39.
- Ligule : membranous structure, continuation of the sheath through the collar.

  Auricle : Horn or claw like appendages projecting from the collar of the leaf.

- The flowering portionof wheat is called Ear or Head but botanically called **Spike.** The central zigzag axis is called Rachis.Spikelet is composed of flowers called florets.
- Flower is closed by lemma (outer bract) and Palea(inner bract)and awn is the extended part of lemma.
- Test wt. : 40g. Shelling % = 60%.

  Harvest Index = Economic yield/Biological yield x 100 = 40-45%
- "Norin" dwarfing gene was isolated from Norin series wheat varieties and "Nor"(Rht) in wheat variety in Japan.

  Rht → Reduced height.
- The first such variety :"Norin 10"(dwarfing gene wheat) 'Norin 10' was brought to USA by S.C.Salamon in 1948.

  Using 'Norin genes Dr. O.A. Vogel developed a dwarf winter wheat variety "Grains"in USA and many other varieties were developed by Dr. Norman E. Borloug, Nobel laureate in 1961-62 at CIMMYT, Mexico.
- In 1963, Govt. of India imported 100 kg of Mexican wheat varieties. Sonora 64,Sonora 63 and Lerma Rojo with the help of Rockfeller foundation and Five varieties by IARI were Lerma Rojo 64 A,S63, Sonora 64, Mayo-64 & S 227.
- After extensive tests, huge quantities of Lerma Rojo 64 A (single gene dwarf variety)and Sonora 64(Double gene dwarf variety) were imported in 1965-1966 from CIMMYT.
- In 1965,both varieties were released for general cultivation in India and brought green revolution (wheat revolution).
- Later on these were replaced by Kalyansona &Sonalika.Later on Kalyansona became susceptible to all rust and sonalika to kernal bunt.
- Source of dwarfing gene : (1) Norin 10 (Japan) (2) Tom thumb (Tibet) (3) Olsen dwarf (South Rhodesia).
- In 1970, triple gene dwarf varieties were released.
- National average yield :22 q/ha.

- Single gene dwarf varieties :Sonalika,UP-262,WL-711,Girija Double Gene dwarf varieties : Kalyansona, Arjun, Janak, UP-215, HD-2204, Pratap.

  Tiple gene dwarf varieties : Hira, Moti, Jawahar, Jyoti, Sangam, HD-1941, HD-1977, UP-301, UP-319.
- ***Phalaris minor*** (weed)is controlled by Tribunil/Dosanex/Isoprot-uron @ 2 kg a.i./ha 32-35 days after sowing (DAS).Dosanex and Isoproturon also control wild Oat ***(Avena fatua)*** At present isopr -turon is ineffective to control phalaris. Pendimethalin @1.0g a.i./ ha 2-3 DAS for Phalaris-sulfosule. Leader @33.3 g/ha or sencor @ 250 g/ha 30 days after sowing is used to control phalaris these days.
- Grain : Straw ratio in Mexican wheat -**1:1.5**, In New plant types of cereals 1 : 1.
- Late sown wheat varieties : Sonora 64, Sarbati Sonora, Sonalika (RR-21), Safed Lerm, UP-301,NP-830.
- Critical stages for Irrigation (i)Crown Root Initiation (CRI) Stage :21 days after sowing (ii) Flowering stage (90 DAS) (iii) Late tillering stage (iv) Milk stage (v) Dough stage.

(a) But when water is available for one irrigation; apply water at CRI Stage.

(b) But when water is available for two irrigation; apply water at (i) CRI Stage and (ii) Boot stage.

(c) But when water is available for three irrigation; apply water at (i) CRI (ii) Boot stage (iii) Milk stage.

(d) But when water is available for four irrigation; apply water at (i) CRI (ii) Late tillering (iii) Late jointings & (iv) Flowering stage.

Maize :

Classification on the basis of character of the kernels (Kipps, 1959)

1. ***Zea mays indurata*** **(**Flint corn**)** : most commonly cultivated in India.
2. ***Zea mays indentata*** **(**Dent Corn**)** :most common in USA.
3. ***Zea mays everta*** (pop corn) :When they are heated the pressure built up within the kernel suddenly results in an explosion and the grain is turned inside out.

4. ***Zea mays saccharata*** **(**Sweet corn**)** :Sweeter than others.
5. ***Zea mays amylacea*** .(Soft corn)
6. ***Zea mays tunicata*** **(**Pod Corn**) :**Primitive type of corn.
7. ***Zea mays ceratinakulesh*** **(**waxy corn**)** :Produces starch similar to tapioca.

- Three types of roots (a) seminal roots (b) crown/coronal root (c) Brace/aerial/Prop roots.
- Maize is monoecious plant/cross pollinated crop. Male inflorescence is called **tassel** and removal of tassel is called detasseling. The style is a very long silky filament,The cluster of which is known as **'Silk'.**
- Maize protein is called **'Zein'**and deficient in tryptophane and Lysine.
- Idea of **Hybrid maize** was first conceived by **E. M. East** and **G. H. Shull** in 1910 by **Single cross technique** (mostly used in USA and China).
- **Double cross technique** for hybrid seed production by **D. F. Jonese** (1920).This double cross technique is mostly used in India.
- All India coordinated Maize Improvement Project was started in 1957. First time in 1961,four **double cross hybrids** were re-leased : **Ganga-l, Ganga 101, Deccan, Ranjeet**. Later on,other double cross hybrids were released –VL54, Himalayan123, Hi–starch, Ganga 2, Ganga 3 and Ganga 5.
- Double Top Crossing/Top crossing

  Top crossing =Single Cross X open pollinated variety.

  (A X B) (C)

  Top cross is produced by crossing a single cross (A x B) with an open pollinated variety and such cross is commonly used for testing the General Combining Ability **(GCA)** of the inbreds.**Trop Cross** varieties are **Ganga2** and **Hi-Starch.**
- Composites are Jawahar, **Vikram**, Kisan, Amber, Sona and **Vijay**. Among composits,Amber has highest yields potential (50-55 q grain/ ha). In 1967 these were released for the first time in the world.

- **Lysine rich** composits (around 10-11% protein and 3.5to 4% lysine) are **Protina, Shakti, and Rattan;** released in 1971, such varities are called **opaque-2 composites.** Quality protein Maize (QPM) are released by using **Opaque - 2 gene e.g. Shaktiman 1, Shaktiman 2, and HQPM1**
- **Critical stages of growth :**

  For Nitrogen application (1) Germination (2) Knee high (3) Tasseling stage.

  For water :Tasseling to silking stage.

Jowar :

- Roots are finer and more fibrous than maize.
- Usually no tillering,prop roots may develop.
- Jowar is poor in lysine but rich in leucine.
- Inflorescence is called Panicle (commonly called Head)
- Thinning is important operation at 12-15 cm (within row).
- First Sorghum hybrid in 1964 :**CSH-1** (95-100 days,Cordinated sorghum Hybrid -1).
- One of the most important male sterile variety : **Combine Kafir 60** (CK-60).
- **Sweet Sorghum** : Sugar content in juice varies from 16-23% brix, Varieties - RSSV59, 46, 24, 45, 57, NSS-219 & 216.

Bajra :

- most drought tolerant crop among cereals & millets.
- Sensitive to water logging and acidic soils.
- First hybrid :**HBI** (Hybrid Bajra -1) released in 1965.

  HB-1= **Tift** 23-A (male sterile) x Bil 3B
- HB-41 first hybrid developed by using local parent,duration 80-85 days.
- Weed control : Atrazine/propazine @ 0.5 kg/ ha as pre-emergence.

Pulses :

The word 'Pulse' is derived from French 'Pottage' or 'poridge'means soup(dal). The term 'legume'from Latin word 'Legne'means to collect or gather refers to collection of Pods.

- Bacterial nodules are found mostly on secondary roots in all legumes.

Low Production Factors in Pulses : Development of sink is more important.

Factors :

1. Flowers drop more.
2. Translocation of stored photosynthates and current synthates is not good.
3. H.I. is very low e. g. 19% in Arhar.
4. Conversion efficiency of sucrose to protein is low and this con version requires more ATP.
5. Filling of Pods is according to hydrodynamic model.Lower devel oping pods (Arhar)transfer ABA to upper developing pods thus causing flower and fruit drop.
6. Grain filling rate is very low.

Physiological determinants of crop growth :

$$\text{yield} = \sum_{k=t} \left[\frac{\text{ECE} \times \text{light interception\%} \times \text{Fertility efficiency}}{\text{Eu \%}} - \text{Respiratory loss} \right]$$

Where Eu% = Energy utilization efficiency i.e. Photosynthesis

t = time of grain development or grain filling percentage

Constraints in Pulse Production in India :

1. **Agro-ecological factor** mostly rainfed,marginal and sub-marginal land, pests and disease.
2. **Biological** :
   (a) Indeterminate growth of plant e.g. Arhar
   (b) Poor partitioning of photosynthates
   (c) Pulses, rich in protein, therefore require more energy and thus prone to insect pest and disease.
3. **Technological :**
   (a) Lack of improved farm implements
   (b) Lack of appropriate post harvest technology to prevent post harvest losses and deteoration in quality.

4. **Organisational/Management constraints :**

(a) Inadequate arrangement for production and distribution of quality seeds.

(b) Untimely supply of required inputs,credits etc.

(c) Lack of transfer of improved technology.

(d) Lack of storage,grading and marketing facilities.

(e) Wide fluctuation in prices.

(f) Research constraints.

5. **Socioeconomic constraints** : Poor crop management, Belief, tradition, education etc. are also limiting factors.

6. **Faulty marketing system.**

- Kharif Pulses : 2n = 22 except horse gram (Kulthi) (2n=24).

Rabi Pulses : 2n = 14 — Lentil / pea / Lathyrus — except chick pea (2n=16)

Kharif Pulses : epigeal except Arhar

Rabi Pulses : Hypogeal except Rajma

- Guar :Legumes crop used for gum manufacturing.
- Cowpea :tolerant to heavy rainfall than other pulses.
- Lathyrus its grain and other parts contain a neuro-toxin called ODAP (βN –oxalyl–2α, β–diamino–propionic acid) which causes lathyrism. i.e. paralysis in the lower limbs. In our country, a variety, Bio L-212 (Ratan) with very low ODAP content is recently released.

Gram :

India is the largest producer of gram in the world sharing 65%and 70%of total global area &prodn. respectively.

- Indian gram is grouped into two groups -

  (a) Desi gram/Brown gram (***Cicer arietinum***) -
  Most widely grown, good branching, 2n = 14&16.

  (b) Kabuli/White gram (***Cicer kabulium***)bold seed, yield poor, taller plant, 2n = 16, branching poor.

- Specific sour taste of gram leaf is due to presence of maleic and oxalic acid.
- Gram seed is sown 8-10 cm deep in soil to escape from wilt disease.

Varieties :

1. Pusa 256 (**BG256**) : most common and best for rainfed condition.
2. C-235 :Widely adopted for dryland,tolerant to ascochyta blight.
3. **Gaurav** : resistant to **ascochyta blight**
4. **Avrodhi** : resistant to **wilt** (and JG-74).
5. **Chaffa** : the first variety,having medium size grain & early maturing (105-110 days) was developed in 1948 through selection in Niphad (Maharashtra).

- Late Planted Chickpea—

Normal sowing of chickpea is affected by seedling wilt in initial stage and excessive vegetative growth owing to high moisture resulting in low yields'That's why normal sowing is modified in delayed planting.When early variety is planted late, it can produce grains by cutting short its vegetative phase and thus plant is escaped from the hazards of seedling wilt or excess vegetative growth.In such type of delayed planting,yield per plant is low.To make up low yield,plant population is increased by 25%(25%more seed rate).

Varieties :Radhey,H-208,BG-261.

Sowing time :First fortnight of Dec.

• Photosynthetic rate falls soon after flowering even in newly formed leaves.It is due to reduction in Nitrogen content of leaves on account of degeneration of nodules just after flowering.Thus due to overlapping of vegetative and reproductive phases there is competition between sink (young pods or flowers)and source (leaves).Therefore during flowering to early stages of pod development,the source capacity is the key factor for pod setting.

Pea : Shelling percentage : 49%.

(a) Garden Pea :***Pisum sativum*** Var. hortenge or Table Pea used mostly in vegetable generally white flowers.

(b) Field Pea (***Pisum Sativum Var. arvense***) : Generally coloured flowers,mature seeds used as dal,also grown as forage or green manuring crop :

Arhar : **Harvest Index : 19% (very low)**

- Carry over effect (legume effect) of Arhar is not so important.
- Earliest Variety : **Prabhat** (110-120 days)

Pigeonpea is often cross-pollinated crop(average outcrossing 20%).Outcrossing is mediated by Bees *(Megachile lanita & Apis florea).*

Hybrid Varieties : Hybrid varieties in Arhar first in India have been developed by using Genetic Male Sterility System.There is a need of rouging of Fertile plants from female lines in it..Therefore effort by developing cytoplasmic genetic male-sterile lines is continuing.

(i) ICPH-8 :Released by ICRISAT;130-132 days duration; yield :20q/ha, released for middle & South India.

(ii) PPH-4 (Punjab) (iii) CoH-1(T.N.) (iv) CoPH

(v) AKPH 4101. AKPH 2022

- Arhar is grown on wide range of soils having p^H 5-8.In acidic soils the growth is adversely affected due to aluminium toxicity or calcium deficiency. It is also sensitive to salinity. Under continuous growing in the same land, the crop growth becomes poorer due to build up of nematodes and inoculum, besides allelopathic effects.

- Area and Production :Maharashtra, U.P., M.P., Karnataka, Gujrat and Andhra Pradesh account for 83.8% of total production comprising 87% area of our country.
- Highest Productivity : (1) Bihar (1115 kg/ha)
  (2) Haryana (1036kg/ha)

Soybean (20% oil) :

- Soybean is the major oilseed crop in the world accounts for 50% of the total area as well as production. It provides approx 60% of vegetable protein & 30% of oil in the world.
- Prophylls : tiny paired simple leaves (less than 1mm in length) present at the base of each lateral branch in Soybean.
- First nodules are visible on 10_{th} days after sowing.
- It is called wonder crop and meat for the poor.
- richest, cheapest and easiest source of best quality proteins and fats.
- Linoleic acid : 56-60%(Soybean) (Unsaturated fat)
  In Sunflower :64%
  In safflower :78%(highest)
- Iodine no. of Soybean oil=140
  Linseed oil =180

Iodine no. is the measure of unsaturation of fatty acid and is based on the fact that halogen addition occurs to the unsaturated bonds until they are completely saturated.It can be determined by Hanus Method expressed as gram of iodine absorbed per 100g of oil.

- Due to presence of enzyme lipoxidase, Soybean is not used as Dal which produces off flavour.
- The climatic requirements for soybean are almost the same as for Maize.
- Soybean : is low in S-containing amino acids viz. methionine (1.2%) and cystine (0.8%). High in lysine (6.2%) and tryptophane (1.4%) [lysine content is very low in cereals]. Proteins in soybean are mostly globulin and partly albumins. Lysine is the major amino acid but some antinutritional substances are also present in the seed like trypsin inhibitor, hemagglutinins and oligosaccharides. 12-14% saturated fatty acids (palmitic and stearic acid),Oleic acid :30-35%,Linolenic acid :5-10%,good amount of vit B.
- Presently cultivated variety is yellow seeded variety.Black seeded variety is not popular due to its susceptibility to diseases, low yield and highly shattering pods.
- Weed control :

Nitrofen (TOK-25)preemergence selective herbicide for all annual grasses and broad leaved weeds.@1.5-2.0kg a.i./ha in 800-1000 lit. water.

Fluchloralin(Basalin) :Pre-plant incorporation @1kga.i./ha

Metribuzin (Sencor) :Pre emergence herbicide effective against annual grasses and broad leave weeds @1kg a.i./ha.

Oilseeds :

Requirement : 38g/capita/day.

Availability :14g/capita/day(1993-94)

- In 2003 - 04, Palm oil & Soyabean were major contributors accounting for 36% & 21% of total consumption of oil in the country. The share of groundnut & rapeseed-mustard were 14% and 13% respectively.
- Sunflower (45-50%oil) :It is introduced in India in 1969 as an oilseed crop from earstwhile USSR whereas Soybean in 1968 (Bragg cv.) from USA.

- Safflower (24-36%oil),used as border crop,and for the preparation of Roghan which is used in the preservation of leather and the production of water proof cloth.
- Sesame/Sesamum :50%oil,18-20% protein.
- Linseed (36%protein,33-47%oil), linseed is known as flax in western countries.In India for flax,it is grown in Palampur (H.P.)

Groundnut (44-50% oil) :

Peg is the gynophore which bends downwards and forces the ovary into the soil.It is only after entering the soil that ovary begins to develop and takes up a horizontal position.

- 8-10 stamens from monoadelphous bundle.Self pollinated crop.
- The biological value of groundnut protein is among the highest of the vegetable proteins and equals that of casein,source of all vit B except B_{12}.
- Shelling% = 70%;Cake :7-8%N; 44–50% oil.
- **Aflatoxin** : Aflatoxin is produced in groundnut (also in Paneer) when infected by *Aspergilus flavus* group of fungi mainly during post-harvest processing & storage under moist condition.
- Technology mission on Oilseeds was set up in 1986 under the directorship of **P.V. Shenoi** to increase the oilseed production. Later on pulses also included in this mission and TMO is changed in TEMPO (Technology Mission on Pulses & Oilseeds).

  SOPA - Soyabean Processors Association in M.P.

  ITC - Indian Tobacco Company (A.P.)

  MAHYCO - Maharashtra Hybrid Company.

To increase the sunflower production,the responsibility to supply hybrid seeds is taken up by MAHYCO and ITC.

- Arachis : greek word means Legume

  hypogea : greek word means below ground.
- Seed rate (recently Agronomists recommended)

  Bunch varieties/Erect : **100-120 kg/ha**

  Plant population :3.33 lakh/ha

  Spacing : 30cm x 10 cm.

spreading type :80-100 kg/ha,

Plant population : 2.22 lakh/ha, spacing : 30cm X 15cm.

Rabi season : In rabi season, the growth is very limited hence bunch/erect type is grown @125-140kg/ha(25%more seed rate) spacing 22.5cm X 10cm

Or 30cm X 7.5cm, population : 4.44lakh/ha.

- **Spreading** type has late maturity than Bunch type and has more yield than of Bunch.
- Bunch type has no dormancy period.
- Three critical stages for irrigation :
  (1) Flowering : 35-40 DAS
  (2) Pegging : 55DAS
  (3) Pod formation/Seed development : 65-70 DAS.

At pod development field should be well drained.

- Varieties :

Kharif Season -

1.erect/Bunch-Spanish type :
M.P.: Jyoti, yield potential 18q/ha
Orissa: Kadiri, Kisan (16q/ha)
T.N.: TMV-11, TMV-12

2. Semi-Spreading/Verginia –Bunch:
TMV-10(T.N.)Kadiri 2 &Kadiri 3(A.P.)
BP-1 & BP-2(Bihar)

3.Spreading type/Verginia runner :
UP :Type-28,type 64,Chandra
Assam:M-13 & M-37
Punjab:Punjab-1
T.N. :TMV-1 & TMV-3

Rabi Season :

A.P.:ICGS-11, 14, Kadiri-2, TMV=2,J-11
T.N.=TMV-2, 7, 12,,JL-24
Guj=J-11, ICGS = 44,

Maharashtra : ICGS-11, TG-17

Orissa :Kissan, AK12 - 24

Karnataka : ICGS-11

ICGS-11 & 44 :yield:35-55 q/ha (highest)

Rabi groundnut gives higher yield than that of kharif due to better management practices and less incidence of insect pest & diseases.

Rapeseed & Mustard :

- Pungency is due to isothiocynate, enzymic hydrolysis product of Glucosinolates (in all **crucifers**)
- Due to presence of toxic glucosinolates in its cake, it is unsuitable source of protein for both human & cattle. It causes goitre and affects growth & development. Harvest Index; 25% The upper limit is fixed at 0.5% by the Govt.
- Pusa Jai Kisan (Bio-902) :First variety of *Brassica juncea* (mustard) developed in the world by Prof.V.L. Chopra et. al. with the help of biotechnology through Somatic hybridisation in India.

  Duration :115-125 days,bold seeded,yield : 19-25 q/ha,

  Oil percentage : 40%, released for irrigated areas of Gujarat, Rajasthan and Maharashtra.
- More than 44% area under rapeseed & mustard is rainfed.In U.P. more than 40% area is under mixed or intecropping.

Varieties :

Rapeseed :

(a) Toria : Type-9, 36, Sangam, M-27, B-54, B-R-23, ITSA (Improved Toria Selection A)

(b) Brown Sarson : BS-70, BSH-1, Pusa Kalyani, Suphla, M-18.

(c) Yellow Sarson : Benoy, YST-151, Type-42, K-88, YS-24, M-3.

Mustard : (Rai/Raya/Laha) : Kranti, Varuna(T-59), Krishna, Rohini, Pusa bold, Vardan, Durgamani, Prakash.

Taramira : T-27(150 day, 8q/ha).

Table : 8.1 Different species of Rapeseed and Mustard

Indian Group	Oil %	Spp.	Area of Cultivation	Local Name	2n
A. Sarson (Rape)	35-48%	*Brassica compestris var. Sarson.B. compestris var. dichotoma*	U.P. Bihar,W.B. Punjab.	Yellow Sarson Brown Sarson	20 20
B.Toria (Rape)	33-46%	*B.c. var.yellow toria* *B.c.var.black toria*	Punjab, U.P.,Assam Punjab,U.P.,Assam	Yellow toria Black toria	20 20
C. Rai (Mustard)	30-38%	*Brassica juncea* *B. juncea* var, *rugosa* *Brassica nigra*	Bihar, U.P.,W.B. -	Indian Mustard Rai/Laha/Raya Pahari Rai Black mustard/ Banarasi rai (Spices)	36 36 16
D. Taramira/Tara	35%	*Eruca sativa*	Origin in South Europe North Africa, Newly evolved spp. and more drought resistant	Common name Rocket Cress	22
E. White mustard	27%	*Brassica alba* *B. hirta*	Grown in earstwhile Russia		

Sunflower : (48-53% oil but 35-43% in India)

- Sunflower is the 3rd most important oilseed crops to soybean and groundnut, in the world. It was introduced into India in 1969 as an oilseed crop. Its oil is the premium due to high level of unsaturated fatty acids like oleic and linoleic acid (90% of the total) absence of linolenic acid, light colour, bland flavour & high smoke point.
- Russian Federation is the largest producer of sunflower in the world. In our country Karnataka, Maharashtra, and Andhra Pradesh are the major producer of sunflower.
- Spraying of borax (0.2% i,e, 2g/l of water) directly to capitulum at ray-floret opening stage increases seed filling, yield and oil content.
- Critical stages for irrigation:

  i) Bud initiation (35-40 DAS)

  ii) Flower opening (55-65 DAS)

  iii) Seed Filling (65-80 DAS)
- High temp above 38^0c during post-anthesis period reduces seed yield and oil content. (Anthesis: means the first opening of a flower in the process of flowering)
- Insecticidal spray during flowering period is harmful because it affects the visit of honeybees. The visit of honeybees is essential for pollination.
- Birds (particularly **parrots**), rats & squirrels cause damage to sunflower when it is raised in isolated and stray-pockets.
- It may be harvested at physiological maturity when back of the head turns to lemon yellow colour and bottom leaves start drying & withering .
- Seeds should be dried well to the moisture content at **9-10%** before storage.

Oilcakes and protein % :

1. Sunflower : 40-44%
2. Safflower : 40-55%
3. Linseed cake :36%(most favourite cattle feed)
4. Sesame cake :37-38%=6-6.2%N
5. Ground nut :45-48%=7-8%N
6. Cotton :43%=6.9%N
7. Tobacco :30-35% crude

Cotton (15-25%oil) :

According to Hutchinson (1947), cotton is grouped into—

(A) Old World Cotton/Desi cotton (n=13) :

1. ***Gossypium arborium*** **:** coarse and short fibre, Length of fibre : 1.25 to 2.10 cm In India its area is 29% of cotton area. Short staple cottons (fibre length less than 19 mm) known as 'Bengals'in trade and used for making quilts and mattresses and for mixing with wool and staple cotton.

2. ***Gossypium herbaceum*** : Fibre length :1.25 to 2.30 cm.Seeds with small fuzz and lint.

(B) New World Cotton/American Cotton (n=26) :

3. ***Gossypium hirsutum*** **:** Generally called **American cotton,** fiber length : **1.80-3.10cm area.** It covers 50% area of cotton area in India. Flowers creamy white when open but soon turn pink.

4. ***Gossypium barbadense*** : called sea Island cotton, or Egyptian cotton or Sudan cotton, fibre length : **3.6 to 5.0** cm (maximum).

   One bale =170kg for cotton.

   1 bale =180 kg for jute.

   Seed cotton = Seed + Lint

   Cotton seed = seed after removing lint.

Sympodial/Cymose : When main axis stops growing and the growth of the branches exceeds the main axis e.g.cotton growing branches.

- Monopodial/racemose :When main axis continues growing and gives off lateral branches which do not exceed itin size e.g.cotton stem branching.

Fig . 8.1

- Application of plant modifiers such as **planofix** (α-NAA) and Cycocel (CCC) at 10 ppm near flowering (80-90 days after sowing) helps in more **bud retention** capacity of plants,brings about an early reproductive growth phase by including **sympodial** branching in plants and also including **drought resistance**.

Topping : Removal of terminal growing point once from each plant at a height of 1-1.2 m(80-90DAS) to protect further terminal growth and to encourage **sympodial branching** and **boll development** by diverting the **energy flow.**

- 'Square'is the appearance of flower bud in cotton.
- **Cotton fibre** is simply an elongation/outgrowth of an **epidermal cell of seed coat**. The long outgrowth forms the 'staple or lint'while shorter outgrowth form **'fuzz'**.One seed has 60000-80000 fibres.
- **Hybrid – 4** : is the first commercial hybrid cotton of the world; developed by Dr. C.T. Patel in 1970 by utilising cheap labour available in rural India for production of hybrid seeds by emascula-tion and pollination.
- **Long staple varieties :**

  Hybrid-4, JKHY-1, MECH-4 all are ***intraspecific*** hybrids ***(G.hirsutum X G.hirsutum)***.

- **Extra long staple varieties :** 40% production of total cotton fibre in India.
  1. **Hybrid-6,** Savita, Surya are intra - specific hybrids (*G.hirsutum X G.hirsutum)*
  2. **Varalaxmi, DCH-32,** TCHB-213, DHB-105, HB-224, all are **inter-specific hybrids *(G.hirsutum X G. barbadense).***
  3. Non-hybrid varieties like Sujata, Suvin of ***Gossypium brabadense (Egyptian)*** and MCU-4, MCU-5, MCU-8, MCU-9 of Gossypium hirsutum (American cotton) are also popular).
- Honeybees are principal pollinators.
- Foliar application of DAP+NAA during flowering and boll-formation stages controls physiological shedding of flowers and bolls and increase the yield of cotton.
- Bad opening of bolls is known as 'tirak'due to growing cotton in soil with saline subsoils,decrease in N-content and increase in tannin content during flowering and boll-formation phase and water starvation as pointed out by Late Prof. Dastur.
- Soil depth of 100-120 cm is ideal for cotton cultivation.
- Fibre Quality : It is judged by

  (a) Staple length : measured by ball's sledge sorter, Bear sorter and fibro graph,first feature.

  (b) Fibre fineness : It is 2nd feature.

  Spinning value if cotton mostly depends on

  (1) staple length and (2) fire fineness.

  Micronaire : Instrument to measure fineness.

  Micronaire = Av. Weight of fibre in micrograms.

  Very fine : When Av. Wt. of fibre is below 3.0 microgram.

  Fine : When Av. Wt. is 3.0-3.9 microgram.

3. Fibre strength : Fibre is very strong when breaking point is more than 95 kg/sq. cm.

4. Ginning % : It is 24 - 43% in different cotton.

$$\text{Ginning \%} = \frac{\text{Wt. of lint}}{\text{Wt. of Seed cotton taken for ginning}} \times 100$$

$$\text{List Index} = \frac{\text{Wt. of 100 seed X Ginning \%}}{100 - \text{Ginning \%}} \times 100$$

5. Maturity of fibre is judged by Arealometer.
6. Nepiness :Sometimes fibre thickness is not uniform and knots present on the fibre are also not distributed uniformly.It causes low grade.
7. No. of knots :good quality cotton-fibre has all the knots properly distributed at equal distance and fibre is more strong.
8. No. of counts : A count is the no. of hanks (each 770.6m or 840 yard long thread) in 450 g lint (one pond). Finer the thread, greater is count. Indian cottons have 22 counts whereas best quality have 80-400.

Jute :

Two types of cultivated varieties—

1. *Corchorus capsularis* : hardy plant, tolerate water logging, more popular (70% area of total jute area), white fibre hence called '**White Jute**'(trade name), leaf is bitter in taste due to corchorin known as 'Tita Pat'Planting time late Feb. to March in low and mid lands.

2. *Corchorus olitorius* : grown on **well drained high lands only**, fibre is fine, softer, stronger, more lustrous,yellow to red, called, tossa (trade name), tasteless leaf hence called 'Mitha Pat', higher yield, planted in April/May in mid and uplands.

- In India Jute growing areas are divided into 8 agroclimatic zones.
- Mesta grows on soils where Jute can not.
- Jute Agricultural Research Institute Barrackpur. Now JARI was changed into CRIJAF (Central Research Institute for jute and Allied fibres) and again CRIJAF renamed as NIRJAFT (National Institute for Research on Jute and Allied Fibre Technology, Kolkata).
- Ideal stage for harvesting :Small pod stage or initiation of Pod formation viz 135-140 DAS.
- Steeping (Soaking) :After 2-4 days of harvesting the plants are shaken for complete leaf shedding and they are tied in bundles of about 20-22 cm(15-20) in diameter.For uniform retting,the bundles should be kept in vertical position for 3-4 days (to ret bottom portion) and then submerged into water in a horizontal position.For proper steeping the bundles should be laid side by side in water and tied together to form a sort of platform usually in 2-3 logs called Jack.

- Retting is a microbiological (microbial) process by which the bast fibre (fibre in bark)gets loosened for easy separation from the woody stalk.

For an ideal retting,the Jack should be kept submerged at least 20 cm below the surface of water because incomplete submergence results in under-retting and produces 'croppy' fibre of extremely low value,while over-retting 'dazed' weak fibre.

Tying of few plants of **Dhaincha or Sunnhemp** in each bundle of jute causes and early retting.In stagnant water add little amount of $(NH_4)_2SO_4$ or bonemeal specially during cold month.

Gently flowing,fairly deep,clear and soft water are congential for an ideal retting. Retting needs around 34^0C water temperature and completes within 10-15 days during July while 18-20 days during August-September and 21-30 days after Sept.(i.e. Oct.-Nov.)

- The fibre is obtained from **Phloem** by retting. Retting consists of disintegration of tissues by micro organism. Disintegration starts from cambium and extends to Phloem and cortex.
- Extraction of Fibre : The fibre is extracted by hands either from individual plants or from a bundle of 10-12 plants by beat-break-jerk method.

  Fibers extracted from individual reed (stem) with fingers is sleek, clean, and free from entanglements while the fiber extracted by bet- break jerk method
- Olitorius being efficient user of N; gives more fibre yield than capsularis in general.

  Capsularis :20q/ha and 4-5 seed/ha.

  Olitorius :27q/ha and 2-3q seed/ha.

Sugarcane :

(a) Tropical cane :

Saccharum officinarum : Thick and juicy canes and good for chewing purpose, indigenous to New Guinea, high sugar content, lowfibre, restricted to tropical areas.

(b) Indian canes :

***(i) Saccharum barberi* :** indigenous to north-eastern India; short and thin stalks; early maturity, low to medium sucrose content.

(ii) *Saccharum sinense* : indigenous to north eastern India; long and thin stalks; low to medium sucrose content and early maturity.

- Sett roots are temporary.Shoot roots provide anchorage to plant.
- **Inflorescence** of a sugarcane is '**open panicle**' and generally called 'arrow'.
- Since 1979, Brazil ranks first in area and production of sugarcane followed by India.

  Brazil>India>Cuba>China.
- Sugar is extracted from.

  (1) Sugarcane, (2) Sugarbeet, (3) Sorghum, (4) Sugar palm.
- Sugar mills (1) U.P. (105), (2) Maharashtra (99), total in India around 400.
- Immature crops give **better germinatio**n since **nitrogenous sub stance & glucose** are higher in such crop. That's why upper 1/3rd part of cane is used for planting.
- Highest producer of sugar from unit area basis :Maharashtra(10.04 tones per ha) followed by T.N. (9.04 tonnes/ha).
- After providing for seeds,50% of sugarcane can be utilized for the production of white sugar, 30% for low purity sugar (jaggery) and 20% for alcohol directly from sugarcane juice alongwith molasses.
- Sugarcane Breeding Institute, Coimbatore (SBI),

  Indian Institute of Sugarcane Research, Lucknow (IISR); Indian Sugar Institute, Kanpur (ISI).
- Three conditions for optimum tillering

  1. High heat requirement : 25-30^{0}c
  2. High moisture requirement :77-88% of field moisture capacity.
  3. Good illumination i. e. bright sunshine and water availability.
- In Brazil, sugarcane is also used as a source of energy, Gasohol is prepared from 80% petrol +20% Alcohol (from sugarcane) which is used in automobiles. In our country, it is also going to be used.
- Higher dose of N decreases sucrose content.

- Four distinct phases for sugarcane.
  1. Germination phase upto 60 DAP(days after planting)
  2. Formative stage:60-130 DAP] water requirement maximum under north India
  3. Grand phase :130-250 DAP
  4. Maturing phase :250-365 DAP.

Methods of Planting :

1. Flat bed planting : shallow Furrow 8-10 cm deep :distance 75-90 cm,generally 3 budded setts,planting by end to end system,Furrow is covered by 5-7 cm soil and field is leveled by planking,popular in north India and some parts of Maharashtra.

2. Furrow method : Deep furrows 10-15 cm in north India and 20 cm in south India;Practised in eastern U.P. and peninsular India particularly in heavy soils.

3. Trench/Java method : 20-25 cm deep trenches at 75-90 cm,give 'U' –shaped trench,common in Java,some coastal areas and in areas where the crop grows very tall and strong wind;setts planted either in these trenches or in small furrows prepared in the centre of trenches by end to end method.Besides these,recent improved planting techniques are :

4. Partha method : in south T.N.,Planting of sugarcane is difficult due to rains in July to Nov. and occasional showers upto middle January which results setts rot and buds fail to germinate.To over-come it Parthasarthy (1961) developed it for waterlogging condition.Fields is divided into ridges and furrows and 3-budded setts are planted at an angle of 45 on ridges leaving at least one bud above the soil.After 5-6 weeks shoots emerge from bud and after another 5-6 weeks when plant attains a higher of 20-25 cm,the setts are pressed horizontally.

5. Spaced transplanting Technique :developed at IISR,Lucknow single budded setts are planted in nursery @20q/ha or 18000 setts/ha After 45-60 days single budded setts are transplanted in main field..

6. Winter Nursery System : at IISR Lucknow,3 budded setts are let close together in the nursery bed in the month of December and covered with a thin layer of soil and setts are submerged by water.Submergence should not be more than 2-3 hours.Floating setts are removed.Nursery is then covered with polythene sheets which allow sunshine to pass through it but keeps off air.Within few hours,dew is formed on the under side of the polythene sheet and on cane begins to drop over the setts.After 5-6 weeks when setts have sprouted,polythene sheets are removed.Such type of setts are called 'Slip setts',and the sett rate is @70q/ha for 3 budded thick variety and 50q/ha for thin variety.

7. Rayungans Method : Indonesian term meaning a developed cane shoot with single sprouted bud.A portion of field is selected for Rayungan production is left at harvesting time.Top of the cane is cutt off which results auxiliary buds begin to sprout. For quick and effective sprouting fertilizer especially Nitrogen in heavy dose is applied and field is irrigated.After 3-4 weeks sprouted buds are separated in a single bud setts and transplanted on ridges.It is costly hence is not commonly adopted in India however is usually used for filling gap.

8. Sablang/sprouting method : Tillers, soon after they develop their own roots, are separated from the mother plant and planted separately. It is successful in Java & Cuba.

9. Tjeblock method : Improvement over Rayungan method because it takes care of proper availability of energy and nutrient to all the buds.Here stalks are cut off at its half length and planted vertically with one node under the soil for rooting.The planted ones and the mother stalks are adequately irrigated and fertilized.Now the upper buds of both Tjeblocks and mother canes,which sprout in due course of time,are planted by cutting them into setts in rayungan.

10. Algin method : Upper most nodes are collected while stripping the canes for crushing;then is planted in wheat field in rows after every 4 rows of wheat at 90cm X 50cm. It is evolved by Allahabad Agriculture Institute,Allahabad.

- Placing of setts in different ways in the field -
  1. End to end method : in this method sett rate is low.
  2. Eye to eye method :
  3. Double row system : For thicker planting and off-season planting.
  4. Single bud planting:Setts having single bud only.

- **Harvesting :**

For proper maturity : Spray Balsario chemical@4.5 kg ha^{-1} in 1000 lit water.The spray of this chemical matures the cane 6-8 weeks earlier.Chemical ripener like Polaris and sodium metasilicate improved the juice sucrose when sprayed on foliage 6 weeks before the scheduled harvest date.

Symptoms for judging the sugarcane maturity—

1. Leaves become yellow.
2. **Arrowing** and plant's growth is stopped.
3. Canes become brittle and break easily at nodes.
4. Canes produce **metallic sounds.**
5. Buds swell out at nodes.
6. Eyes start sprouting.
7. Brix schachrometer/Hand refractrometer :when Brix% reading of middle portion stalk reaches **16-18%** (Milliable canes)
8. When glucose content is less **than 0.5%** tested by Fehlings solution.

Under high temperature,Sucrose gets converted into glucose and quality of the produce becomes poor).

Recovery :

Juice% by Ordinary Crusher	:	50-65%
Juice% by Power crusher	:	60-70%
Juice% by Vacuum Crusher	:	70-75%
Jaggery from juice	:	9-10% (max. 15-18%)
Jaggery/Gur recovery from juice	:	**10%**
Crystallised sugar from gur	:	**62.5%**
Sucrose content in canne	:	13-24%
Sugar from juice	:	6-10%
Molasses	:	3.5%
Rab	:	18-20% of juice

- **Comparision between sugarcane and sugarbeet :**

	Particulars	Sugarbeet	Sugarcane
1	Duration	5-6 months	10-12 months
2	Crops in a year	2	1
3	Recovery percentage	10-12%	6-8%
4	Purity	84%	10-12%
5	Sucrose %	18-21	14-17
6	Suitability to saline soil	Suitable	Unsuitable

Extraction of sugar from sugarbeet is by Diffusion process.Sugar-beet provides 40% of the world sugar.

Tobacco :

Indian Tobacco has two spp.

(a) *Nicotiana tabacum* : Plants Height : 150-250 cm,large and narrow leaf,texture of leaf is fine,nicotin content :**0.5-5.5%**,used for **smoking and chewing purpose;** grown on light and high lands;seeds rate is higher than rustica.

(b) *Nicotiana rustica* : Plant Height 90-120 cm,bushy;nicotine content is **3.5-8.0%**;used for **hookah, chewing** and **snuff purposes** (for inhaling); grown on heavy,low lying soils..

- Central Tobacco Research Institute, Rajah-mundri, A.P.(1947).
- Sodic soils are unfit for tobacco production because the plants absorb a lot of chloride ions (Cl^-)which results in a poor burning quality of leaves and a mild acidic soils (p^H=5 to 6) are always better for the production of superior quality leaves.
- Nicotine is produced mainly in **plant-roots** and is carried through stem to leaves where it is stored.
- In Gujrat,tobacco is mainly grown in sandy to sandy loams called 'goradu'soils under irrigated conditions.In Orissa natu tobacco is known as Pikka tobacco.

Areas :

Indian Tobacco :

- → **Flue cured Virginia (FCV) :** cigarette tobacco, 30% area and 20% production of total.
- → **Non-Virginia types :** 70% area of total tobacco area, Bidi, Natu, chewing, hookah, cigar and cheroot,burley and snuff tobacco.

Flue cured Virginia tobacco : major exportable type :mainly grown in (1)AndhraPradesh (100,000ha) and (2) Karnataka (20,000ha).

Bidi Tobacco is most important non-virginia type : its 70% area and 80% production in Gujarat, Other states are : Karnataka, Maharashtra, and Andhra.

Bidi tobacco Research station Anand released two varieties viz. var. Anand 119 and Anand 2.

Chewing tobacco : U.P.,T.N.,W.B., Gujarat, Bihar and Orissa.

U.P and W.B. :grow *N. Rustica*

T.N.- grow, *N.tabacum*

Rest states grow both types.Natu tobacco is grown mainly in Andhra.Hookah (mostly rustica) :W.B.,Bihar & U.P.cigar and cheroot :Andhra,T.N. and W.B.,cigar & Filler tobacco:T.N. and W.B.

Cigar wrapper : W.B.

Burley tobacco :East Godavari and Visakhapatnam district of A.P

Snuff :T.N.,A.P.,Kerala,W.B.,U.P.,Gujarat & Punjab.Tamil Nadu produces 40% of total snuff tobacco.

Planting : Seeds are mixed with soil,sand or ash before sowing or broad-casting because seeds are very small in size usually 10000-12000 per gram seed. Seed rate is 20-30 g. seeds per 100 m^2 area in nursery for one ha i. e. 2-3 kg seed/ha.

Time for nursery sowing :**2nd Fortnight of August**

Transplanting Age : 4-5 leaves stage

- → 7-9 weeks in tabacum
- → 5-6 weeks in rustica

Topping : Removal of flower heads either alone or with few upper or top leaves from the plants to improve the size,body,and quality of leaves in most of the tobacco except wrapper tobacco.Thus due to topping the energy and nutrients absorbed by the plants are diverted to the leaves rather than flower heads.It helps in the full development of the top leaves or otherwise they will remain relatively shorter when the lower leaves are going to mature.It gives a uniform quality product and prevents excessive coarseness in the leaves.

Types	No. of leaves left on the plant
Cigarette&cheroot	=8-9
Wrapper (Cigar)	=10-12
Hookah,chewing &Bidi	=10-14

- **De-Suckering** : After topping,axillary buds grow;Removal of such lateral branches or suckers or auxiliary buds is called Desuckering.

The main aim of topping and de-suckering is to **divert energy and nutrients** from flower head to leaves.

- **Priming** : Removal of lower leaves which come in contact with soil and lose their commercial value.Such lower leaves are called sand leaves.
- **Harvesting** : two common methods -
  1. **Priming :** Removal of matured leaves.Entire harvest needs 5-6 priming used in cigarette and wrapper tobacco.
  2. **Stalk— Cut method** :used in Hookah,Bidi,cigar cheroot and chewing tobacco.Bidi tobacco:When top leaves develop **red rusty** spots called **spangles.**

Cigar and cheroot :when leaf is puckered and become brittle and yellowish green.

Chewing : Leaves develop pronounced puckering.

Hookah : Yellowish brown spots of puckering.

- **Curing** : is essentially a drying process where by most of the moisture of leaf is removed to impart required colour,texture and aroma to the final product.

Four types of curing are prevalent in India—

1. Air curing/shade curing :practiced for wrapper tobacco in W.B. and Lanka and for lanka tobacco in A.P.

There are three main methods of air curing—

(a) Ground-curing : Hookah tobacco in U.P. chewing tobacco in Karnataka.

(b) Rack-curing :

T.N.	Cigarette, cigar, cheroot, chewing-snuff.
Maharashtra	Bidi and chewing.
Andhra	Cheroot and bidi tobacco.
Assam	Hookah.

(c) Pit curing : not very common but mainly in Punjab, T.N., Maharashtra & A.P.

2. Sun curing :cigar and chewing,Bidi tobacco.
3. Fire curing/smoke curing :chewing tobacco in T.N.
4. Flue curing : cigarette curing process consists of three stages—

(i) **Yellowing of leaves** : temp 32-35°C and high humidity 85-90% for 30-40 hrs.

(ii) **Fixation of colour** : Any delay in drying of leaves, after yellowing is over, result in development of a brownish tinge on the yellow leaves due to oxidation of tannin like substances called sponging. A sudden rise in temp. when leaves are still wet,results in a bluish black discolouration called scalding.

iii) **Drying** : at 160°F to dry veins and mid ribs.

❑❑❑

9 Seed Technology

Seed is defined either Agronomically or Botanically.Agronomically a seed or seed material or propagule is living organ of the crop in rudi-mentary from that is used for propagation or in other words,any part of the crop from which a new crop will grow.

Botanically seed is a fertilized ovule consisting of intact embryo,stored food and seed coat which is viable and has got the capacity to germinate.

Agronomically germination means the capacity of seeds to give rise to normal sprouts within a definite period fixed for each crop under optimum field conditions.

- **Four essential factors for germination of seeds -**
  - (i) Capacity of seed to germinate
  - (ii) Moisture
  - (iii) Temperature
  - (iv) Oxygen-supply.
- **Factors governing growth of plants—**
  - (i) Parental feature (ii) Air (iii) Moisture (iv) Temperature (v) Light (vi) Minerals.

Q. Which one of the following is a essential factor for germination of seed -

(i) Light (ii) Minerals (iii) genetic make up (iv) Air.

Ans. (iv) Air means oxygen also.

Field germination is always lower than the germination of seeds tested in Laboratory.The mortality of seedlings after germination in the field frequently depends upon entomological,phytopathological,edaphological and meteorological factor as well as toxic effects of organic secretions and applied chemicals and storage period of the seed i. e. ageing.

- Factors affecting the emergence of seedlings are—
  1. Deep sowing or depth of soil cover over the seed.
  2. Inadequate or excess soil moisture.
  3. Poor aeration.
  4 Higher soil compaction or impermeable layer of soil or outer materials.
  5. Low temp.
  6. Rapid desiccation of soil.
  7. Longer time period.
  8. Injurious level of salt content.
  9. Poor seed capacity.
  10. Detrimental physical, chemical and biological soil conditions.
- Germination of seeds depends upon their natural peculiarities and biological conditions e. g. for the purpose of seed production, sunflower, mustard and flax can be harvested 15-20 days earlier and coriander 10 days earlier than for market purposes.

Foliar feeding of N at the grain filling stage improves germination,initial growth and growth vigour of seedlings and weakens dormancy.

Chemical ripener e. g. Magnesium chlorate @ 20 kg/ha can be applied on sunflower 40 days after flowering to reduce the moisture content of seeds twice without affecting the sowing quality.

- **Seed Index** : Weight of 100 seeds (in case of bold seeds like maize).

  Test weight : Weight of 1000 seeds (in case of small seeds like rice).

  Seed quality Index : is the vigour of seed germination.

Real Value of seed :is the percentage of germinability of percentage of purity of seed lot of a crop plant.

$$RV = \frac{\text{Purty \% X Germination \%}}{100}$$

RV is expressed in percentage and also known as **Utility percentage of seed.**

Conditions affecting the real value of seeds are the method of production,the method of handling and the method of storage.Seeds having a real value lower than 70% are usually not preferred for sowing purposes because of poor germination and purity values.

$$\text{Purty \%} = \frac{\text{Wt. of pure seed}}{\text{Wt.of bulk seed}} \text{X } 100$$

(i.e. seed + imprities)

$$\text{Impurity \% (or Dockage)} = \frac{\text{Wt. of Impurities in bulk sample}}{\text{Wt. of bulk sample}} \text{X } 100$$

Impurity percentage is called **Dockage.**

Genetic purity :

Genitic purity is required 100% as far as impurities by seeds of other varieties of same crop is concerned but in case of Impurity by seeds of other crop species;it is permitted upto 0.1% only.

Isolation Distance: Isolation means to keep apart. Isolation distance is the specified distance from potentia contaminants, where an acceptable level of contamination is expected. Isolation of seed crop is essential to avoid genetic & physical impurities. Genetic purity i.e. varietal purity may be deteriorated by cross - pollination from the plants of nearby plots. Physical impurities may occur due to physical or mechanical admixture with other crop species or other varieties of the same species from the nearby plots. Therefore the protection from these sources of contamination is necessary for maintaining genetic and physical purity i.e. good quality of the seed. Isolation distance is affected by mode of pollination (i.e. self-pollination or cross pollination), pollination activity as well as direction and velocity of wind. Essentially self- pollinated crops like rice, wheat & soybean have only 3 metre as isolation distance. The increase in natural cross pollination percentage requires increasing isolation distance e.g.

Natural cooss pollination Percentage	**Isolation distance in metre (m)**
0-10%	: 100 - 200
11-25	: 200 - 300
26- 50	: 300 - 500
51-75	: 500 - 1000
76- 100	: >1000 (more than 1000 m)

Minimum isolation requirements for some crops

Crops	Minimum isolation distance (m)		Remarks
	Foundation Seed	Certified Seed	
Field crops viz. rice, Wheat, Barley, Oats	3	3	150 m iolation for loose smut susceptible wheat for both classes; from fields with loose smut infection more than 0.1% for foundation and 0.5% for certified seed
Pearl millet hybrid seed parent pollinator	1,000	200	Isolation from other varieties;the same variety not conforming to varietal-purity requirements to varietal-purity requirements for certification
Cotton hybrid	50	30	Isolation will be 5 m from the varieties of other species (different ploidy level) and 5m isolation in certified classes between blocks of parental lines of the same hybrid
Soybean,groundn-ut	3	3	Isolation from other varieties,the same variety not conforming to varietal purity requirements for certification
Rapseed and mustard	50 for self-compatible,-100 for self-incompatible	25 for self compatible 50 for self-incom-patible	Other varieties of the same species,the same variety not conforming to varietal purity requirements for certification,rocket salad and other Brassica spp.
Castor	300	150	Other variety;the same variety not conforming to varietal purity requirements for certification
Okra(bhindi)	400	200	Isolation from *Abelmoschus manihot, A.moschatus* (also are objectionable weeds)
Cabbage	1,600	1,000	Isolation from *Brassica oleracea* varieties,brussels sprouts,cauliflower,collards and kales, knol-khol
Onion Bulb Seed production	 5 1,000	 5 500	Other variety,the same variety not conforming to varietal purity requirements for certification
Tomato	50	25	Other variety,the same variety not conforming to varietal purity requirements for certification

Courtesy : Handbook of Agriculture (ICAR)

Types of pure seeds :

There are four classes recognized by the International Crop Improvement Association.

1. Breeder/Nucleus seed : Such seeds are produced directly under the supervision of plant breeder and possessed all the required genetic characters. Such seeds have high genetic value and are costly due to very little in quantity.It may be produced by the breeder's concerned Institution when breeder is not available.Breeder seed is used to produce foundation seed. In case of self pollinated species,mass selection may be regularly practiced to retain the genetic purity of the variety.

2. Foundation seed : Produced from breeder seed and is the source of all other certified seed classes either directly or through registered seed.It is also known as **Mother seed**. Production of foundation seed is the responsibility of National Seeds Corporation.It is produced on Government farms,at experiment stations,Agriculture Universities or on cultivators field under strict supervision of research scientists and experts from NSC or SSC.It has specific genetic identity and purity.White tag denotes foundation seed

3. Registered seed : Produced from FS or Registered seed itself.It is the parent of cetified seed or RS.It has satisfactory genetic identity and purity and is usually produced by progressive farmers under technical guidance and supervision from SSC.It is often omitted and certified seed is produced directly from FS.It is general practice in India.It has purple tag.

4. Certified seed : Progeny of RS,FS or CS itself.It has satisfactory genetic identity and purity.It is approved and certified by the state seed certifying agency,annually produced by progressive farmers according to standard seed production practices.It has blue tag.It is available for general distribution to farmers for commercial crop production.It is generally produced by SSC but NSC also undertakes its production,if required.Two classes of CS are produced i.e. F_1 and F_2 .Recetification is not permitted from F_2 generation (second) of seeds.

Other types of seeds in Agronomic use are -

(a) Improved seed : is a better seed substitute for pure seed but is not so good with respect to genetic and physical factors.It has at least 10-15% more genetic potential than the local seed and are resistant to pest and diseases,well adapted to agro-climatic conditions of the locality,high in response to better condition of growth.

(b) Hybrid seed.

(c) Composite seeds.

(d) Mutant seeds.

Viability Test :

Seed's viability is defined as the capacity of the seed to remain capable of germination for some specific period of time.When the seeds are dormant or very slow in germination.the rapid test is extremely useful.

1. Respiration test : Respiration is an indication of life but such tests are complicated.

2. Electrical conductance method : Seeds are soaked in distilled water and electrical conductance of such water is tested.The increase in conductance is roughly proportional to the percentage of dead tissues.The increase in the conductivity is due to leaching of metabolites from dead seeds.

3. Potassium permaganate method : It is a qualitative test to find out whether seeds are viable.Increase in the proportion of dead seed ,increase the discoiouration of the solution since leaching from living cells is very less.

4. Indigocarmine method : Portion of dead seeds is determined by counting the number of stained.

5. Embryo Culture Method : Takes about 7-10 days,judged by counting the number of germinated embryos.

6. Tetrazolium Chloride Test : Known as Biological test.Seeds are soaked in 0.5-2% solution of 2,3,5 **triphenyl tetrazolium chloride.** The viable or living seeds take bright colouration and the colouration becomes more intense in the embryo.It can be made quicker by cutting kernels, using vacuum and maintaining a temperature of 40-45^{0}c, it takes only 4-5 minutes to complete a test of 100 kemels.

7. Grodex Test : Grodex is a seed germination indicator and is a brand name of triphenyl tetrazolum bromide in powder form.

Seed Moisture :.

Germination occurs when seed moisture is above 40-60%.

The moisture determination in seeds by—

1. Traditional bite test
2. Electrical moisture meters
3. Infra-red moisture meters
4. Oil distillation method.
5. The lab. Oven method.
6. Karl Fisher reagent method.

Seed Dormancy :

Dormancy is the arrested development and reversible rest period of plant organs either of a seed or of any vegetative part.The formation of dormant structures is commonly associated with the suspension of metabolic,synthetic and morphogenetic activities that are associated with the minimum physiological and a minimum moisture content.During this period,there is very poor or total suspencion of respiration or rather anaerobic respiration with higher respiration quotient.

Dormancy is due to lack or inactivity of hydrolytic enzymes.**Gibberellins** are the predominant germination agents found in the germination phase during the food reserve degradation stage.**Cytokinis** exert their influence later on the initiation of cell proliferation and expansion. Red light (660mu) promotes and blue specially far-red light (735mu) inhibits germination.

Dormancy can be broken artificially by various special treatments which may be grouped into two groups.

(A) Physical Treatment :

1. Heat treatment at 40-45^0 C for different duration.
2. Low temp. treatment at 2-8^{0}C for 12-24 hours but seeds should

be presoaked for 36 hours before they are given the low temperature treatment.

3. Altemate heating and cooling for several times.
4. Alternate drying and wetting for several times.
5. Exposure for 24 hours of water soaked seeds to red light for 1-2 hrs. at 15-25^{0}C temp.
6. Dehusking or removal of seed coats (scarification) by rubbing to make it permeable to water.

(B) Chemicals :

(a) **Inorganic Chemicals :**

(i) by acid treatment : Dilute solutions of HNO_3, HCl or H_2SO_4(0.1-0.5%) for different durations in minutes.

(ii) **KNO_3** (1-3%) : **Strongest** and used for immediate dormancy break after harvesting; NH_4NO_3(1-3%). H_2O_2, H_3BO_4 etc.

(iii) by gases : by increasing O_2 concentration.

(b) **Organic Chemicals** :

(i) Non–hormonal :**Thiourea**, Ascorbic acid.

(ii) Hormonal :GA3(1-1000 ppm)— Commonly used hormone to break dormancy ; Others are Kinetin (1-100ppm), Ethylene (Ethrel solution of 100-300 ppm).

- **Dehumidification** : Removal of water – vapour from the air in storage. Silica gel is the most common desiccant used as chemical dehumidifier.
- **Tempering :** Process of bringing grains or other products to a desired moisture or temperature for processing.
- **Matrical** is the heterogeneity in the quality due to location of the seeds in the inflorescene e.g. difference in flowering pattern.
- Indian seeds act passed in 1966.

* **Seed Treatment** : Besides breaking seed dormancy,seed treatment is also done before sowing the seeds for the following purposes—

1. To protect from seed borne and soil borne diseases and insect pests damage.

2. To promote germination
3. To hasten nitrogen fixation capacity in case of pulses.
4. For easy sowing.

(A) Treatment for Diseases and Insects control :

(a) Physical treatment :

(i) Hot water treatment : seeds are kept in hot water at a certain temp. for certain period.

Later on, after cooling in cold water such seeds are dried in shade. This method is successful in controlling following diseases—

Loose smust of Wheat (54^{0}C for 10 minutes)

Loose smust of Barley(54^{0}C for 13 minutes)

Alternaria blight of Wheat (38oC for 10 minutes)

Leaf spot of Til (54^{0}C for one hour)

(ii) **Solar treatment** : After soaking the seeds in water for some hours, seeds are dried in scorching sun in the month of May-June by keeping on cemented floor or metal-sheet.This method is used to control loose smust of Wheat and Barley.

(b) Chemical treatment : Diferent fungicide e.g. Agrosan G.N., Ceresan, Captan, Thiram etc. are used for the same.

(B) Seed Innoculation in Legumes :

Before sowing the legumes crops in the new area,the legume seeds are to be imoculated with **Rhizobium** culture.

Seed Plot technique of Potato :

Objective : To produce seeds free from virus Y,A,X and S.These viruses are transmitted by aphids.Aphid population is low in hills during April to August while it is low in the plains during October to early January.

Nucleus seed of potato is produced in the hills during April to August when aphid population is low.It is brought to plains and is stored in cold – storage for planting the seed crop in October. The seed production practices are.

1. Start with disease free seed stocks.
2. Select suitable location i.e. aphids free during crop growth.
3. Adopt lands isolation and other requirement of certified seed production.
4. Rogueing and inspection of crops.
5. **Good crop management -**
   (a) Timely planting i.e. first week of Oct.
   (b) Raising the crop on not too rich soil i.e.no judicious manuring and heavy fertilization.
   (c) Seeds are planted closely and use of pre-sprouted tubers.
   (d) Restricted irrigation after the crop has been tuberised well, with hold irrigation 10-15 days before.Restrict the irrigation by middle of December and gradually by end of December.
   (e) Haulms should be dried up by the middle of January.The haulms should cut or wilt before aphid build up population by the end of Jan. If the crop is still green, kill the haulms with a 2% solution of $CuSO_4$ or cut green haulms to prevent transmission of viruses to tubers.
   (f) Tubers are allowed to remain in the soil upto the end of Feb. or early March or may be spread thinly after harvesting in a dark place to allow thickening of skin of potato.
   (g) Tubers are harvested,graded and kept in cold storage to serve as seed for the next year's commercial crop.One ha of seed crop poduces enough seed for 10-15 ha of commercial crop.

National Seed Corporation (NSC) :

NSC was registerd in 1963 having two main objectives—

1. To promote development of a seed industry in India.
2. To produce and supply foundation seeds of various crops.

State Seed Corporation (SSC)

SSC's are chiefly concerned with production and supply of certified seed.

To reduce the work load of NSC and to stimulate a faster growth of industry,SSCs were established in view of great success of and the impact made by the tarai development corporation (TDC) , Pantnagar.

State Seed Certification Agency : Is responsible for seed certification,SSCA makes field inspections and conducts seed tests required for seed certification.

The seed Act, was enacted by the parliament on 29.12.1966 and the seeds Rule were notified in1968. This Act came into force throughout the country on 2nd Oct. 1969. Seed was declared as an essential commodity under the Essential Commodities Act. (1955)

❐❐❐

10 Weed Problems

The Term 'weed' was firstly used by Jethrotull.Weed is such undesirable plant grown in association with crop which snatches major parts of nutrients,water,light,place and CO_2 available to the crop.Due to tough competition crop suffers adversely and the production is reduced.Following are the main points -

(i) Unwanted plant

(ii) Plant growing where it is not required i.e. a plant out of place.

(iii) Extremely noxious,useless,unwanted or poisonous plant.

(iv) Any plant or vegetation excluding fungi interfering with the objectives or requirements of people.

Thus weed may be defined as unwanted and undesirable plant which interfere with the utilisation of land and water resource and adversely affect human welfare.

Criteria of to be a weed :

Whether a plant is weed or not depends upon

(i) Characteristics and habit of plant

(ii) Relative position

(iii) Time of occurrence of plant.

Therefore all plants may become weeds in a particular situation.

Classification of weed :

A. On the basis of site of predominance :

A_1) Obligate weeds : such weeds are grown only in association with man and his agriculture.It is never found in wild form e.g. Chenopodium, Anagallis etc.

A_2) Facultative weeds :such weeds are grown both as wild and in cultivated habitats e.g. *Argemone mexicana, Euphorbia hirta.*

B. On the basis of Dependence on Host :

B_1) Total Root parasite/Holo root parasite e.g. **Orobanche**

B_2) Total stem parasite/Holo stem parasite e.g. **cuscuta**

B_3) Semi-root parasite e.g. **striga**

B_4) Semi-stem parasite e.g. Loranthus (birdvine)

On the basis of Botanical or Morphological characters :

C_1) Grasses :Cylindrical and hollow stem having nodes and internodes,Leaf emerges from node,fibrous root,graminae family e.g. Echinochloa,Phalaris etc Most of the weeds belong to this group.

C_2) Sedges :Triangular stem,no node,very large internode and leaf at top,Cyperaceae family,no branch,usually three leaves at top,pith present e.g. Cyperus (Motha).

C_3) Broad leaf :broad leaf alternately arranged on stem,cylindrical stem,usually tap root e.g. all dicot weeds like Camellina,*Chenopodium album* (Bathua).

D. On the basis of season :

D_1) Kharif Weeds :e.g.

i) *Amaranthus viridis (junglee chaulai)*
ii) *Boerhavia diffusa (Vishkhopra)*
iii) *Cassia occidentalis (Kasaundhi)*
iv) *Cyperus rotundus (Motha)*
v) *Euphorbia thymiloia(Chhoti duddhi)*
vi) *Euphorbia dracunculoides(Titali)*
vii) *Digera arvensis (Lahsua)*
viii) *Datura festoosa.*

D2) Rabi weeds : e.g.

i) *Anagallis arvenis (Krishna nil)*
ii) *Argemone mexicana(Satyanashi)*
iii) *Convolvulus arvenis (Hiran Khuri)*
iv) *Chenopodium album (Bathua)*
v) *Desmodium trifolium(Tinpatia)*
vi) *Cuscuta reflexa*
vii) *Asphodelus tenuifolius (Vanpyaji)*

E. Other basis :

E_1) Relative weed e.g. rice in wheat field.

E_2) Absolute weed e.g. Such weeds are due to its characteristics e.g. *Cyperus rotundus.*

E_3) Rogue :The off type crop variety in the field of certain crop variety is called rogue.

E_4) Mimicry weeds :Such weeds are similar in external morphology with the crop e.g. Phalaris in wheat field and wild rice in the rice field.

E_5) Volunteer weeds :Such weeds are grown from the fallen seeds of previous or preceeding crop in the field.

E_6) Noxious weeds :Such weeds are declared undesirable,extremely noxious and poisonous by the law e.g.Parthenium.

Crop – weed Competition :

Weeds are the naturally grown plant which are able to extract nutrients & water from the soil and are component enough to harvest light energy.In the growing plants, leaf area index (LAI) and root density increase which leads to mutual interference in the utilisation of growth factors. The population of weeds crossing over the threshold level increases the competition for growth factors resulting in reduction of crop production.The weed and crop compete for such common growth factors whose availability is less.In the normal cases the weed competes with the crop mainly for nutrients,water,light,place and CO_2 but in some instances weeds show the allelopathic effect on crops.

(i) For Nutrients : Naturally grown weeds absorb more nutrients from the soil than the crop.Nutrient analysis of weed and crop shows that the concentration of nutrients is far more in weeds than the associated crop.In the beginning of the crop growth, weed absorbs excess nutrients because majority of the annual weeds complete its life cycle within 50-60 days where as the crops take 100-130 days.It is experimentally proved that in the highly infested rice field with Echinochloa,there is the uptake of 60-80% of the available nitrogen from the soil by weeds itself.In some cases weed takes up nutrients directly from the host plant.Some weeds activate denitrification,for example,Nutsedge releases a type of organic scented chemicals in the soil which increases the population of dentrifying bacteria and thus leaves very less amount of nitrogen for the crop.

(ii) For Water : There is a tough competition between crop and weed for water in unirrigated and dryland areas. In such area,scarcity of soil water is the most common feature, The water uptake is comparatively more by weeds.The amount of water required to produce unit amount of dry matter is called transpiration ratio.The transpiration ratio of weed is comparatively high. It means in comparison to crop,to produce one tonne biomass of weed requires more water than that of crop.Under dryland areas,soil water is absorbed quickly by weeds depriving the crop of water and the crop suffers adversely.The weed is capable to extract soil moisture even available at above 0.3 atmospheric metric tension.Therefore weed thrives well at permanent wilting point of the crop (PWP for most of the crop is 15 atm),.The transpiration ratio of *Cynodon dactylon* is around 813 compared to 450 for Bajra and 430 for sorghum.Amaranthus depletes water from deep soil layers and maintains its higher leaf water potential and turgor potential than cotton through out the growth period.Water hyacinth transpires 140-170% more water in comparison to normal evaporation from the open water surface.Therefore weed control is utmost important in dryland or barani agriculture.

(iii) For Light : The vast growth and rapid biomass production of weeds cause shading and smothering effect on the associated crop.Majority of photosynthetically active radiations (PAR)is harvested by weeds which results in reducing photosynthesis in the crop.Thus it reduces the life of the smothered and shaded crop.In the groundnut,weeds like *Celosia argentia* and *Amaranthus viridis* cause shading effect and interfere in harvesting of light.It is reported that *Amaranthus hybridis* reduced the photon flux density by 90% reaching the cotton canopy at mid-day.

(iv) For CO_2 : When weed infestation is more,competition for CO_2 may set in.Most of the weeds are C_4 plants therefore weeds deplete CO_2 to a very low concentration in comparison to crops.

(v) Allelopathic effect : Some of the weeds releases some chemicals which have allelopathic effect on crops.Such weeds are *Agropyron repens, Sorghum halepense,Lantana camara, Abutilon theophrasti, Cyperus rotundus, Euphorbia maculata, Ambrosia psitostachya* etc. The extracts from the rhizomes of Ambrosia reduce germination and seedling growth of several crops like wheat,oats,tomato and lucerne.Exudates of roots of wild oat at 2 and 4 leaf stage are toxic to wheat crop.The extract from *Abutilon theophrasti* decreased water status of soyabean and this induced water shortage in soyabean,leads to stomatal closure and breaking down of chlorophyll.

The critical period for crop=weed competition is around 30 days for most of the crops.The average annual loss due to weeds is 30-40% in different crops.The rapid and vast growth of the crop reduces the length of this critical period.Such crops are sorghum, maize, sunflower and cowpea etc.The length of the critical period for dwarf varieties is shorter and for taller traditional varieties is longer.

Characteristics of weed :

1. They thrive well even under adverse conditions.
2. They spread vegetatively.
3. Able to regenerate lost parts.
4. Compete with cultivated plants.
5. Produce enormous number of seeds.
6. Its seed remain viable for many years.
7. All seeds of weed mature simultaneously.
8. Some weed seeds have hairy appendages.
9. Weeds have allelopathic effects on crops.
10. It produces seeds in flushes.
11. Seed production through apomixis,spore etc.
12. Weed seeds have dormancy either inherited or induced.

Multiplication and Dissemination :

Weeds are prolifc with high fecundity. It multiply profusely by—

(i) Bearing seeds or

(ii) Propagules or by

(iii) Both

Weeds are disseminated by three ways -

(i) Weed seeds either fall near the mother plant or

(ii) Travel a long distance and short distance or

(iii) Travel with agricultural crops.

Most of weeds travel by various agencies -

(a) Structural modification like saccatie fruits,winged,parachute fruits,plumed fruits,hook or spiny appendages.

(b) Introduction from outside like Johnson grass *(Sorghum halepense)*.

(c) Impurities in crop seeds and in feed stuff.

(d) Threshing and cultivating machines spread rhizomes.

(e) Manuring with partially decomposed FYM & compost.

(f) Wind e.g. canada thistle seeds disseminate by wind.

(g) Water/irrigation.

(h) Animals e.g.xanthium.

(i) Human activities e.g. *Phalaris minor*.

Weed Control :

→ Preventive method

→ Curative method — Eradication / Control

(A) Preventive method :

Has two dimensions -

(i) **Time :** to prevent the infestation prior to weed germination.

(ii) **Space :** to prevent the introduction or spread to new areas.

Preventive method has following measures -

(a) Crop management practices are :

(i) Use of vigorous and fast growing varieties.

(ii) Proper placement of fertilizer.

(iii) Better irrigation practices.

(iv) Proper crop rotation.

(v) Higher plant population.

(vi) Effective prevention of weed seed production both in cropped and non-cropped area.

(b) Use of weed free crop seeds :Following procedures are adopted :

(i) Clearing and testing.

(ii) Separating crop seeds by separators.

(iii) Employing clean agl.equipments.

(iv) Using well decomposed weed free FYM and compost.

(v) Adopting measures to prevent carrying of weed seeds and propagules.

(c) Seed certification.

(d) Weed Laws :There is no weed law in India except Karnataka which declared *Parthenium hysterophorus* as a noxious weed.

(B) Curative/Remedial Method :

(B1)Eradication methods; are

(i) Destroying the spp.at the initial stage of introduction and before it produces any propagules or enforces its regenerative capacity i.e. at an early growth stage.

(ii) Degenerating the buried dormant but viable seeds by fumigation, flooding, heating & other methods.

(B2) Control Methods

→ Cultural methods.

→ Biological methods.

→ Chemical methods.

(a) Cultural methods :

→ Mechanical/Physical methods.

→ Cropping and competitive/Ecological method

(i) Mechanical or physical methods—

(i) Hand weeding :Two to four hand weedings for most of the field crops.The interval between two hand weedings is 15-20 days.

(ii) Dredging :With the help of mechanical force to remove weeds along with their roots and rhizome.

(iii) Chaining :Floating aquatic weeds are removed by chaining. Other physical methods are – Hand hoeing, Digging, Spudding, Tillage, Mowing, Mulching, Flooding, Clipping, Burning, Chilling.

(ii) Cropping and competitive/crop husbandry or ecological method :

(i) Selective stimulation of crops.

(ii) Bushing cultivation :Ploughing of field in the standing crops to control weeds & reduce over crowding.

(iii) Stale seed bed technique :weeds are allowed to germinate by wetting or rain and they are tilled before sowing rice.

(iv) Crop rotation.

(v) Summer fallowing.

(vi) Competitive cropping.

(vii) Clean cultivation.

(viii) Suitable time and method of planting crops.

(b) Biological method :–by employing crop plants, parasites, predators and pathogens

(i) Cropping and competition.

(ii) Use of parasites ,predators & pathogens;Before the release of such agents "starvation tests " are conducted regarding the safety of non-target flora in and around the area under a specific habitat.

Insects :1st attempt to control weeds by insects in 1920s to control ***Lantana camara*** a prickly shrub introduced in Hawaiian islands around 1860.

In Australia :Lantana camara was controlled by two beetles viz. *Octotoma scabripennis and Vroplata giraldi.*

In Argentina :To control opuntia, *Cactoblastis cactorum* and *Dactylopius opunita* (from USA) were used.

Sri Chandrashekhar Lohumi discovered a bug (1975) that destroys the flowering weed Lantana in Nainital.

Prickly pear weed (Opuntia) by *Dactylopius indicus* and *D.tomentous* in Maharashtra *and* T.N.

Parthenium (Congress weed)was successfully controlled by ***Zygo- gramma biocolorata*** when it was introduced in 1983 from Mexico by IIHR Bangalore.

Alligator weed (Alternanthera philoxeroides)aquatic weed by *Agasicles hvgronhvla.*

Fungi : *Water hyacinth (Eichhornia crassipes)by Rhizoctina blight.*

Eichhomia crassipes by Alternaria eichhomia. (Water hyacinth was first introduced in India as ornamental plant in 1896 from Brazil.)

Opuntia sp.by Fusarium.

Repeated grazing also control weeds.Ducks destroy floating weeds.Pigs devour underground storage tissues.

Limitations of Biological control :

(i) The weed must be highly specialised.

(ii) It must, in its native habitat be subjected to control by insect that will thrive in the new environment in which it has established itself.

(c) Chemical Control :

- The earliest attempt in India was made to control weeds by herbicides in 1937 in Punjab for controlling *Carthamus oxycantha* by using sodium arsenite.
- Credit for introducing 2,4-D as a herbicide goes to Marth and Mitchell of USA in 1944.2,4-D was first time tested in India in 1946.
- It was Paraquat which introduced the chemical concept in weed control in India particularly in tea plantation.
- Maximum herbicide is used in Tea & coffee because there only way is to use herbicide.

Classification of Herbicides :

1. On the basis of chemical structure :

→ **Inorganic :** First chemical used, Arsenic, Sodium sulphuric acid, Sodium arsenate, Sodium chlorate, Broax, Copper sulphate, Copper nitrate etc.

→ **Organic Hebicides :** 16 to 17 group

Group	Herbicides
1. Aliphatics	**Dalapon, TCA,** Acrolein CH_3Br.
2. Amides & Acetamides	**all - chlor** like Butachlor, Alachlor (Lasso), Propachlor, **Propanil**
3. Benjoics	2, 3, 6-TBA, Dicamba, Tricamba, Chloramben
4. Bipyridiliums	**Paraquat, Diquat** (contact)
5. Carbamates	Propham, barban, dichlormate
6. Thiocarbamates	Butylate, **Thiobencarb** or benthiocarb
7. Dithio carbamates	C D E C, Metham
8. Nitriles	Bromoxynil, Dichlobenil.
9. Dinitro anilines	**Fluchloralin (Basalin), Pendimethalin,** Trifluralin, Nitralin.
10. Phenols	Dinoseb, D N O C, P C P
11. Phenoxys	2,4,5-T; MCPA, 2,4 - DB; Dichlorprop.
12. Triazines	**Atrazine, Propazine,** Simazine (Soil applied)
13. Ureas	All-ron (like Diuron, Mouron, **Isoproturon**)
14. Uracils	Bromacil, Terbacil.
15. Diphenyl ethers	**Nitrofen** (Toke-25)
16. Others	Picloram, Pyrazon.

Common and Trade name of Herbicides :⟶

S. No.	Common name	Trade name	Sl. No.	Common name	Trade name
1	Acetachlor	Harness	16	Cinosulfusam	Setoff
2	Acetachlor Safener	Surpass	17	Chlorimun	Classic,Kloben
3	Acifluorfen	Blazer	18	Chlosulform	Glean,Telar
4	Alachlor	Lasso, Crop-star, Judge	19	Cinmethylin	Argold,Cinch
5	Ametryn	Evik, Gesapex	20	Clrthodin	Select
6	Anilifos	Aniloguard	21	Clodinafop	Topik
7	Atrazine	Aatrex, Gesaprim, Conquest, Atrataf	22	Clomazon	Command
8	Barban	Carbyne	23	Clopyraid	Reclaim, Lontrel
9	Benefin	Balan,Benfluralin	24	Dalapin	Dowpon,Hexapon
10	Bensuffurm	Londax	25	Diallat	Avadex
11	Bensulide	Prefar,Betasan,Pre San G	26	Dicamba	Banvel,Trooper, Clarity
12	Bentazon	Basagran,Pledgd	27	Dichlobend	Casoron,Dyclomec, Norosac
13	Butachlor	Machete, Aimchior,Lambast	28	Diclofop	Hoelon,Illoxan
14	Butylate	Sutan	29	Difenzoausal	Avenge
15	CDAA	Randox	30	Diquat	Aquaclde, Weedtrine D,Reglone

S. No.	Common name	Trade name	Sl. No.	Common name	Trade name
31	Dithiopyr	Dimension	50	Linuron	Lorex,Linex, Afalan
32	Diuron	Karmex	51	MCPA	Chiptox, Rhomene, Class, Agrozone
33	DSMA	Ansar184, Ansar8100	52	MCPB	Thristol, Cantrol
34	ETPC	Eptam	53	Mecoprop	Mecomec, Methoxone M,Vipex
35	Fenoxaprop	Whip, Bugle, Option, Whip Super, Puma Super	54	Metolachlor	Dual
36	Fluazifop-butyl	Fusilade	55	Metoxuron	Dosanex
37	Fluchloralin	Basalin	56	Metribuzin	Sencor, Lexone
38	Fluometuron	Cotoran, Meturon	57	Metsulfuron	Ally,Escort
39	Fluoroxypyr	Satane, Starrane	58	Metsulfuron-m-ethyl	Algrip
40	Glufosinate	Basta, Liberty, Ignite, Finale, Rely	59	Metsulfuron+-Chlorimuron	Almix
41	Glyphosate	Roundup, Accord, Honcho, Ranger, Glycel	60	Molinate	Ordram,Hydram
42	Halosulfuron	Permit	61	MSMA	Ansar 529,Drexar,Acme
43	Haloxyfop	Galant, Verdict, Focus	62	Napropamide	Devrinol
44	Imazethapyr	Pursuit, Hammer	63	Nitrofen	Tok
45	Ioxynil	Totril	64	Nicosulfuron	Accent
46	Isopropalin	Paarlan	65	Norflurazon	Zorial,Evital, Predict
47	Isoproturon	Arelon	66	Oryzalin	surflan,Ryzelan
48	Isoxaben	Gallery	67	Oxadiargyl	Raft, Topstar
49	Lactofen	Cobra	68	Oxadiazon	Chipco-Ronstar, Ronstar

S. No.	Common name	Trade name	Sl. No.	Common name	Trade name
69	Oxyfluorfen	Goal	84	Quizalofop-p	Assure, Targa super
70	Paraquat	Gramoxone, Cyclone, Weedol, Herb-axone	85	Rimsulfuron	Titus
71	Pebulate	Tillam, PEBC	86	Sethoxydim	Poast, Vantage, Torpedo
72	Pendimethalin	Prowl, Stomp, Herbadox	87	Simazine	Aquazine, Princep, Gesatop
73	Phenmedipha-m	Spin-aid, Betanal	88	Sulfometuron	Oust
74	Picloram	Tordon, Borolin, Grazon	89	Sulfosate	Touchdown
75	Pretilachlor	Rifit	90	Sulfosulfuron	Leader
76	Primisulfuron	Beacon	91	Benthiocarb Thiobencarb	Bolero, Abolish, Saturn
77	Prometon	Pramitol, Gesafram	92	Tralkoxydim	Achieve, Grasp
78	Prometryn	Caparol, Gowanprometryne, Cotton pro, Gesagard	93	Triallate	Far-Go, Avadex BW
79	Pronamide	Kerb	94	Triasulfuron	Logran, Amber
80	Propachlor	Ramrod, Bexton	95	Tribenuron	Express, Granstar
81	Propanil	Stam, StamF-34, Proster, Wham DF, Riverside	96	Triclopyr	Garlon, Turflon
82	Pyrazosulfuron	Clincher	97	Trifluralin	Treflan, Tri-4, Gowan, Trilin, Heritage, Advance
83	Quinclorac	Facet	98	2,4-D	2,4-D Weedar, Plantgard Weedtrol,Weedmar, Fernoxone

Courtesy : Handbook of Agriculture (ICAR)

2. On Selectivity :

(a) Selective herbicide : Such herbicide kills only target weeds e.g. Simazine,Atrazine,2,4-D, MCPA, Butachlor, Pendimethalin, Fluchloralin, Isoproturon etc.

(b) Non-selective herbicide : Kills all vegetation (crop+weed both) when comes in contact e.g. **Diquat** and **Paraquat**.However selectivity depends on dosage of herbicide.

3. On Translocation :

(a) Systemic herbicide : Such herbicides move within the weed either through xylem or phloem and thus affect the whole system like photosynthesis & respiration.Most of the systematic herbicides are selective at recommended dose e.g. Propanil, 2,4-D, Atrazine, Simazine.

(b) Contact herbicide : kills the vegetation either weeds or crops when it comes in contact.It kills the part of the plant that is in contact with herbicide e.g. **Paraquat** & **Diquat**.

4. On the basis of time of application :

(a) Pre-plant applied : such herbicides are applied before planting of crop in field e. g. **Fluchloralin** & **Alachlor**.

(b) Pre-emergence : applied before emergence of weeds but post-emergence of crops e.g. only selective herbicides.

(c) Post-emergence : applied after the emergence of weed (and also crop)e.g. 2,4-D,Propanil,Diquat,Paraquat,Dalapon.

5. On the basis of Method of application :

(a) Foliage : It may be either contact or translocated.

(b) Soil applied : either selective or nonselective.

(c) Aquatic application : e.g. Copper sulphate,2,4-D.

- **Synergistic Effect**

(i) Atrazine +2,4-D (low conc.)

(ii) Paraquat (low)+Pentachlor.

(iii) Atrazine +Alachlor (widely used in corn).

* **Antagonistic effect :** Generally contact +systemic herbicide combination show antagonistic effect.

(i) Dalapon+Atrazine

(ii) TCA+2,4-D.

(iii) TCA+MCPA.

	Short name	Full name
1.	**2,4-D**	: 2, 4-dichlorophenoxy acetic acid
2.	2, 4-DB	: 2, 4-dichloro phenoxy butyric acid
3.	2,4,5-T	: 2, 4, 5 trichlorophenoxy acetic acid
4.	TCA	: Trichloro acetic acid
5.	Propanil (Stam - F34, Rogue)	: 3,4-dichloropropionanilide
6.	MCPA	: 2-methyl, 4-chlorophenoxy acetic acid
7.	MCPB	: 2-methyl, 4-chlorophenoxy butyric acid
8.	Dalapon	: 2, 2- dichloropropionic acid (Sod. Salt)
9.	IPC (propham)	: Isopropyl N - phenyl carbamate

- National Research, centre for weed science at Jabalpur (1988). 1st Director: Dr. Vishnu Mohan Bhan.
- **Active ingredient (a.i.) :** Chemical in commercial product that is directly responsible for the herbicidal activity is called active ingredient. Generally expressed as % by weight or by volume.

$$\text{The amount of commercial Product to be required} = \frac{\text{Recommended doze}}{\text{a.i.}} \times 100$$

- **Acid Equivalent (a.e.) :** Some herbicides are active organic acids like phenoxy acetic acid,picloram& chloramben.But many of these are generally supplied in the form of their salts and esters e.g. instead of 2, 4- D.in acid form,sodium salt or amine salt or ester form is used.The theoretical yield of parent acid from such a herbicide formulation is called its acid equivalent.The acid equivalent of sodium salt of 2,4-D is 92.5%.It means 2,4-D is 92.5% in sodium salt of 2,4-D and a.e. is less than a.i.

Q. A product has 80% a.e.;To apply 2 kg a.e./ha What will be the quantity of formulation required.

Solution :

∵ To provide 80 kg a.e. 100 kg formulation is required.

∴ To provide 2 kg a.e.100/80 x 2 = 2.5 kg/ha

Q. One has to apply 0.5kg a.i./ha of 80% atrazine then quantity of atrazine required would be

(a) 0.40kg, (b) 0.48kg (c) 0.52kg (d) 0.62kg/ha

Solution :

$\because$ 80kg atrazine is available in its 100 kg formulation.

$\therefore$ 0.5 kg atrazine is available in its $\frac{100}{80}$ x 0.5=0.62kg/ha.

- **Orobanche** (Broom rape):Total root parasite on **solanaceous** plants like tobacco,chilli,Brinjal,Potato.To control it,a long crop rotation with tobacco crop once in 3 years and preceded by a chilli crop (as a trap)will reduce the loss due to parasite.
- **Striga** (witch weed):Semi-root parasite found in **sorghum,maize, sugarcane,sunflower.**

* **Herbicides for weed control in different crops.**

Crops	**Herbicides**	**Dosage (kg a.i./ha)**	**Time of application**	**Remarks**
Wheat	2,4-D	0.5	30-35 DAS	2,4 -D does not control grassy weeds leaf
	Isoproturon	1.0	30-35 DAS	
	Pendimethalin	1-1.5	PE	
	Clodinatop	0.06	30-35 DAS	Do not control broad weeds
	Sulfosulfuron	0.025	30-35 DAS	
	Fenoxaprop	0.12	30-35 DAS	
Sugarcane	Simazine or Atrazine	2.0	PE	Spray atrazine or simazine with high volume spray
	Glyphosate	2.0	POE	To be applied as directed spray, controls all weeds
Barley	2,4-D	0.5	30-35 DAS	Does not control grasses
Rice(transplanted)	Butachlor	1.5	All at 3-4 DAT	Do not control sedges
	Fluchloralin	0.65		
	Benthiocarb	1.5		
	Oxadiazon	0.6		
	Anilophos	0.4		
	Pendimethalin	1.5		

Pearl millet	Atrazine	0.5	PE	
Cotton	Fluchloralin Diuron Paraquat Pendimethalin	0.65 0.5 0.6 0.75	PPI PE POE (directed spray) PE	
Sorghum	Atrazine	1.0-1.5	PE	
Maize(corn)	Atrazine Simazine Alachlor	1.0-1.5 0.5 1.5	PE PE PE	
Pulses (green gram, black gram, pigeon pea,lentil, chickpea, cowpea)	Fluchloralin Pendimethalin Clodinafop Quizalofop	0.75 1.0 0.06 0.05	PPI PE Post Post	Control most grasses and some diots For control of annual grasses only
Soyabean	Alachlor Pendimethalin Metolachlor Fluchloralin Trifflurallin	2.0 1.5 1.5 1.0 1.0	PE PE PE PPI PPI	
Oilseeds (groundnut,sesame,mustard,sunflower,linseed)	Fluchloralin Pendimethalin Alachlor Isoproturon Clodinafop Quizalofop	1.0 1.5 1.5 0.75 0.06 0.05	PPI PE PE POE POE POE	Only on mustard crop. For control of annual grasses

DAS → Days after sowing; DAT → days after transplanting;

PE → Pre-emergence; POE → Post-emergence

PPI → Pre-plant incorporation

Courtesy : Handbook of Agriculture (ICAR)

❑❑❑

11 Assessment of Agronomic Interactions

(a) Assessment of competition and yield advantage

1. LER (Land Equivalent Ratio): It denotes relative land area under sole crop required to produce the same yield as obtained under a mixed or an inter cropping system at the same level of management.It is the ratio of land required by pure crop to produce the same yield as that of inter-crop.

$$LER = La + Lb = \frac{Ya}{Sa} + \frac{Yb}{Sb}$$

Where La → LER of crop 'a'

Lb → LER 'of crop 'b'

Ya and Yb → Yield of individual crops 'a' & 'b' respectively in mixture.

Sa and Sb → Yield of individual crops 'a' & 'b' respectively in pure stand.

Let La = 0.70 and Lb = 0. 40

∴ LER = 1.10

LER = 1.10 means to produce the combined mixture yield by growing in pure stands would require 10% more land i.e. the mixture gives 10% yield advantage.

Thus it gives a better picture of the competitive abilities of the component crops.It also gives actual yield advantage of intercropping.

In another words LER is the unit to measure the production efficiency of different intercropping system by converting the production in terms of land acreage.LER gives an accurate assessment of the biological efficiency of intercropping.

When LER > 1. intercropping is beneficial.

2. Relative Crowding Coefficient (RCC): proposed by de Wit (1960).

It is used in replacement series of intercropping.It indicates whether a species or crop,when grown in mixed population,has produced more or less yield than expected in pure stand.

In 50 : 50 mixture -

$$Kab\ (RCC) = \frac{\text{Mixture yield of a}}{\text{pure stand yield of a - mixture yield of a}}$$

$$= \frac{Yab}{Yaa - Yab}$$

Yab → mixture yield of a crop grown with b

Yba → mixture yield of b crop grown with a

Yaa → yield in pure stand of crop a

Ybb → yield in pure stand of crop b

Zab → proportion of sown spp. a in mixture with b

Zba → proportion of sown spp. b in mixture with a.

For all mixture :

$$Kab = \frac{Yab \times Zba}{(Yaa - Yab)\ Zab}$$

K > 1; Means yield advantage (more yield than expected)

K = 1; No difference

K < 1; Yield disadvantage(less yield than expected)

Crowding coefficient and LER give the yield advantage but only LER gives the magnitude of advantage. Therefore LER is preferred to assess the competition effects and yield advantage in intercropping situations.

3. Aggressivity: Proposed by Mc Gilchrist (1965). It is the mixture of how much the relative yield increase in component a is greater than that for component b.

$$Aab = \frac{\text{Mixture yield of a}}{\text{Expected yield of a}} - \frac{\text{Mixture yield of b}}{\text{Expected yield of b}}$$

$$Aab = \frac{Yab}{Yaa \times Zab} - \frac{Yba}{Ybb \times Zba}$$

Aab = 0 means component crops are equally competitive.

Aab = negative means dominated.

Aab = bigger value either positive or negative means bigger difference in competitive abilities.Agressivity and crowding coefficient can be used only in replacement series whereas LER in both replacement and additive series.

4. Competition Index: By Donald (1963).

It is measure to find out the yield of various crops when grown together as well as separately.It indicates the yield per plant of different crops in mixture and their respective pure stand on an unit area basis.

If the yield of any crop,grown together is less than its respective yield in pure stand then it is harmful association but on increased yield means positive benefit.

$$CI = \frac{(Yaa - Yab) \times (Ybb - Yba)}{Yaa \times Yab}$$

5. Competition Ratio: by Willey and Rao (1980)

$$CRa = \frac{Yab}{Yaa \times Zab} \div \frac{Yba}{Ybb \times Zba}$$

$$= \left(\frac{Yab}{Yaa} \div \frac{Yba}{Ybb}\right) \times \frac{Zba}{Zab} = \frac{LERa}{LERb} \times \frac{Zba}{Zab}$$

It is simply the ratio of individual LERs of the two component crops,but correcting for the proportion in which they were initially sown.

6. Competition coefficient: Ratio of the Relative crowding coefficient (RCC) of any given spp. In the mixture.

$$CC = \frac{\text{RCC of a given spp.}}{\text{Total RCC of all crops in mixture}}$$

It is used to find out the relative crowding from which maximum yield can be obtained without any adverse effect on any of the species.

(b) Assessment of land use and productivity in multiple cropping :

1. Multiple Cropping Index: by Dalrymple (1971)

It measure the sum of areas planted to different crops and harvested in a single year divided by total cultivated area times 100.

$$MCI = \frac{\sum_{i=1}^{n} ai}{A} \times 100$$

Where n = Total no. of crops

ai = area occupied by i^{th} crop

A = total land area.

2. Cropping Intensity Index: by Menegay et al (1978)

It assesses a farmer's actual land use in area and time relationships for each crop or group of crops compared to the total available land area and time, including land temporarily available for production.

$$CII = \frac{\sum_{i=1}^{Nc} ai\ ti}{A_0 T + \sum_{j=1}^{M} Aj\ Tj}$$

Where

Nc = Total no. of crops grown during time period T

ai = area occupied by i^{th} crop.

ti = duration occupied by i^{th} crop (months that crop occu pied area ai)

T = time period (usually one year)

Ao = Total cultivated land area available for use during T.

M = Total no. of fields temporarily available during T.

Aj = land area of jth field.

Tj = Time period when Aj is available.

Efficient cropping zone is judged by **CII** and **LER**.

3. Cropping intensity/Intensity of cropping :

$$CI = \frac{\text{Total cropping area}}{\text{Net cultivated area}} \times 100$$

$$= \frac{\text{Area under Kharif + Rabi + Zaid Crops}}{\text{Area under actual cultivation}}$$

4. Specific Crop Intensity Index: by Menegay et al (1978)

It is derivative of CII and determines the amount of area-time denoted to each crop or group of crops compared to the total time available.

$$SCII = \frac{\sum_{K=1}^{N_k} a_k t_k}{A_0 T + \sum_{j=1}^{M} A_j T_j}$$

N_k = Total no. of crops during T

A_k = area occupied by K^{th} crop

T_k = duration of Kth crop.

It is used for vegetable intensity index,rice intensity index,field crops intensity index etc.

5. Relative Cropping Intensity Index: It is the modification of CII and determines the amount of area-time allotted to one crop or group of crops relative to the area-time actually used in the production of all the crops.

$$RCII = \frac{\sum_{K=1}^{N_k} a_k t_k}{\sum_{i=1}^{N_c} A_k T_i}$$

It is used for classifying farmers viz. when relative vegetable intensity index is more than 50%,then farmer will be called vegetable grower.

6. Cultivated land Utilization Index: by Chuang (1973).

It is calculated by summing the products of land area planted to each crop, multiplied by the actual duration of that crop and divided by the total cultivated land area times 365 days.

$$= \frac{\sum_{i=1}^{n} a_i d_i}{A \times 365}$$

d_i = days by i^{th} crop occupied.

A = Total cultivated Area.

Diversity Index: by Strout (1975) and Wang and Yu.

It measures the multiplicity of crops or farm products which are planted in a year by computing the reciprocal of sum of squares of the share of gross revenue recieved from each individual farm enterprise in a single year,

Q.1 In the following farms, which farm is mostly specialised.

Crop	**Income (Rs) From Different Farms**		
	A	**B**	**C**
Sugarcane	30,000	----	10,000
Cotton	10,000	20,000	20,000
Wheat	40,000	20,000	10,000
Jowar	20,000	10,000	40,000
Potato	----	50,000	----
Total	1,00000	1,00000	1,00000

Solution :

Crop	**Shares of individual crops in different Farms**					
	Farm 'A'		**'B'**		**'C'**	
	Share	Square of its share	Share	Square of its share	Share	Square of its share
Sugarcane	0.3	0.09	---	----	0.125	0.0156
Cotton	0.1	0.01	0.2	0.4	0.25	0.0625
Wheat	0.4	0.16	0.2	0.4	0.125	0.0156
Jowar	0.2	0.04	0.1	0.01	050	0.2500
Potato	---	----	0.5	0.25	---	----
Total	1.0	0.30	1.00	0.34	1.0	0.3437

Diversity Index :

$$\text{Farm A} = \frac{1.00}{0.30} = 3.33$$

$$\text{Farm B} = \frac{1.00}{0.34} = 2.94$$

$$\text{Farm C} = \frac{1.00}{0.3437} = 2.909$$

Therefore, Farm 'A' is more diversified and Farm 'C' is more specialised.

Lower is the diversity Index, Higher is the specialisation.

Rotational Intensity :

$$RI = \frac{\text{no. of crops grown in a field}}{\text{Years of rotation}} X\ 100$$

e.g.

Maize – Potato – Onion : 1 Year.

$$RI = \frac{3}{1} X\ 100 = 300\%$$

Similary :

1. Maize - Sugarcane : 2 Years
$$RI = \frac{2 X 100}{2} = 100\%$$
2. Maize - Sugarcane - Ratoon : 3 Years, RI = 100%
3. Paddy - Pea - Potato - Gram : 2 Years, RI = 200%
Because pea & Gram are grown in same season.
4. Maize - Early Potato - Late Potato - Chillis : 1 Years, RI = 400%
5. Jowar + Guar - Berseem + Oats : 1 Year.
$RI = \frac{2 X 100}{1} = 200\%$ '+' Sign means mixed cropping
6. Jowar + Arhar - Fallow - wheat : 2 Year.
Here Jowar & Arhar are sown together but Arhar is taken as Rabi crop. Therefore rotation takes 2 Years.
$$RI = \frac{3}{2} X\ 100 = 150\%$$
7. Jowar + Arhar - wheat : 1 Year. RI = 200%
8. Sanai – Sugarcane : 2 Years, $RI = \frac{1 X 100}{2} = 50\%$
↓
(For green manuring)

* Crop taken as green manuring is not considerd as crop grown.

9. Maize for cobs – Early Potato – Late Potato – Sugarcane – Ratoon – Barseem

$$RI = \frac{6 X 100}{3} = 200\%$$

10. Paddy – Pea – Sugarcane – Ratoon – Lobia / Wheat :

$$RI = \frac{5 X 100}{3} = 166.6\%$$

11. Fallow – Sugarcane – Ratoon

$$RI = \frac{2 X 100}{3} = 66.6\%$$

12. Paddy – Pea – Sugarcane – Ratoon – Lobia – Wheat : 4 Years.

How does this rotation take 4 years?

	Crop		Sowing	Harvesting
1.	Paddy	:	June 2003	Oct. 2003
2.	Pea	:	Oct./Nov.2003	Jan./Feb. 2004
3.	Sugarcane	:	Feb. 2004	Feb. 2005
4.	Ratoon	:	Feb. 2005	Feb. 2006
5.	Lobia	:	June/July 2006	Oct. 2006
6.	Wheat	:	Nov. 2006	April. 2007

Here	June 2003 – May 2004	1Year
	June 2004 – May 2005	1Year
	June 2005 – May 2006	1Year
	June 2006 – May 2007	1Year
	Total	4 Years

$$RI = \frac{6 X 100}{4} = 150\%$$

Agriculture Year : 1 June to 31st May

❑❑❑

12 Agro – Forestry and Social Forestry

Agro-Forestry/Forest Gardening :

Agro-Forestry is primarily a system where agriculture and forestry are practiced either simultaneously or separately on the same unit of land and has affinities with taungya system of regenerating forests which in Burmese means cultivation of trees and crops.Agriculture in forest is called agro-forestry.Dr. King and Chandler defined it as a sustainable land management system which increases the use of the land combines the production of crop (including tree crops)and forest plants and /or animals—simultaneously or sequentially,on the same unit of land and applies management practices that are compatible with the cultural practices of the local population.It is localized in its concept.

Main Feature :

The main feature of the multistoreyed production of different species of economic plants with suitable plant geometry.Agro-forestry is 6-F programme because its provide i.e. **Food** for man, **feed** for livestocks, **Fibre** for clothing, **Fuel** for village and urban homes, **Furniture** timber and **Funds** for the poor forest villagers and is govt. development programmes.

Neccessity of agroforestry :

(i) Continuous supply of fuel and fodder.

(ii) Ecological balance.

(iii) Fresh oxygen through photosynthesis.

(iv) Acts as windbreaks.

(v) Creates employment opportunity.

Objectives of agroforestry :

(i) Diversified and or more sustainable production at a higher level from the available resources.

(ii) Economically superior land use with intensive agroforestry systems on fertile soil of small and marginal farmers.

(iii) production of basic needs of farmers namely, food, fruit, fodder, fuel etc from their own piece of land and

(iv) Enhancement of land productivity and sustainability in lands affected by low soil fertility,high erosion and high soil degradation.

In hilly areas,the practice of agro-forestry has been developed.

(a) To get material for cheap and light type of constructions.

(b) To meet the basic requirements of fuelwood in the locality.

(c) To get fodder for livestock and green manure for agricultural crops and

(d) To get protection from cool breeze to agricultural crops.

Components of agroforestry :

The essential aim of the components of agro-forestry is to conserve and improve the site and to optimise the combined production.The different types and component of agro-forestry are—

1. **Agri-Silviculture:** The growing of agricultural crops along with the forest crops is known as agri-silviculture. Agri-silviculture = Agl.crops + Forest crops (silviculture).
2. **Sylvo-Pastoral systems:** Such land management systems in which forests are managed for the production of wood as well as for rearing of domesticated animals.

   Sylvo - Pastoral system : silviculture + Pasture managements.
3. **Agro-sylvo-pastoral systems :**

   Agl.crops + Forest crops + Pasture management.
4. **Agri. horti - Silviculture System :**

   Agri crops + horticulture + silviculture .

5. **Multipurpose forest tree production systems :**

Forestry for the multipurpose benefits or uses .

6. Boundry Plantation (trees is on boundry + crops)
7. Block Plantation (Block of trees + Block of crops)
8. Energy Plantation(trees + crops during initial years)
9. Alley cropping (perennial hedges + crops)
10. Agri-horticulture (fruit trees + crops)
11. Silvi-olericulture (trees + vegetables)
12. Horti-pasture (fruit trees + pasture or animal)
13. Horti - olericulture (fruit trees + vegetables)
14. Forage forestry (forage trees + pasture)
15. Shelter-belts (trees + crops)
16. Wind-breaks (trees + crops)
17. Live fence (shrubs and under - trees on boundary)
18. Silvi or Horti- sericulture (trees or fruit trees+sericulture)
19. Horti-apiculture (fruit trees + honeybee)
20. Aqua - forestry (trees + fishes)

Agroforestry Practice	Brief Description(Of arrangement of components)	Major groups of components	Agroecological adaptability
Agrisilvicultural Systems (Crops-including shrub/vine/tree crops-and trees)			
Improved fallow	Woody Species planted and left to grow during the 'fallow phase'	w: fast-growing preferably leguminous h : common agricultural crops	In shifting cultivation areas
Taungya	Combined stand of woody and agricultural species during early stages of establishment of plantations	w: usually plantation forestry spp. h : common agricultural crops	All ecological regions (where taungya is in practices); several improvements possible

Alley cropping (hedge-row Intercropping)	Woody species in hedges: agricultural species in alleys in between hedges; microzonal or strip arrangement	w : fast-growing, leguminous, that coppice vigorously h: common agricultural crops	Subhumid to humid areas with high human population pressure and fragile (productive but easily degradable) soils
Multilayer tree gardens	Multi species, multilayer dense plant associations with no organized planting arrangements	w : different woody components of varying form and growth habits h : usually absent; shade tolerant ones sometimes present	Areas with fertile soils, good availability of labour, and high human population pressure
Multipurpose trees on crop lands	Trees scattered haphazardly or according to some systemetic patterns on bunds,terraces or plot/field boundaries	w : multipurpose trees and other fruit trees h : common agricultural crops	In all ecological regions esp,in subsistence farming; also commonly integerated with animals
Plantation crop combination	(i) Integerated multistorey (mixed,dense)m-ixtures of plantation crops (ii)Mixtures of plantation crops in alternate of other regular arrangement (iii)Shade trees for plantation crops;shade trees scattered (iv)Intercropping with agricultural crops	w;plantation crops like coffee, cacao,coco-nut,etc,and fruit trees,esp in (i) fuelwood/fodder spp.,esp in (iii) h;usually present in (iv),and to some extent in (i);shade-tolerant species	In humid lowlands or tropical humid/subhumid highlands(depending on the plantation crops concerned) usually in small holder subsistence system

Home gardens	Intimate,multistorey combination of various trees and crops around homesteads	w;fruit trees predominate;also other woody species,vines,etc. h : shade tolerant agricultural species	In all ecological regions,esp,in areas of high population density
Trees in soil conservation and reclamation	Trees on bunds,terraces,-raisers,etc with or without grass strips; trees for soil reclamation	w : multipurpose and/or fruit trees h : common agricultural species	In sloping areas,esp in highlands, reclamation of degraded,acid alkali soils,and sand dune stabilization
Shelterbelts and windbreaks,live hedges	Trees around farmland/plots	w : combination of tall growing spreading types h : agricultural crops of the locality	In wind prone areas
Fuelwood production	Interplanting firewood species on or around agricultural lands	w : firewood species h : agricultural crops of the locality	In all ecological regions
Silvopastoral systems (trees+pasture and/or animals)			
Trees on range land or pastures	Trees scattered iirregularly or arranged according to some systematic pattern	w : multipurpose;of fodder value f : present a : present	Extensive grazing areas
Protein banks	Production of protein rich tree fodder on farm/rangelands for cut-and-carry fodder production	w : leguminous fodder trees h : present	Usually in areas with high person : land ratio

Plantation crops with pasture and animals	Ezample;cattle under coconuts in south-east Asia and the south pacific	w : plantation crops f : present a : present	In areas with less pressure on plantation crop lands
Agrosilvopastoral system (trees+crops+pasture/animals)			
Homegardens involving animals	Intimate,multistorey combination of various trees and crops, and animals, around homesteads	w : fruit trees predominate;also other woody species a : present	In all ecological regions with high density of human population
Multipurpose woody hedge rows	Woody hedges for browse, mulch, green manure,soil conservation etc	w : fast-growing and coppicing fodder shrubs and trees h : (similar to alley cropping and soil conversation)	Humid to subhumid areas with hilly and sloping terrain
Apiculture with trees	Trees for honey production	w : honey producing(other components may be present)	Depending on the feasibility of apiculture
Aquaforestry	Trees lining fish ponds,tree leaves being used as 'forage'for fish	w : trees and shrubs preferred by fish(other components may be present)	Lowlands
Multipurpose woodlands	For various purposes (wood,fodder,soil protection, soil reclamation etc.)	w : multipurpose species : special location-specific species(other components may be present)	Various

Note : W : woody, h : herbaceovs, f : fooder for grazing a : animals

Courtesy : Handbook of Agriculture (ICAR)

Social Forestry :

Social forestry means plantation of such trees which are useful for community development.The word 'Social forestry' was first time used by Westoby in 1968.He defined as "social forestry is a forestry which aims at producing flow of protection and recreational benefits for the community".According to the National commission on Agriculture (1976),Social forestry denotes programmes for raising plants and trees for supply of fire wood,fodder, and small timber for the community.

Principles of Social Forestry :

1. **Principle of democracy :** social forestry implies the culturing of trees by the people,for the people and of the people.

2. **Principle of forest area extension :** It aims to increase the forest area by rehabilitating wastelands while producing biomass both for industrial and local uses.

3. **Principle of Poverty – Eradication :** It is single largest development strategy to eradicate poverty by providing the job.

4. **Principle of employment :** It is a continous process discouraging the migration of labour to the urban habitats.

5. **Principle of Govt. Based programme:** It is primarily a govt. based programme. Sometimes social forestry is also tagged as 'sick land for sick people'.

Objectives :

(a) Ecological and environmental balance and purity.

(b) Easy and abundant availability of fuel and firewood for the families, fodder for their domestics and timber wood for domestic use.

(c) Forestring of agro-based industries in rural areas for absorption of people in gainful employment.

Components of Social Forestry Programmes :

1. **Farm Forestry :** Objectives of encouraging farmers to plant and raise trees on their own plot of land through free or subsidised supply of seedlings.In dry areas,trees are grown around the farm.

2. Rural Forestry : For the benefit of the community as a whole through massive plantations along roadside and canal banks,around tanks and ponds and on fallow and uncultivable lands. It is also called **Extension Forestry** as this results in extending forests beyond the existing boundaries.

3. Urban forestry : Forestry in the urban areas i.e. on the useless land near govt. buildings. schools, colleges and universities, roadsides, hospitals, recreation gardens, municipal areas etc.Community woodlands are planted by particular communities themselves on land of their own or on that pooled by themselves and benefits of which are shared by them equally.

Need for more forests :

By the sixth plan,only about 22.7%of the total land area is under forests and not more than 40-50 % of this has good forest cover.But for environmentally sound country,ideal forest area should be 33%.Every year country has been loosing 1.5 mha of forests and about 12000 m tonnes of top soil due to erosion.

With a vigorous programme of social forestry,India may land into a new green revolution through which it will be possible to increase wood in the villages.At the same time it will help the poor to cross the poverty line.

❑❑❑

13 Water Management
(Irrigation and Drainage)

Irrigation is the artificial application of water to land for growing crops or trees.According to Israelsen and Hansen (1962),the artificial application of water for the purpose of supplying moisture essential to plant growth is called Irrigation.

- **Free energy or gibbs free energy :** The free energy of a saturated soil is zero but in unsaturated soil,it is less than 0, hence it's value is negative.
- **Matric Potential :** The portion of total water potential that is attributable to the solid colloidal matrix of the soil system. It is negative pressure potential which results from the capillary and adsorptive forces emanating from the soil matrix.
- **Capillary potential :** Buckingham (1907) introduced the concept of 'capillary potential' to define the energy with which water is held by soil.
- **Gravitational potential :** Due to gravitational force field and is dependent on the elavation.
- **Seepage :** Horizontal flow of water in channel is called seepage. Water loss from the irrigation channel or canal is mainly due to seepage.
- **Percolation :** Vertical movement of water in the soil i.e.downward flow of water in the soil is called percolation.It occurs in water saturated soil.
- **Infiltration :** Downward movement of water from the upper layer of soil is called Infiltration. It occurs in unsaturated soil.

- **Leaching :** Downward movement of nutrients and salts from the root zone with the water is called Leaching.

P^F Value :

P means potenz meaning power at 10. P^F value was first time introduced and defined by Schofield (1935). P^F is the scale like P^H through which we measure the force with which water is retained in capillary or soil. F denotes 'force'. It is an exponential expression of a free-energy diference (based on the height of a water column above free water level in cm). It is defined as the logrithm to the base 10 of the numerical value of the negative pressure of the soil moisture expressed in centimeter of water.

$P^F = \log 10^h$ Where h = soil moisture in cm of water.

1 atmosphere (atm) =1036 cm of water or 76.39 cm of Hg.

1 bar = 1023 cm of water column $=10^6$ dynes/cm^2

1 millibar = 10^{-3} bar.

$P^F = 2$ means the force with which water is retained,is equivalent to the weight of a column of water 100 cm. in height. The scale of values ranges from 'O' (which corresponds to a pressure of $\frac{1}{1000}$ atm) to '7'(corresponds to 10,000 atm).

- Cumec means cubic metre per second = $m^3/s = m^3s^{-1}$

  Cusec means cubic feet per second = $ft^3/s = ft^3\,s^{-1}$

  1000 cubic centimeter = $1000cm^3$ = 1 litre

  1 cubic metre $= 1m^3 = 1m \times 1m \times 1m$

  $= 100cm \times 100cm \times 100cm$

  $= 100 \times 100 \times 100 \times cm^3$

  $= 1000 \times 1000cm^3 = 1000$ litre

$\therefore 1\ m^3 = 1000$ litre

- Moisture equivalent : an amount of water retained by a sample of initially saturated soil material after being subjected to a centrifugal force of 1000 times that of gravity for a definite period of time, usually half an hour.In medium textured soils, the values of field capacity and moisture equivalent are nearly equal.

FC ≅ ME in medium textured soil.

FC > ME in sandy soil

FC < ME in very clayey soil, field capacity is less than moisture equivalent.

- Permanent wilting point (PWP) was originally proposed by **Briggs** and **Shantz** in 1912.They utilized dwarf sunflower *(Helianthus anus)* as indicator plant.
- Higher the clay content of a soil,higher its hygroscopic coefficient. Organic matter also enhances it.
- **Wilting Coefficient :**

$$WC = \frac{\text{Hygroscopic Coeff}}{0.68} = \frac{\text{Moisture equivalent}}{1.8}$$

The percentage of moisture in rootzone at the permanent wilting of plants is called the wilting coefficient or critical moisture point (CMP).

- Moisture content (% by volume) = Moisture content (% by weight) X BD

  Let bulk density (B. D.) of soil = 1.6g/cc and moisture content on dry weight basis = 20%.

  Then moisture content by volume = 20 X 1.6 = 32%

  32% means 32 cm per 100 cm depth of soil.

  = 32 cm per metre depth of soil

- Soil moisture tension in a salt free soil at field capacity increases when texture changes from sandy to clay. In sandy soil =0.06, loamy sand =0.1, silt loam =0.3, clay =0.6.
- **Moisture movement under saturated conditions**

(a) Poisseuille's Law : expresses the flow of water in a narrow tube. According to it rate of flow of water in sandy soil is more than that of loam and least in the clay.

Sandy > loam > clay.

(b) Darcy's law : expresses water movement in soil. The quantity of water passing a unit cross section of soil is proportional to the gradient of hydraulic head.

Considering both the above laws :The conductivity of water does not depend upon the total amount of pore spaces.

Measurement of Soil Moisture Content :

Direct measurement of soil moisture content in the soil—

(a) Gravimetric method : soil is dried at 105-110 ^{0} c and before and after the drying, weight of the soil is taken. And from the difference, moisture percentage is calculated.

(b) Infra-red moisture balance method :This method is used either for individual tests or series of test, in practically all organic and in or-ganic materials directly. Infrared lamp is used for drying with moving air.

Indirect method -

(a) Look at the soil and feel it.

(b) Neutron Scattering method : rapid tests,in situ measurement of soil moisture, direct measurement of moisture content by volume.

Principle : Measurement of the no. of H^+ nuclei that are present in a unit volume of soil.

(c) **Tensiometer :** Two types – (i) Tensiometer with mercury manometer, used for research. (ii) With vacuum guage for practical field use. Tensiometer (or Irrometer) measure Soil moisture tension directly. It is used in coarse soil and the practical limit about 0.8 bars. Hence it can not be useful under all soil moisture conditions. Tensiometer consists of porous ceramic cup which may be chocked in clay soils and this cup is sensitive to salts to a certain extent. Therefore it is not suitable for clay soils & salty soils. Soil moisture tension in a salt free soil at field capacity ranges from less than 0.1 to 0.7 atm depending on soil texture.

(d) **Electrical resistance method :** Osmotic pressure is determined by using vapour pressure psychrometer. Resistance blocks are made up of various materials like gypsum, nylon, fibre glass, plaster of paris or a combination of these. Nylon and fibre glass units are more sensitive in the higher ranges of soil moisture. Plaster of paris blocks function most effectively between 1 to 15 atm, and Nylon block upto 2 atm and fibre glass units over entire range of available water. A combination of fibre glass+plaster of paris provides sensitivity in both wet and dry range and provides good contact between soil and the unit. Gypsum block or plaster of paris resistance units first prepared by **Bouyoucos and Mick** (1940) hence electrical resistance instrument is also called Bouyoucos moisture meter. It is especially designed on Wheat-stone bridge principle.

Resistance blocks like tensiometer can be used to schedule irrigation, to irrigate green house beds, to determine rooting zone and soil properties and to evaluate fluctuation in water table. Unlike tensiometers resistance blocks do not work properly at higher moisture levels. Their efficiency is better at lower moisture levels upto wilting point. Resistance blocks do not give precise result in saline conditions because the soluble salts interfere with conductance of the blocks. Thus nylon, plaster of paris and fibre-glass blocks should never be used while gypsum blocks may be used in saline conditions upto some extent. Resistance units are used in **fine textured soils.**

Potential Evapo-transpiration (PET) :

The concept of PET was given by Thornthwaite (1948) who defined it as evapo-transpiration from a large vegetation covered land surface with adequate moisture at all times,According to Dalton (1882) :Evaporation is a function of the difference in the vapour pressure of the water and the vapour pressure of the air.

$$E = f(V.P._{water} - V.P._{air})$$

- Solar energy arrives at the upper limit of the earth's atmosphere @ about 2 calories per minute per square centimeter.

Measurement of E.T. :

1. **Lysimeters :** It permit the specific measurement of ET, rainfall and water uptake by plants. Lysimeter are tanks, buried in the ground to measure the percolation of water through soils. Lysimeter are the most dependable means of directly measuring the evapo-transpiration rate, but their installation must meet four requirements :

(i) Lysimeters itself should be very large and deep to reduce the boundary effect and to avoid resticted root development. For short crops lysimeter should be at least one cubic meter in volume.

(ii) The physical conditions within the lysimeter must be comparable to those outside.

(iii) Lysimeter will not be representative of the surrounding area if the crop in the Lysimeter is either taller, shorter, denser or thinner, or if lysimeter is on the periphery of non-cropped area. The effective area of a lysimeter is defined as the lysimeter Evapotranspiration per unit area of the surrounding field.

(iv) Each lysimeter should have a guard ring area around it maintained under the same crop & moisture conditions in order to minimise the clothesline effect.

Lysimeter is of two types viz. - Non-weighing type lysimeter and weighing type.

Non-Weighing type Lysimeter is called drainage lysimeter. It operates on the principle that **E.T.** is equal to the amount of rainfall and irrigation water added to the system minus leaching.

ET = Rainfall + irrigation water – Leaching.

The drainage lysimeter is useful only in determining the PET rate and should be irrigated every four or five days,unless rainfall intervenes.

Weighing type Lysimeter : gives more accurate estimation of ET, capable of measuring ET for a short period,diurnal pattern of ET, the phenomenon of midday wilt, short term variation of energy partition and the relationship between transpiration and soil moisture tension. It measures the water balances i.e. water added water retained by soil and water lost through all sources – evaporation, transpiration and deep percolation. Effective rainfall is also measured. Therefore weighing type lysimeter are the most direct and accurate instrument for the determination of E.T.

2. Evaporimeters : used for estimating the PET of various crops. Most commonly used evaporimeters in India are (i) **US Weather** Bureau class. A pan and (ii) Sunken screen evaporimeter.

(i) **USWB class I Pan :** Most widely used evaporation pan and also written as USPAN (United States pan).Pans have higher rates of evaporation than a large free water surface.

∴ Evaporation from large water surface area

= Reading of Pan X Pan Coefficient

And pan coefficient = 0.7 approx.

ET = Pan evaporation X Crop factor.

(ii) **Sunken screen Evaporimeter :** Dastane and Sharma (1968)at IARI developed a sunken screen evaporimeter to simplify the mea suring device of E.T They observed

$$\frac{\text{Evaporation}}{\text{ET}} = 0.95 \text{ to } 1.05 \text{ as compared to } 1.20 \text{ to } 1.50$$

In USWB pan evaporimeter.

Consumptive Use of Water : It is used to designate the losses due to evapotranspiration and the water used by plant for its metabolic activities.Since water used in the actual metabolic process is less than 1% of ET i.e. insignificant,the term consumptive use =Evapotranspiration.

CU = ET + water used in metabolic activities

- Amount of water to be applied (in cm)

$$\frac{\text{Moisture}_{FC} - \text{Moisture}_{WP}}{100} \times \text{B.D.(g/cc)} \times \text{Depth of soil (cm)}$$

Where Moisture $_{FC}$ means Moisture content at Field capacity

Moisture $_{WP}$ = Moisture content at Wilting Point.

B.D.=Bulk density in g/cc

Water use Efficiency :

The water utilisation by the crop is generally described as water use efficiency (kg/ha-cm)or q/ha. cm .It may be defined in two ways .

(a) Crop water use efficiency : It is the ratio of crop yield (y) to the amount of water depleted by the crop in the process of evapo transpiration.

$$\text{WUE} = \frac{\text{Y}}{\text{ET}}$$

(b) Field water use efficiency : is the ratio of crop yield (y) to the total amount of water used in the field (WR).

$$WUE = \frac{Y}{WR} \quad \text{Where WR} \rightarrow \text{water required or used}$$

Increasing WUE in Unirrigated areas or low rainfall areas : Principal objective is to conserve and utilise a large portion of the available rainfall for crop production. To increase the productive utilisation of the atmospheric precipitation naturally reaching a given area, the following practices are required to adopt—

1. Maximum absorption of the precipitation into the ground i.e. elimination of surface runoff. The following practices are :
   (a) Runoff control.
   (b) Tillage
   (c) Increasing infiltration.
   (d) Water harvesting and recycling of water
2. Decreasing the agl. unproductive loss of moisture namely evaporation, consumption of water by weeds and losses by sub-surface flow and deep percolation.The following practices are :
   (a) Weed control.
   (b) Timely harvest of rainy season crops.
   (c) Mulching.
   (d) Tillage.
3. Increasing the moisture reserve of the plants :The practices which increase effective utilization of rainwater and stored soil moisture :
   (a) Growing suitable crops and their varieties.
   (b) By adopting suitable cropping sequence.
   (c) Increasing root depth of crops by removal of hard pan and fertilization.
   (d) Growing intercrops of different rooting depths.
   (e) Seeding technique and plant stand :Proper placement of seeds and fertilizers become important in winter season when crops are to grow on stored soil moisture.

Irrigation Efficiency : The percentage of applied irrigation water stored in the soil and made available for consumptive use by the crop. When the water is measured at its entry to a farm,it is called Farm Irrigation Eficiency; When measured at the field, known as Field Irrigation Efficiency, and when measured at the point of diversion, it is called as Project Efficiency.

Irrigation Efficiency (Ei)

$$= \frac{\text{Water stored in the soil for crop growth}}{\text{Water applied as irrigation}} \times 100$$

$$E_i = \frac{W_{et} + W_i - Re}{W_i} \times 100$$

Where, W_{et} = Volume of irrigation water in a specified area transpired by plant, and evaporated from soil.

W_i = Volume of water necessary for leaching (salt control)

R_e = Vol. of effective rainfall.

$$\Rightarrow \quad E_i = \frac{E_s}{100} \times \frac{E_c}{100} \times \frac{E_a}{100} \times 100$$

Where, E_s = Reservoir storage efficiency.

E_c = Water conveyance efficiency.

E_a = Water application efficiency.

Where, W_f = Water delivered to field.

Water conveyance efficiency $(E_c) = \frac{W_f}{W_d} \times 100$

W_d = Water delivered from source.

Water application efficiency $(E_a) = \frac{W_s}{W_f} \times 100$

Where, $W_n \rightarrow$ Water needed in the root zone prior to irrigation.

Water Distribution efficiency $(E_d) = \left(1 - \frac{\overline{Y}}{\overline{d}}\right) \times 100$

Where, $\overline{d}$ = Average depth of water stored.

$\overline{y}$ = Average numerical deviation from $\overline{d}$

Q. Suppose, average depths of water penetration at five places are 75, 80, 97, 100, 95 cm respectively. What will be the water distribution efficiency of the field?

Solution :

$$\overline{d} = \frac{75 + 80 + 97 + 100 + 95}{5} = 89.4 \text{ cm}$$

$$\overline{y} = \frac{|89.4-75| + |89.4-80| + |97-89.4| + |100-89.4| + |95-89.4|}{5}$$

$= 9.52$ cm

$$\therefore E_d = \left(1 - \frac{9.52}{89.4}\right) X\ 100$$

- Irrigation period : The number of days that can be allowed for applying one irrigation to a given area during the peak consumptive use period of the crop that is irrigated.

$$\text{Irrigation Period} = \frac{\text{Net irrigation requirement}}{\text{Peak use rate}}$$

- Intensity of Irrigation : The area that is proposed to be irrigated every year or every season is known as the intensity of irrigation, usually expressed as the percentage of the total cultivable area.
- Crop ratio : The ratio of area under different crops of a particular canal or entire project is called crop ratio. Normally a winter crop like wheat requires about $\frac{1}{4}$ th water requirement of a Kharif crop like rice. If there is only rice in Kharif then crop ratio of Rabi to Kharif will be 4 :1.
- Moisture Regime :The percentage of moisture in the soil at atmospheric pressure is known as moisture regime.
- Duty of Water : A form of expression for the quantity of water required for irrigation to bring a crop to maturity.
- Base : The period of irrigation which crop requires for full maturity.

- Delta (Δ) : The total depth of water required by a crop during the entire period the crop is in the field.

$$\text{Duty of Water} = \frac{8.64 \text{ X Base}}{\text{Delta } (\Delta)}$$

Duty of Water :

→ Grosss duty of water : Area commanded by the flow of water as measured at the source of supply; It includes wastage in channel in addition to what is used for measuring crops.

→ Net duty of Water : Area commanded by water delivered at field. It includes losses of water in the field.

The difference between gross and net duty of water gives efficiency of distributaries.

- **Weirs :** Used to measure the flow in an irrigation channel or the discharge of a well or canal outlet at the source.A weir is a notch of regular form through which irrigation stream is made to flow.It may be rectangular, trapezoidal (cipoletti weirs) or triangular (90^0 – V notch). **'V' notch** weir is commonly used to measure small; and medium size streams.
- **Parshall Flume (Venturi flume) :** Parshall (1926) designed a device to measure water flow known as venturi flume which was gradually refined by Parshall (1950) himself as Parshall flume which is most commonly used now for water flow measurement.Most commonly used devices at farms are (1)Parshall flume and (2)'V'notches.

Methods of Irrigation :

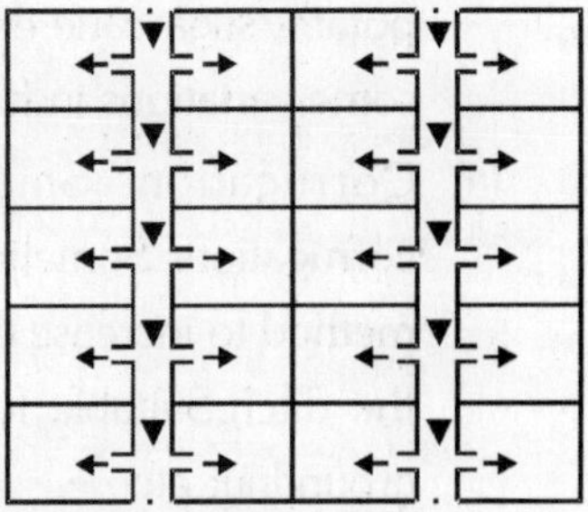

(A) Surface Irrigation :

1. **Flooding :** exclusively used for lowland rice though for other crops also, minimum labour intensive.

2. **Check Basin Method :** Most com-

mon method spreading time of water is low to minimise percolation losses,suitable for close growing crops.The size of check basins ranges from 4m X 3m to 6m X 5m depending upon the stream size and soil texture.

Advantage : Water application uniformly.

Disadvantages :

(a) Labour requirement is more for field layout and irrigation.

(b) Wastage if land is more under channels and bunds.

(c) Interculture or cultivation is difficult due to obstruction by bunds.

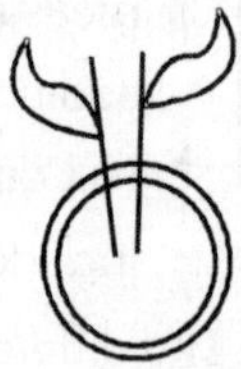

3. **Ring basin method :** Here basins around the trees are made, suitable for fruit trees. Basins are generally round in shape and occasionally square in shape.

4. **Border strip method :** Field is laid out into long, narrow strips bordering with small bunds. Length of strips varies from 30 to 300m and width from 3 to 15m. However the most common size is 30 to 50 m in length and 3 to 5 m in width.Slope range is 0.1-1.0%, suitable for close growing crops & medium to heavy textured soils but not suitable for sandy soils.

Demerit : Distribution of water is not uniform.

Merits : Large irrigation streams can be efficiently used.

5. **Furrow method :** suitable for crops grown with ridges and fu rows e.g. sorghum, maize, cotton, tobacco, brinjal, tomato, potato, sugarcane etc.Furrow length ranges 30-300 m.There are some variations in furrow method depending upon situations.

(a) Corrugation :Small and shallow furrows are known as corrugations.Sometimes prepared in the case of the border-strip method to increase efficiency and uniformity of overflooding from the ditch.Suitable for close growing crops like Wheat, setaria, groundnut etc.

(b) Short Furrows :Short furrows of 5-6 m length and 5-6 rows are grouped into a basin used to irrigate kitchen gardens with small streams.

Corrugation and short furrows are characterised under Every furrow Irrigation (EFI).

(c) Skip Irrigation/Alternate furrow Irrigation(SI/AFI) :Water is applied to one side of each crop row.Therefore water applied is reduced by 25-35% compared to EFI and crop yield may be reduced by 2-16%.

(d) Wide Spread Furow Irrigation (WSFI) : Furrows of more than 2-5 m apart having 2 or more than two crops rows are irrigated.The advance of water is slow in WSFI compared to EFI.The water applied is about half of the water applied in EFI.

(e) Within Row Irrigation (WRI) It slightly differs from every furrow irrigation in method of planting the crop.Such method aims at reducing the amount of water applied for each irrigation.Here, a presowing irrigation is given and seeds planted in the furrow itself. Subsequent irrigation is given to a shallow depth in the furrow. Water required for this type of irrigation is 40% less than conventional furrow irrigation.

Merits of furrow Irrigation. :

(i) Water in the furrows contacts only one-half to one-fourth of land surface thereby reducing evaporation losses.

(ii) Crust problem is avoided altogether in it.

(iii) Intercultivation is possible and labour requirement for irrigation is less.

6. Surge Irrigation : Intermittent application of water to the field surface under gravity flow which results in a series of 'on' and 'off' modes of constant or variable time spans.

Merits :

(i) Infiltration uniformity is increased.

(ii) Deep percolation is reduced compared to continuous water application due to intermittent wetting and dewatering process.

7. Cablegation : Automatic method,It saves water and labour.Cablegation is a form of gated-pipe system.To automate the system, the plug is allowed to move downslope through the pipe at a controlled rate.

(B) Subsurface/Subsoil Irrigation :

It is through underground perforated pipes,through deep trenches at 15-30 m intervals water gradually wets root zone through capillary movement.

Merits :

(i) Evaporation loss is less due to dry surface

(ii) Less weed problem.

(iii) Deep trenches may be used for drainage.

(iv) Suitable where water table is shallow.

Demerits :

(i) Deep percolation through trenches.

(ii) Maintenance of pipe lines is difficult.

Such method is practiced in Kerala for coconut gardens and in Kashmir for vegetables.

(C) Sprinkler Irrigation :

Water is applied as a spray or as rain over the crop.Rate of water delivery is more than 1000 lit/hr. It operates at pressure of more than 2.5 bar and throw water as a spray upto the distance of more than 10 metre.

Merits :

(i) No conveyance losses [The loss of water in surface irrigation is 15-20 % in well-irrigated areas and 30-35% loss in canal and tank irrigated areas.]

(ii) Uniform application of water.

(iii) Irrigable area is increased by 1.5 to 2 times with the same amount of water.

(iv) Saving of water from 25 to 50% for different crops.

(v) Saving of 10-16% land.

(vi) No risk of runoff and erosion,inundation.

(vii) Suitable for sandy soils,shallow soils where land leveling is difficult, steep slopes and Rolling topography and undulating land.

(viii) Suitable for the areas where is water and labour scarcity.

(ix) Suitable for saline soils to leach salts.

(x) Sprinkling of water before frost occurrence helps in maintaining higher temp. and minimizing frost damage.

Demerits :

(i) Not used under high wind velocity.

(ii) Not suitable for areas with hot dry winds.

(iii) Power requirements are usually high since it operates at a pressure of 0.5-1.0 kg/cm^2.

(iv) High initial cost.

(v) More spreading of diseases.

(D) Drip/Trickle Irrigation :

Discovered in Israel, Discharge rate of water per dripper is generally 1 – 4 lit/hr. and the water diffuses by capillary action in the soil. It consists of main pipeline, submains, laterals and emitters. Unlike sprinkler heads, emitters release water without any pressure. The discharge of most emitters ranges between less than 1lit/ha to 15-20 lit/ha. The no.of emitters per unit area depends on plant spacing, soil characteristics, root development and discharge of emitters. The lateral is looped around the basin of the plants so as to get more no.of emitters for each plant. The irrigation interval is 1-4 days.It saves 50-70% water and even more.

Merits :

(i) Well suited to areas of **acute water shortage.**

(ii) Deep percolation,surface runoff,evaporation losses are minimum.

(iii) Water is maintained at field capacity all through.

(iv) Salt concenteration is less even in salty soils due to high moisture content.

(v) No land leveling is necessary.

(vi) Herbicides and fertilizers are applied to the crop by solublising into the irrigation water i.e. **Herbigation** (Hebicide+Irrigation) and **Fertigation** (Fertilizer application through irrigation).

(vii) Energy requirement is less.

(viii) Less disease and weed infestation

Demerits :

(i) Clogging of emitters.

(ii) Damage to lateral system due to rodents and other animals.

(iii) Salt accumulation near plants due to lack of sufficient water for leaching.

(iv) Inadequate root development.

(v) Frequent installation and retrieval of the system for annual crops.

(vi) High initial cost.

(vii) Due to limited wetting zone,the entire soil volume is not exploited, However it is overcome by high moisture in the root zone.

Most suitable for **widely spaced crops ; orchard trees and vegetables** like apples, mango, lemon, tomato, brinjal, cotton etc.

Typhoon system of Drip irrigation : is an efficient drip irrigation for sugarcane. It uses less than 50% of water needed to raise the crop by conventional irrigation methods;developed in Israel for irrigating row planted crops. It is used in Hawaiian sugar plantations.

Micro-Irrigation :

Characteristics : Water is applied at.....

(i) a low rate.

(ii) over a long period of time.

(iii) at frequent intervals.

(iv) directly into the plant's root zone.

(v) Via a low-pressure delivery system.

The following types of irrigation are categorised under Micro-irrigation—

(a) Drip irrigation.

(b) Micro-jet irrigation :Water leaves the jets at a pressure of nearly one bar, throws water at the distance of 1-4 m,water discharge is 5-160 lit/hr.

(c) Micro sprinkler :water discharge 28-223 lit/hr; operating pressure: 0.8 to 4 bar and throw distance :0.9 to 4 m, suitable for forest trees.

(d) Bubbler irrigation :Vertical risers are anchored to stakes or posts and their heights are so adjusted that they deliver water at the rate desired. Water bubbles out of open vertical tubes practiced in mango,coconuts etc.

- Pudding : Irrigation before sowing of crop is known as Pudding .

Drainage :

Agricultural drainage is the removal of excess water known as free water or gravitational water from the surface or below the surface of the farm land so as to create favourable soil conditions for plant growth.

Causes of water logging :

(a) Natural

(i) Poor natural drainage of subsoil.

(ii) submergence under floods.

(b) Artificial

(i) High intensity of irrigated agriculture irrespective of the soil and subsoil.

(ii) Heavy seepage of losses from unlined canals,distribution and farm wate courses.

(iii) Enclosing irrigated fields with embankments and chocking up natural drainage.

(iv) Hydraulic pressures from upper saturated areas at higher elevations.

(v) Non-maintenance of natural drainage or blocking of natural drainage.

Methods : — Surface drainage
— Sub-surface drainage

(A) Surface Drainage :Simplest and most common in India by digging open drains at suitable intervals and depth.

Merits :

(i) Cheap

(ii) Defects in the open drainage can be seen easily and rectified

(iii) Requires less available fall or grade for adequate outlet

Demerits :

(i) Some land is wasted for open drains

(ii) Causes hinderance to field preparation and intercultivation.

(iii) Periodical desilting is necessary.

(iv) Heavy weed growth.

(v) It may be damaged by rodents and farm animals.

There are several methods and systems for surface drainage having three functional parts viz. Collection,Disposal and Outlet systems.

(a) Drainage of flat areas (slope less than 2%). For flat areas,two pro cesses are adopted—

1. Smoothing/grading/forming :means land with continuous slope i.e. elevated areas are cut off and excess soil is spread over lower areas so that surface becomes even with uniform slope.Excess run off is collected and conveyed into the field ditches.

2. Field ditches system :

(a) Bedding system : Small furrows are formed at known intervals parallel to the slope for draining out water. Those furrows are called dead furrows and land between these furrows are called beds.

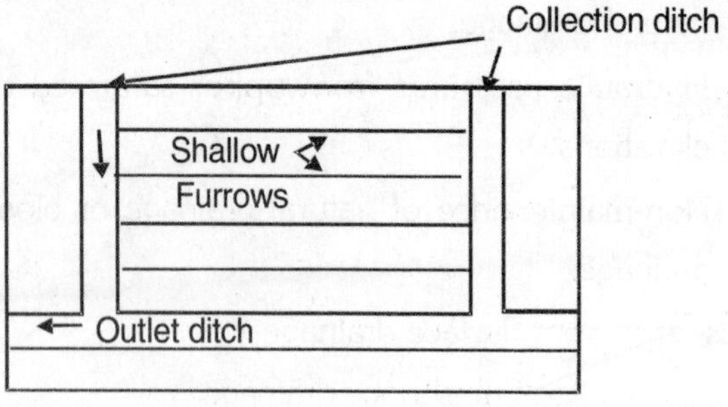

Fig. 13.4

(b) Parallel field ditch : Similar to bedding system except for deep drains and uneven interval between drains. Most effective and suited for both irrigated & rainfed.

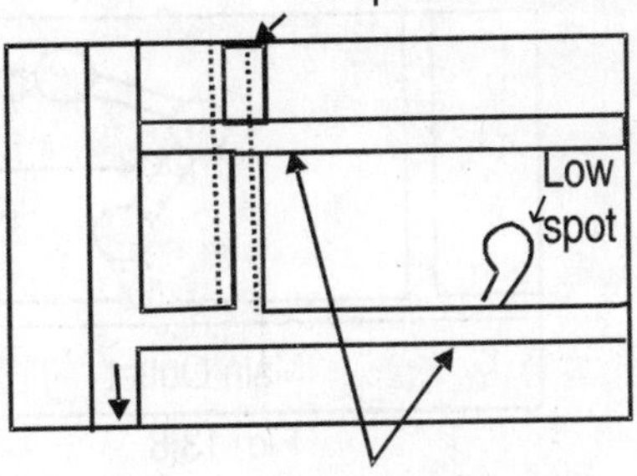

Fig. 13.5

(c) **Parallel open field ditch** : Deeper in depth and steeper side slopes than parallel field ditch hence is the name 'open';also called diversion ditch system; It is used in both surface and sub-surface drainage.

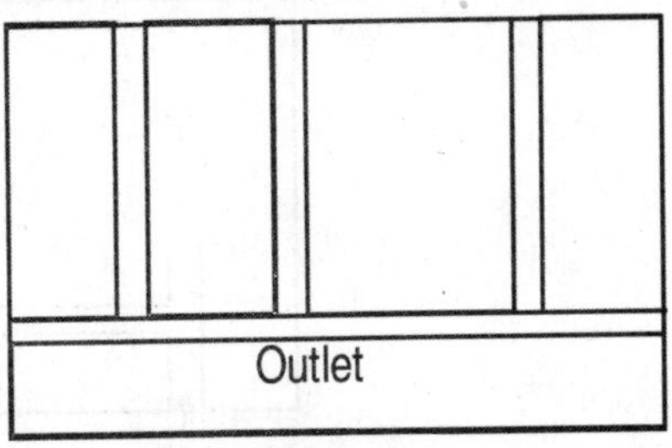

Parallel Ditches
Fig. 13.6

(d) **Broad Bed and Furrow system** (BBF) : Field is laid out into 120-150 cm width beds and 45 cm wide furrows across the slope. About 0.5 % slope is provided for free drainage.Crops are sown on beds having two or more than two rows each.Beds are 15 cm raised. BBF method is widely practiced in **groundnut** in **clay** soil.

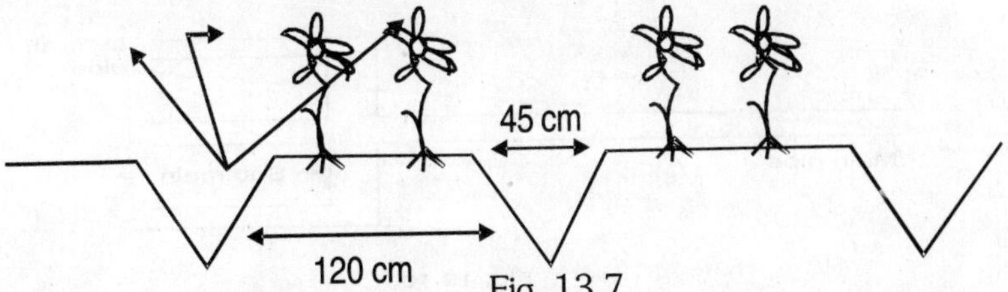

Fig. 13.7

(b) Drainage of Ponded Areas :Micro ponds or depressions are con nected by means of shallow channels or ditches.

Drainage of water is made by random field ditch system.

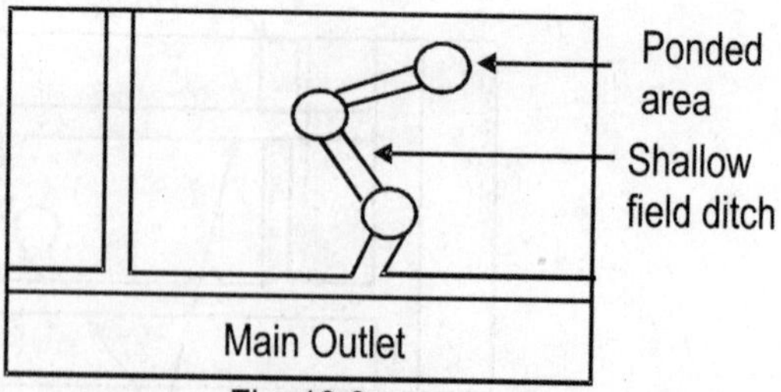

Fig. 13.8

(c) Drainage of sloping areas (slope > 2 %) by interception system or cross slope ditch system.

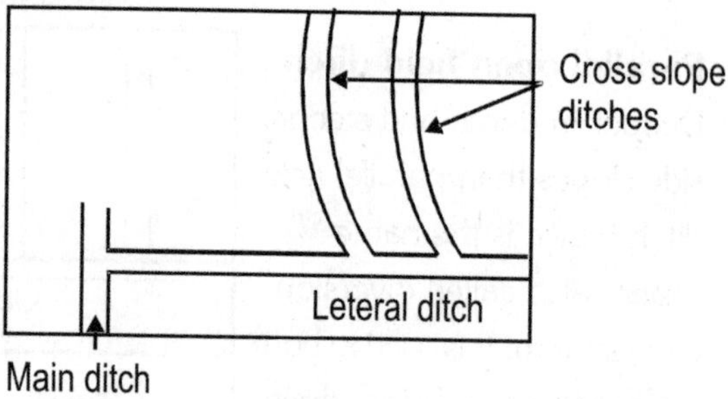

Fig. 13.9

(B) **Sub-surface drainage** : Purpose is to lower down the ground water level below root zone.These are the following methods—

1. Tile Drains including perforated pipes.It has following types of ar rangements of tiles or pipes :

(a) Natural or Random

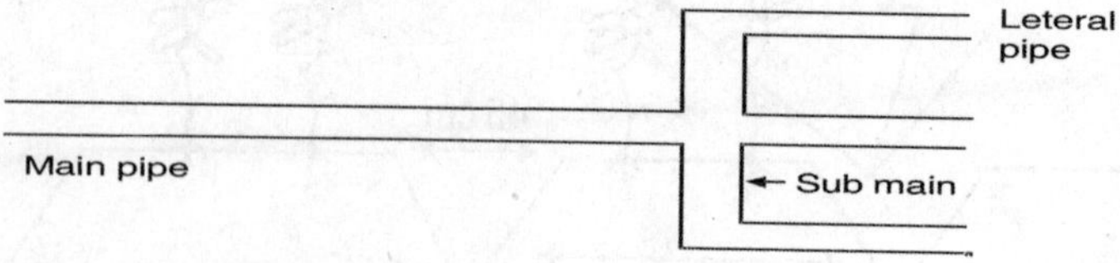

Fig. 13.10

(b) Herring bone type :Here there is no submain

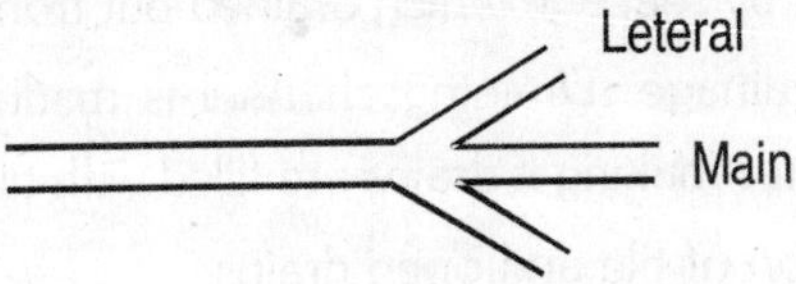

Fig. 13.11

(c) Gridiron type :Tiles arrangements from one side only.

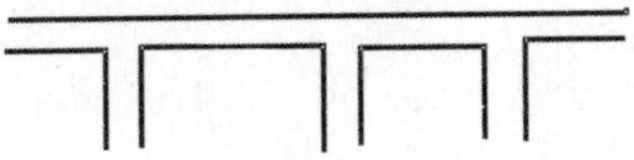

Fig. 13.12

(d) Cut off or Interceptor : It intercepts seepage moving down a slope. The interceptors should be usually placed at about the upper bound ary of the wet area.

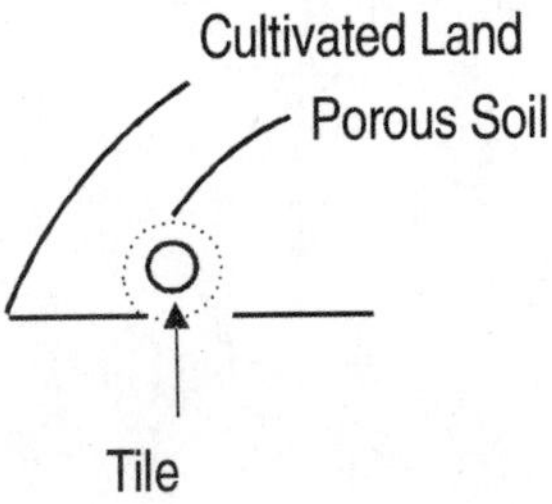

Fig. 13.13

2. Mole drainage : A mole is the egg- shaped drain which is made in the clay soil with the help of Mole plough. Once made mole drain remains functional for 3-5 years and especially suitable for clay soils.
3. Vertical Drainage :Drainage by wells is called vertical drainage. Wells are digged out to collect water from the nearby areas through seepage and collected water is pumped out.

4. Deep open Drainage :Deep open drains are made in which water is collected by seepage which drained out from the field.
5. Burried Drainage :Draining channel is made below the ground surface. After making it drains are filled with tiles,fibres or plastics.
6. Combination of tile and open drains.

❑❑❑

14 Rocks and its Weathering

Depending upon the mode of formation,rocks are of three main classes viz.

1. Igneous rock :

Examples : **Granite, basalt, diorite, gabbro, syenite, Grano-diorite, Peridotite.**

The original molten material solidified and formed Igneous rocks.Depending upon the mode of formation,such rocks are of two types.

(a) Plutonic rocks : Originally formed inside the earth by the solidification of the magma and pushed to the surface by various earth movements in pre-geological periods.Cooling inside the earth was a slow process.During cooling crystals developed and the rocks that formed were **crystalline.** When cooling was very slow,large crystals formed.Basic plutonic rocks are Diorite,granite,syenite,Gabbro.

(b) Volcanic rocks : The cooling of lava takes place on the surface of the earth fairly rapidly and the solidified material,known as volcanic rock;and is glassy in structure e.g. andesite,trachyte and basalt.Basalt is prominent in Deccan trap area.

2. Sedimentary rocks :

Limestone, Dolomite, Sandstone, Shale, Conglomerate.

Such rocks are derived from igneous rocks hence secondary formations and are formed by the consolidation of fragmentary rock materials and the products of their decomposition deposited by water, glacier,wind and gravity (colluvial soils).

Sedimentary rocks cover 4/5 th of the extent of rocks visible on the surface of the earth,yet they may not constitute more than 5% of the total bulk,the igneous rocks make up the rest.

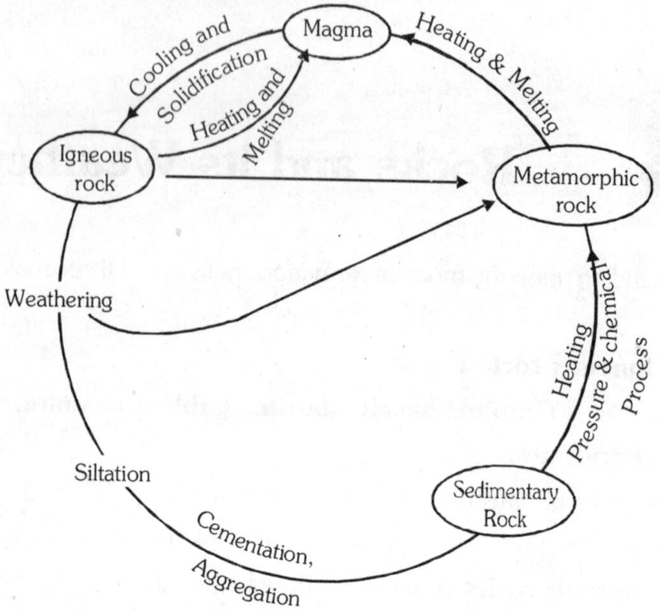

Fig. Rock Formation Cycle

3. Metamorphic rocks :

e.g. **Gneiss, marble, schists, slate, calcite and quartzite.**

Such rocks are formed from the Igneous or Sedimentary rocks by the action of intense heat and high pressure or both resulting in considerable change in the texture and mineral composition ; gneiss from granite, quartzite from quartz or sandstone, **marble from Limestone** and slate from shale.

Rock minerals :

Minerals are natural occuring substances with definite chemical composition and physical features e.g. Quartz (SiO_2) and orthoclase ($KAlSi_3O_8$). Rocks are mixture of minerals.

Quartz : Silica or SiO_2. It forms double hexa gonal pyramid crystals; **chief constituent of sandy fraction,** rocks containing free silica in abundance, that is not combined with bases are called 'acid rocks'. In granite (acid rock), quartz is present in a pure form as a prominent constituent. Basic rocks like Basalt and augite contain only a little, if any of free silica. The acid rocks give rise to the sandier types of soils under moderate rainfall and Basic rocks to the clayey types of soils.

Feldspars (Felspars) : Formula $K_2O, Al_2O_3, 6SiO_2$. Such minerals constitute about **48% of earth's crust** whereas **Quartz** constitute 36% and micas 10% of earth's crust.Potash feldspar forms the orthoclase feldspar group.Albite or soda feldspar and anorthite or lime feldspar constitute the plagioclase group.Potash feldspar is a chief constituent of acid igneous rocks and lime soda –feldspar of basic igneous rocks.Feldspar weather easily and give rise to clay on hydrolysis.

Micas : Potash mica is white,clear and transparent and known as muscovite mica. It is double silicate of K and alumina with a part of potash replaced by hydrogen :$K(OH)_2 Al_2Si_3O_{10}$.

Magnesium mica is called Biotic mica;resembles muscovite mica but black in colour : $K(Mg . Fe)_3 (OH)_2 AlSi_3O_{10}$.

White mica is more resistant to weathering than black mica. Micas are more resistant to weathering than Feldspar and other silicates.

- **Olivine :** Ferro-magnesium silicate $(FeMg)_2 SiO_4$. Two hydrated forms of olivine are talc and serpentine,is a hydrated silicate of Mg.
- **Tourmaline :** Is Boro-alumino silicate.

Sources of plant nutrients :

Nutrient		Source
N	→	Organic matter (O.M.)
C	→	Carbamate
P	→	Apatite, Fe/Al Phosphate, OrganicMatter.
K	→	Micas, Feldspar, **Orthoclase,** Microline.
Mg	→	**Dolomite,** Muscovite, Biotite, Olivine **Hornblende, Brucite,** Serpentine, Talc, Vermiculite, Glauconite.

B	→	**Tourmaline** (Source of Boron)
Mo	→	Olivine
Cu	→	Chalcopyrite, Olivine, Homblende, Augite, Biotite.
Zn	→	Sphalerite, Olivine, Homblende.
Cl	→	Apatite.
Mn	→	**Pyrolusite**
TiO_2	→	**Rutile** (Titanium oxide)
Ba	→	Baryte.

Primary and Secondry Minerals :

Primary minerals dominate the coarser fractions of soil viz. **sands**, gravel, stone whereas **secondary** minerals and clay minerals are most prominent in the fine materials especially in **Clays.**

Examples of primary minerals : Quartz, Muscovite, Orthoclase, Biotite, Microcline, Plagioclase, albite, Hornblende, Augite, Anothrite, Olivine.

Examples of secondary minerals : Geothite, Haematite, Gibsite, Clay minerals, Dolomite, calcite, Gypsum.

- Very slowly Weathered / Most resistant to weathering minerals are
  - (i) **Quartz** (most resistant)
  - (ii) **Muscovite** (more resistant)
- Moderate resistant/slowly weathered minerals :
  - (iii) Feldspar (orthoclase) (iv) Biotite.
- Easily weathered / Least resistant minerals :
  - (v) Augite (vi) Hornblende
  - (vii) Olivine (viii) Calcite (Least)
- Classification of rocks on the basis of silica content :
  - (a) **Acid rocks :** Such rocks have 65-70% silica e.g. Sandstone, granite, gneiss, rhyolite, Pegmatite.
  - (b) **Basic rocks :** contain 40-55% silica e.g. **limestone, Basalt,**

diabase, **gabbro**.

(c) **Intermediate rocks :** contain 55-65% Silica e.g. diorite, sy enite, andesite.

Weathering of Rocks :

Weathering is a **geological** process and essentially destructive in nature which leads to the formation of simple compounds from the solid rocks of earth's crust.

Rock containing complex minerals weathers more easily .Basic Igneous rocks weather more than acid igneous rocks.Sedimentary rocks are more resistant to weathering than igneous and metamorphic rocks.Limestone weathers more easily than sandstone.

There is the mechanical breakdown of the rocks to fragments,which constitutes the physical or mechanical weathering of rocks.Then chemical change take place in the material.Finally there is addition of organic matter,the invasion of material by various organisms and thus rock is changed into soil.The mechanical breakdown of the rocks is referred to as weathering of rocks and the chemical and biological change taking place may be taken to constitute the "Formation of Soil".

There are 3 types of Weathering—

(A) **Physical weathering :** Without accompanying any chemical change by various agents :

(i) Heat and Cold : Different minerals have different coefficient of expansion and contraction and thus peeling and flaking off of surface layer take place.

(ii) Freezing and Thawing : When water freezes,its volume increases by about 9% and the force exerted is 150 tons per square foot (1 ton =1016 kg : 1 tonne = 1000kg).

(iii) Glaciers : Ice has greater grinding power and force exerted by ice is 40 ib/inch ^{2} for every 100 ft. thickness.

(iv) Erosion by streams : The transporting power of moving water varies as 6th power of its velocity.Streams grind the materials into powder form.

(v) Waves and

(vi) Winds laden with sand particles cause abrasive action on the rocks.

Soils in early stage of development,desert soils and the soils of arctic and alpine regions are formed mainly due to the result of physical weathering.Soils evolved mainly as a result of physical weathering are called "Skeletal soils".

(B) Chemical Weathering : Involves two phases viz.

(i) Disappearance of certain minerals and

(ii) Formation of secondary products.

Chemical weathering is minimum under **desert** conditions due to the absence of water and under arctic and alpine conditions due to low temperature.

1. Simple solution : $H_2O + CO_2 \rightarrow H_2CO_3$

During raining, lightening and thundering;ammonia, nitrous oxide, nitrogen peroxideare dissolved in rain water. Carbonic acid has solubilising effect.

2. Carbonation : $\underset{\text{Calcite}}{CaCO_3} + H_2CO_3 \rightarrow \underset{\text{bicarbonate}}{Ca(HCO_3)_2} \downarrow$ leached

The soluble bicarbonate is leached to the lower layers where CO_2 is liberated and the insoluble $CaCO_3$ is precipitated.

The precipitation is around the $CaCO_3$ granules formed earlier,which serve as nuclei and promote further accretions and growth in the form of nodules.It is common phenomenon in the tropic and the lime nodules are given the name of 'Kankar'.Kankar nodules are found in most red soils and black cotton soils in south India,at 1 or 2 ft. from the surface in red soils and deeper at 3-4 ft. in black soils.Kankar is deposited in thick layers,is used as lime for construction of buildings.

3. Oxidation : mainly affects the Fe^{2+} iron.

$$2Fe\,SiO_3 + 3H_2O + O \rightarrow Fe(OH)_3 + 2SiO_2$$

Iron pyrite and marcasite produce hydrated ferric oxide and sulphuric acid on Oxidation.

$$2FeS_2 + 7H_2O + 15O \rightarrow Fe(OH)_3 + 4H_2SO_4$$

If $CaCO_3$ is present,it reacts with H_2SO_4 to give Gypsum.

4. Reduction : Under anaerobic condition

$$Fe_2O_3 + H_2 \rightarrow 2FeO + H_2O$$

Heamatite.

5. Hydration : Occurs mostly in humid regions

$$2Fe_2O_3 + 3H_2O \rightarrow 2Fe_2O_3.3H_2O$$

Heamatite (Red) Limonite (Yellow)

Volumes increase due to hydration.

6. Deposition : It occurs usually in the belt of cementation.When Fe/Al moves as colloids and comes in contact with electrolytes,it gets deposited due to intermittent heating and drying.

7. Hydrolysis : $2H_2O \rightarrow H^+ + OH^-$

At ordinary temp one mole of H_2O contains only 10^{-7} moles of $[H^+]$

$$KAlSi_3O_8 + HOH \rightarrow HAlSi_3O_8 + KOH$$

Orthoclase Acid silicate clay.

(C) Biological Weathering : Essentially it is physical and chemical weathering by biological agencies" - Said by Joffe.

Lower forms of plant life like mosses and lickens first appear on the broken and physically comminuted rock fragments.

Roots help in widening the cracks and crevices,produce CO_2.Worms burrow and invert the soil.

Development of mineral profile :

Is mainly due to the movement of water in soil,There are 3 possibilities

1. Under humid conditions,as a result of leaching certain constituents are either deposited in lower horizons or are completely removed in drainage.
2. Under arid condition,the deposition of material may occur at various depths as classification takes place.
3. In impervious subsoil layer,water can move only laterally over to the horizon of impedence.

Chemical composition of earth's crust (outer) :

	Element	**Percentage**
1.	**Oxygen**	49.2%
2.	**Si**	25.67%
3.	Al	7.5%
4.	Fe	**4.71%**
5.	Ca	3.39%
6.	Mg	1.93%

❐❐❐

15 Soil Formation

Soil is formed from weathering of rocks and minerals.Soil formation is a slow process.Formation of one inch soil needs 800-1000 years.

Factors of soil formation :

Dokuchaev : "Soil is the result of the combined activity and reciprocal influence of parent material,plant and animal organisms,climate,age of land and topography".

S = f (cl,o,r,p,t,....)

S = Soil formation,	r =relief/topography.
Cl = climate	p = parent material
o = organisms	t =time or age of land.

According to **Jenny** : "Soil property is determined by the relative influence of these factors".

The factors of soil formation may be kept under three groups—

(a) Passive factors :

1. Parent material
2. Relief or Topography

(b) Active factors :

3. Climate
4. Organisms

(c) Neutral factors :

5. Time or Age of land.

1. Parent material : is the unconsolidated mass from which the solum develops.The physical,chemical and mineralogical properties of parent material have significant effect on soil formation. These parent materials are grouped into three classes -

(i) Residual parent material :Such parent materials are remained at same place for a longer time.Soil formation takes long time from this parent material.Igneous rock,sedimentary rock and metamorphic rock are the examples of this category.

(ii) Transported parent material :The breakages and pieces of rocks and minerals are transported from one place to the other by the agencies viz. water,ice,air,gravity etc.

(iii) Biological parent material :Decomposed or partially decomposed biological matters are termed as biological parent material.

Chemical composition of soil and physically weathered soils has some resemblance to that of parent material,texture determines the depth of soil profile.Chemical composition of soils depend upon the types of rocks and their mineralogical composition.

2 . Topography/relief : Influences through its effect on : drainage, runoff, soil erosion and micro-climate i.e. exposure of land surface to sun and wind.The depth of soil solum influenced by the degree of slope.

The soils of the upper slope are less clayey,lower in P^H, soluble salts & organic matter contents.Lighter in colour and well drained in comparision to the soils at lower slope.Soil of the valley is generally deep and clayey.

Variation in aspects and elevation influence the climatic factor.Vegetation differs if slope faces a certain direction.

3. Climate : Includes precipitation , temp, humidity, and wind.

Rainfall and temperature are most important factors.Water acts both physically and chemically and moulds the morphological,chemical and physical features of soil profile.Intensity,distribution and duration of rainfall are also important factors.

Soil properties affected by climatic conditions are P^H,base saturation of exchange complex,organic matter content,clay mineralogical composition amount and nature of clay.

Evaporation,transpiration and humidity modify the effect of precipitation by either reducing or increasing the quantity of water available for percolation and reaction.

In scanty rainfall regions,salts are accumulated at the soil surface. But in high rainfall regions, total salts are leached into the lower horizon due to which soil becomes acidic. High temperature fastens the weathering process.Soil formation process is activated under favourable temperature and rainfall.

4. Organism (biosphere) : Flora and fauna.Flora includes plant roots, micro organisms, forest and grasses. Fauna means earthworm, ants, termite, rodents etc.

Flora and Fauna plays active role in the formation of soil.soil formed in the forest consists of relatively higher biological matter above the 'A' horizon which is called 'O' horizon.Undecomposed biological matter is prominently found in the forest.Fauna plays an active role in the mixing and churming of such matters in the soil.

5. Time and age of land : Age is the span of period from inception of soil development to present stage.Soil formation is a very slow process.Conditons that hasten the rate of soil development are warm and humid climate,forest vegetation,permeable and unconsolidated parent material,low in lime content,flat topography with good drainage.

Time factor has no relevance after the soil reaches its maturity,If the soil solum is removed by erosion or deposited over by a fresh transported parent material,a new cycle of soil formation takes place over the buried soil.

Soil Forming Processes :

Basic process according to Simonson (1959) are -

(i) Addition of water,organic matter and minerals to soil.

(ii) Losses of water,O.M. and minerals from soil.

(iii) Translocation means movement in solution (leaching) and move ment in suspension of clay,O.M and hydrous oxide.

(iv) Transformation of mineral and organic matter and formation of definite layers.

Fundamental Processes are :

1. Humification :Process of decomposition of O.M. leading to the formation of humus.

2. Eluviation : Means **washing out,** Eluviation is the process of **removal** of constituents in suspension or solution by the percolating water **from the upper layers** to lower layers. It refers to movement and removal of material in solution from entire solum.

3. Illuviation : Means **Washing in,** It is the process of **deposition** of soil materials in **the lower layers** i.e. immobilization and accumulation of the eluviated constituents.

4. Horizonation : The development of all the horizons in soil is horizonation.In the favourable conditions,development of different horizons in soil takes about 250 years but in adverse condition,it takes even more than 1000 years.

Specific Pedogenic Processes :

(a) Zonal Pedogenic Processes :

5. Calcification and gypsification : Calcification means deposition of $CaCO_3$ in soil profile, In arid and semi arid regions, following reactions occur.

$$CO_2 + H_2O \rightarrow H_2CO_3$$

$$H_2CO_3 + Ca \rightarrow Ca(HCO_3)_2 \text{ soluble}$$

$$Ca(HCO_3)_2 \xrightarrow[Co_2\uparrow]{temp\downarrow} CaCO_3 \downarrow + H_2O + CO_2$$

(insoluble)

6. Decalcification : Removal of $CaCO_3$ from soil by leaching.

$$CaCO_3 + H_2O + CO_2 \xrightarrow[Co_2\uparrow]{temp\downarrow} Ca(HCO_3)_2 \text{ soluble}$$

7. Podozolization : Calcification tends to concentrate $CaCO_3$ in the lower part of the B-horizon ,where as Podzolization **leaches** the **entire** solum.Apart from Ca,the other bases are also removed and the whole soil becomes **distinctly acidic** as the **Sesquioxide (Fe& Al) move out**, the A-horizon gives a bleached grey or ashy appearance hence the term podzol (Russian term pod=under,zola =ash like)and sesquioxide may accumulate in B –horizon.

A typical and mature podzol profile exhibits strongly contasting B-horizon with 3 distinct layers i.e..

(i). Precipitated humus

(ii) A reddish brown layer due to deposition of sesquioxides and

(iii). A yellowish horizon which gradually merge into the parent material. The podzols are low in fertility and are mainly used for forestry and pastures.

Podzolization is a type of eluviation in which humus and sesquioxides become mobile, leach out from the upper horizons and are deposited in the lower horizons.

8. Laterization : Contrast to podzolization . Later means brick or tile. It refers specifically to a particular cemented horizon in certain soils which when dried, become very hard like a brick. Such soil (in tropics), when massively impregnated with **sesquioxides** (Fe & Al oxides) to the extent of **70-80%** of the total mass, are called laterites or latosols (oxisols).

Laterization is the process that removes silica, instead of sesquioxides from the upper layers and there by leaving sesquioxides to concentrate in the solum. Laterites/ Latosols / Oxisols are rich in sesquioxides and poor in silica. It is mainly used for shifting agriculture, low intensity grazing and plantation of coffee, banana, pine apple and coconut.

Favourable conditions are warm and humid climate with 2000-2500 mm rainfall and continuous high temp, throughout the year.

Calcination, Podzolization and Laterisation are **zonal** soil forming processes where soils produced are under normal prevailing conditions of climate & Vegetation.

(b) Intrazonal Process :

In the Intrazonal soils, the profile characteristics are more influenced by certain local conditions such as relief or parent material than the climate and vegetation, These processes are occured primarily in arid and semi – arid regions except gleization.

9. Gleization : Development of a glei (a Russian word meaning blue, grey or green clay) in the lower part of the soil profile above the parent material due to poor drainage and water- logging. Such soils are called **hydromorphic soil**. It is not dependent on climate but often on drainage conditions. Due to unaerobic condition, Fe is reduced with the formation of mottles and concretions of Fe & Mn.

10. Salination : Accumulation of soluble salts in soil. Factors affecting are-

(i). arid/ semi- arid climate associated with shallow and brackish ground-water.

(ii). Lower topography

(iii). Poor drainage

(iv). Old lake bottom which on drying leaves salts at its surface.

(v). Alluvial deposits along sea coasts

(vi). Saline irrigation water.

11. Solonization / Alkalization : is accumulation of high exchangeable Na^+ ions and p^H more than 8.5 Soil colloids are dispersed and tend to move downward results in poor physical condition. The O.M. forms black Organo- clay coatings and hence called black alkali soils.

12. Solodization or Dealkalization : Means removal of Na^+ ion from the soil .

$$2NaX + CaSO_4 \rightarrow Na_2SO_4 + CaX$$

13. Pedoturbation : Process of mixing of soils . It is of three types.

(a). Faunal pedoturbation : by animals

(b). Floral Pedoturbation : by plants

(c) **Argillipedoturbation** : by churning process caused by swell-shrink clays as observed in **deep black cotton soils** of central India i.e. **Vertisols.**

❑❑❑

16 Soil and its Physical Properties

Generally soil is the porous, powdery and unconsolidated outer layer of the earth's crust which is formed by weathering of minerals and decomposition of organic substances. Soil is dynamic, three- dimensional (having length, breadth and depth) piece of landscape with a three phase (solid, liquid and gaseous) system. Soil is the basis and stratum of life hence someone rightly abbreviated the term 'Soil' as

S → Stratum / Soul
O → Of
I → Infinite
L → Life

There are two basic concepts of soil –

1. Pedology : Pedology = Pedon + logos = greek word
↓ ↓
Soil/earth discourse/Study

Therefore to study the origin, classification and description of soil is known as pedology . According to pedology, Soil is a **natural body** or entity and is a biochemically weathered and synthesized product of nature. Pedology does not focus primarily on the immediate practical utilisation of the soil. A pedologist studies the origin of soil, classifies the soil and describes the soil in its natural environment. Therefore it is useful to highway and construction engineers.

2. Edaphology : Edaphos + logos = greek word
↓ ↓
Soil/ground discourse

To study the soil from the stand point of higher plants is known as edaphology: According to Edaphology soil is **natural habitat** for plants. Edaphology studies the various properties of soil in relation to plant production. The food and fibre production is the main objective of edaphology.

- **Regolith may be defined as**
  (i). The unconsolidated product of rock by weathering.
  (ii). Soil material above the earth's surface i.e. bedrock
  (iii). Loose earth materials above solid rock.

Such weathered material is either the inherent product of the underlying rock or is deposited on other bedrock by transporting agents i.e. wind, water or ice.

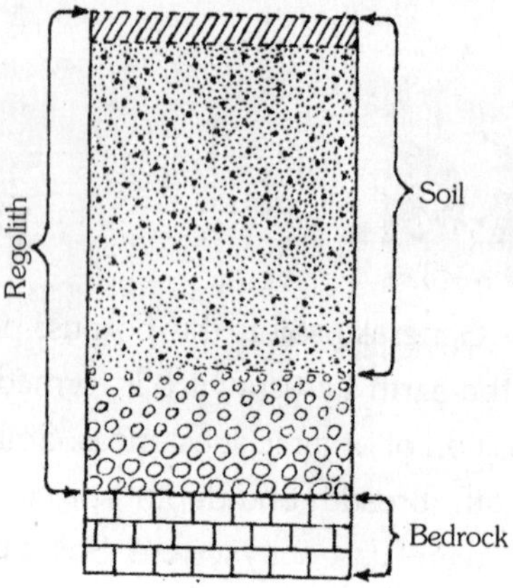

Fig. 16.1

- **Regosol** : Any soil of the azonal order without definite genetic horizons and developing from or on deep, unconsolidated, soft mineral deposits such as sands, loess or glacial drift.
- **Organic soil** : Such soil having at least 20% organic matter (by weight) in low clay content soil and at least 30% in high clay content (upto 60%) soil, is called organic soil.
- **Mineral soil** : Such soil have usually less than 20% organic matter but may contain organic surface layer upto 30 cm thick and shows the properties determined by mineral matter is called mineral soils. Mineral soils are the biochemically and physically weathered upper portion of the regolith.

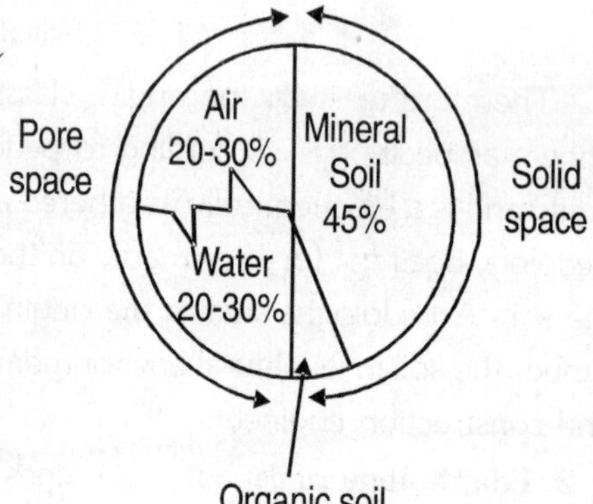

Fig. 16.2

There are four major components of soil = mineral soil, organic soil , Air and water.

The air and water in a soil are extremely variable and their portions determine in large degree the soil's suitability for plant growth.

Solid Space = Mineral soil (45%) + Organic soil(5%)

Pore Space = Air (20-30%) + Water (30-20%)

- **Law of Minimum :** Von Liebig(1840) : By the deficiency or absence of one essential constituent all the others being present,the soil is rendered barren for all those crops to the life of which that one constituent is essential.

  The level of plant production can be no greater than that allowed by the most limiting of the essential plant growth factors.

Soil profile :A vertical section of the soil through all its horizons and extending into the parent material and the individual layers are regarded as Horizons.These horizons above the parent material are collectively reffered to as the Solum (latin word solum meaning soil).The layers resulting from soil forming processes are grouped under five heads :O,A,E,B & C and all these horizons in combination are called Master Horizons.

'O' horizon : Organic horizons above mineral soil, occurring commonly in forest areas.Such 'O' horizon is **visible in virgin soil** and **absent in arable soils.**

'A' horizon :Topmost mineral horizon, containing a strong admixture of humified O.M. which tends to impart a darker colour than that of the lower horizon.

'E' or 'A_2' : Horizon of **Maximum Eluviation** of clay, Fe and Al oxides and a corresponding concenteration of resistant minerals such as quartz in sand.

AB/EB : Transition layer between A(or E) and B with properties more nearly like those of A(or E) than of underlying B (formally called A_3). Sometimes it is absent.

'B' horizon : Horizon of Maximum accumulation (Illuviation) of materials such as Fe and Al oxides and silicate clays (**Illuvial horizon**) and also $CaCO_3$, $CaSO_4$ and other salts in arid zones.Organic matter content is generally higher than that of E.

'C' horizon : Unconsolidated material underlying the solum (A&B) zone of least weathering,accumulation of Ca, Mg carbonates, Cementation, sometimes high bulk density.

- When a virgin soil is put under cultivation,the upper horizons become the furrow slice.

Physical Properties of soil :

The physical properties of a soil e.g. water holding capacity, permeability to wate, aeration, plasticity and nutrient supplying ability, are influenced by the size, proportion, arrangement and mineral composition of the soil particles.

Soil Texture : Concerned with the size of mineral particles means relative proportions of sand,silt and clay.The size of particles in mineral soil is not subject to ready change (i.e. by cultural practices).Thus it is considered as a permanent feature and a **basic property** of a soil.Mechanical analysis of soil separates i.e. the percentage of sand,silt and clay is done by **Hydrometric method.**

Coarse fragments and fine earth :

Stone	:	More than 250 mm in diameter.
Cobbles/Flags	:	250-75 mm in diameter.
Gravel	:	2-75 mm in diameter.Stones are generally rock fragments larger than 20 mm in diameter.
Gravel	:	2-20 mm in diameter.
Fine earth (sand,silt & clay)	:	less than 2 mm in diameter.
Sand	:	Coarse sand :0.2-2 mm in diameter (dominant in Quartz mineral)
Fine sand	:	0.02-0.2 mm (Quartz & Feldspar)
Silt	:	0.02-0.002 mm in diameter. Because of an adhering film of clay,they exhibit some plas

ticity, cohesion, adhesion and adsorption; Dominant minerals in silt are **Quartz and Feldspar.** Both sand and silt particles are approximatly spherical and cubical in shape.

Clay : **Less than 0.002 mm in diameter.**

Clay particles commonly are platy in shape and highly plastic when moist.It exhibits properties like flocculation,deflocculation and plasticity.

They have highest surface area since surface area is inversely related to size.Clay behaves like a weak acid which is neutralised by bases thus serving as store house for several nutrients.-

Classification of soil particles (International system)

Particle	Diameter in mm.
Stone	>20
Gravel	2-20
Fine earth	<2
Coarse sand	0.2-2
Fine sand	0.02-0.2
Silt	0.002-0.02
Clay	<0.002

Textural classes :

The proportion of predominant size of fractions of sand,silt and clay is the basis for classification of soil texture.If the soil contains more than 85% sand,soil is called sandy soil.When >80 % silt,then silty soil and when > 40% clay then clay soil.

Textural classes of soils

Textural class	Range (%) of		
	Sand	Silt	clay
Sand	85-100	0-15	0-10
Loamy sand	70-90	0-30	0-15
Sandy loam	43-80	0-50	0-20
Loam	23-52	28-50	7-27

Silt loam	0-50	50-88	0-27
Silt	0-20	88-100	0-12
Sandy clay loam	45-80	0-28	20-55
Clay loam	20-45	15-53	27-40
Silty clay loam	0-20	40-73	27-40
Sandy clay	40 -65	0-20	35-45
Silty clay	0-20	40-60	40-60
Clay	0-40	0-40	40-60

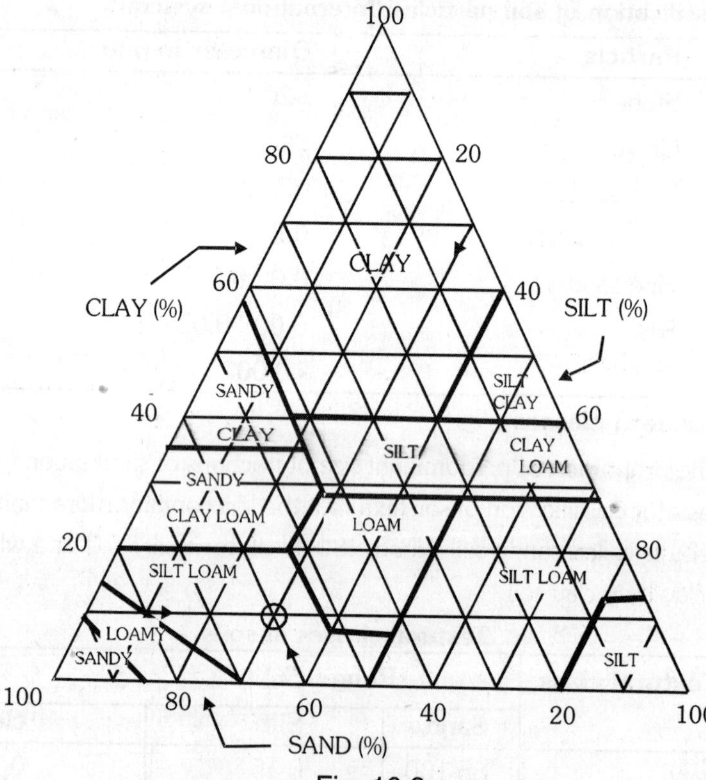

Fig. 16.3

Three broad and fundamental groups of soil textural classes are sands, loams and clays. An ideal loam is defined as the mixture of sand, silt and clay particles,that exhibits light and heavy properties in about equal proportions.Roughly it is a half and half mixture on the basis of properties.Loamy soil are best for agricultural production because they retain more water and nutrients than sandy soils and have better drainage,aeration and tillage properties than clay soils.

Soil texture influences soils physical and chemical properties like water holding capacity,nutrient retention,nutrient fixation,nutrient availability,drainage,strength,compressibility and thermal regime.

The clay imparts cohesion and stickness and fine texture to the soil and induces resistance to the passage of implements or in other words it makes the soil 'heavy'.Comparatively sandy soils offer little resistence to implements and are called 'light soil'.The term heavy and light refer to the degree of resistance offered by soil to the passage of implements,but not to its weight in fact the heavy clay soil weigh less than an equal volume of the light sandysoil.

Rice, Cotton, Sorghum, Coriander are grown on heavy textural soils which include clay loam,silty clay loam,silty clay and clay.

Medium textured soils like **loams**,silt loams,silts and **sandy loams** are suitable for **most of the crops.**

Sandy, loamy sand, sandy loam and sandy clay are light textured soils and these are suitable for groundnut, potato, tobacco, pearl millet and leguminous fodder crops.

Soil Structure :

The primary soil particles viz sand,silt and clay are usually grouped together in the form of aggregates. The arrangement of primary soil particles and their aggregates into certain defined patterns is called soil structure. Individual aggregates are known as peds or secondary units . Soils that do not have aggregates with naturally preserved boundaries(peds) are considered to be structureless. Two forms of structureless condition are recognised.

(i). Single grain : Particles are easily distinguished . Particles are unattached to each other and each particle acts individually. Movement of the air and capillary water is highest. Such structure is usually found in sandy soil and upper surface of podsols.

(ii). Massive : Individual particles adhere closely to each other but the mass lacks planes of weakness. When clayey soils are ploughed under wet condition, its individual particles adhere closely to each other and soil becomes massive or puddled. Such puddled soil retains water due to lack of pore spaces and thus is suitable for paddy.

Soil conditions and characteristics such as water movement, heat transfer, aeration, bulk density, amount and size of pore space (porosity), availability of plant nutrients and growth of microorganisms are much influenced by soil structure. In fact, the important physical changes imposed by the farmer i.e. ploughing, cultivating, drainage, liming and manuring his land are structural rather than textural.

Soil Structure Formation :

The aggregate formation is influenced by the amount and nature of colloidal clay. The nature of adsorbed cations also influences it. Ca^{++} and H^+ bring about better aggregation than Mg^{2+} and K^+. Wetting of clay particles with a liquid like water is required to form aggregates because water molecules show dipole movement . Sesquioxides i.e. iron and aluminium oxides, act as cementing agents for binding sand and silt particles to form aggregates.

Aggregates formed with sesquioxides are more stable than those formed by silicate clays. Colloidal organic matter is more effective in forming aggregates than clay.

Types of soil structure:

It refers to shape and arrangements of aggregates. Soil particles differ in size, shape and orientation. In general three broad categories of soil structure are recognised viz. Single grained, Massive and Aggregated. Between single grain and massive two extremes, an intermediate condition in which the soil particles are associated in quasi- stable small clods are known as aggregates or peds.

Shapes of Soil Structure :

The shapes of aggregates observable in the field-

Fig. 16.4

1. **Platy :** Peds are arranged in relatively thin horizontal layers, plates, leaflets or lenses. It occurs in recently deposited clay soils, surface layers of virgin soils but also in subsoil horizons i.e. 'B'.

2. **Prism like :** Vertically oriented aggregates or pillars. Occure in 'B' horizon of clayey soils in arid and semi- arid regions. When the tops are flat, these vertical aggregates are called prismatic and when rounded they are known as columnar. The size of prism like structure is upto 15 cm in diameter.

Prismatic (Level tops)

Columnar (Rounded tops)

Fig. 16.5

3. **Block like :** Original aggregates have been reduced to block, irregularly six – faced, cube like blocks of soil, 2-8 cm in size, common in heavy subsoils particularly of humid regions and in upper part of 'B' horizon.In cube like block structure, edges are sharp and rectangular faces distinct.

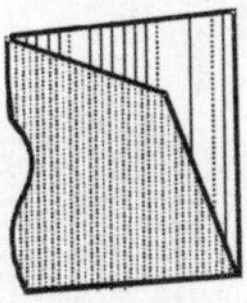

Blocky
Cube Like

Blocky
(subangular)

Fig. 16.6

4. **Spheroidal** : Rounded aggregates or peds not more than 2cm in diameter (generally) often found in a loose condition in the 'A'

horizon . Such units are ordinarily called granules and when granules are especially porous the term 'crumb' is used . It is the char-

Granular (Porous) Crumb (Porous)

Fig. 16.7

acteristics of furrow slice and subject to wide and rapid change. For the farmers, crumb and granular structure is favourable for plant growth. Formation of crumbs is more pronounced in clayey than in sandy soils. Clay humus and lime in the soil promote the formation of crumbs.

The shapes, sizes and densities of aggregates generally vary within the profile. The aggregates of B-horizon are bigger due to weight of the top layers & less shrinkage and expansion activity as there are less fluctuations of soil moisture. The profile in semi-arid region contain granulated A-horizon with a prismatic B-horizon . In humid temperate regions, the A-horizon contains graunulated aggregates but B-horizons have platy or blocky aggregates.

Other classification of soil structure :

According to the soil – aggregates, soil is grouped into two structural heads -

(A) Soils having definite cleavage :

(a) Structure having Vertical cleavage :

- (i) Prismatic &
- (ii) Columnar structure

(b) Soil structure of horizontal cleavage :

- (ii) Platy
- (ii) Laminar (very thin plates)

(c) Soil structure of both vertical and horizontal cleavage in mixed form :

(i) Blocky structure

(ii) Nut-like structure :When particle aggregates of blocky structure are somewhere smaller,such soil str. is known as nut-like str. and found in B - horizon

(iii) Granular

(iv) Crumb structure

(B) Soils having indefinite cleavage :

(i) Single-grain structure :

(ii) Massive structure.

Large masses of soil aggregates are called 'clods'.

Soil Structure classes :According to the size of the aggregates,five size classes are very fine, medium,coarse and very coarse.

Soil Str. Grades : Refer to degree of aggregation i.e. degree of inter aggregate adhesion and to aggregate stability.Foure grades are—

(i) Structurless — Single grain / — Massive.

(ii) Weak : Poorly formed peds.

(iii) Moderate : Moderately durable peds and well formed.

(iv) Strong : Durable peds and quite evident in undisturbed soil.

Factors affecting soil structure :

1. Soil management : A good soil management (proper tillage) with a proper crop rotation system has the effect of maintaining the soil in a good state of aggregation.

2. Adsorbed cations : Na^+ and K^+ ions disperse the soil, while ca^{2+}, Ba^{2+} have favourable effects on the aggregation. Similarly the presence of soluble salts favour flocculation.

3. Micro - Organisms : Filamentous fungi and microbial decomposition products of O.M. have a binding effect on soil particles,burrowing activities of insects, excrements of earthworms change the soil structure.

4. Variation of soil moisture : Drying of soil forms cracks and big clods; poorly drained soils have unfavourable structure.

5. Influence of O.M : Binding effect,water holding,porosity are increased by adding O.M. to soil.Cementing action of humus is equivalent to certain inorganic compounds like oxides of Fe which provide most of the long-term aggregate stability.

Polyvalent inorganic cations e.g. Ba^{2+} ,Ca^{2+},Mg^{2+} and Al^{3+} cause flocculation which act as bridges between O.M. and soil clays.There is only one practical way to improve the soil structure is to add O.M. to the soil.

Particle Density of Mineral Soil :

It is the mass (or weight) of a unit volume of soil solids. It is denoted as P.D. or D_p and was previously known as Real specific gravity. P.D. of the most mineral soils : 2.60-2.75g/cm^3 or Mg/m^3 (mega gram per cubic metre. 1Mg =10^6gram). For general calculation D_p of average arable mineral surface soil is about 2.65g/cm^3

D_p of O.M. is 1.1–1.4g/cm^3

Particle Density depends on the chemical composition and crystal structure of the mineral particle. The size of the particles of a given mineral and the arrangement of the soil solids have no effect on Dp.

Bulk Density (Apparent Specific gravity) :

It is the mass (or Wt) of a unit volume of dry soil including both solids and spaces (i.e.pores). Fine textured soils like clay, silt loam etc. have lower B.D. than sandy soil due to high pore spaces. Soils with high proporation of pore space to solids have lower B.D.(or D_b). Increase in pore space, decreases the B.D.

B.D. of general soil for calculation is 1.33g/cm^3(Just half of P.D.). Deerper the profile, higher the B.D. the reasons are low content of O.M., less aggregation, root penetration and compaction caused by overlying layers weight. Cropping increases the B.D. of the top soils in all cases while pore space decreases proportionately.

$$B.D. = \frac{\text{Weight of soil}}{\text{Volume of solids and pores}}$$

Pore Space / Solid Space :

Let Ws = Wt. of oven dry soil (solids)

Vs = Volume of oven dry soil (solids)

Vp = Volume of Pores.

Vs + Vp = Total soil volume.

By definition :

$$P.D = \frac{Ws}{Vs} \quad \therefore Ws = PD \times Vs \text{ ....(i)}$$

$$B.D = \frac{Ws}{Vs + Vp} \quad \therefore Ws = BD \times (Vs+Vp) \text{ ......(ii)}$$

From (i) and (ii) equations.

$$PD \times Vs = B.D \times (Vs + Vp)$$

$$\therefore \frac{Vs}{(Vs + Vp)} = \frac{BD}{P.D}$$

$$\text{Since } \frac{Vs}{(Vs + Vp)} \times 100 = \text{percent solid spaces}$$

$$\therefore \boxed{\%\ \text{Solid Spaces} = \frac{BD}{P.D} \times 100}$$

% Pore Spaces = 100 - % solid spaces

$$\therefore \boxed{\%\ \text{Pore Spaces} = 100 - \left(\frac{BD}{P.D} \times 100\right)}$$

Soil type	**Total pore space in percentage**
Clay	50-60%
Loamy	30-50%
Sandy	20-30%

In sandy soils, there are 25000 pores/m^2 whereas in clay, 25 X 10^6 pores/square meter.

- Pores less than about 0.06 mm in diameter is known as Micropores whereas more than 0.06 mm in diameter is known as macropores. Cropping reduces macropores space for ready movement of air by about one half.

Soil consistence :

Refers to the resistance of soil materials to deformation or rupture. It depends upon the degree and kind of forces which attract one molecule to another . It varies with size, shape and arrangement of soil particles and the nature of the water films around it, nature and amout of inorganic and organic colloids. When water dries from soil particles, soils develop greater cohesive force and mechanical strength.

Soil colour :

Is due to either mineral matter or O.M. and mostly to both. Organic Matter imparts black to dark grey tinges, iron compounds for red, brown and yellow tinges: and silica lime and other salts give light, white and grey tinges. Red colour is related to unhydrated ferric oxide although manganese dioxide and partially hydrated iron oxide may also contribute to it. The yellow colour is due to oxides of iron ($2Fe_2O_3 \cdot 3H_2O$). Yellow colour indicates more moist soils than red colour. Soil colour directly influences soil temperature.

Soil Air :

Gaseous phase of the soil. Under moist field conditions macropores generally constitute the air space. Ordinarily, the occupation of nearly one-third of the pore space in the soil by air and two – third by water constitutes the most favourable condition for plant growth. Plant growth suffers when oxygen diffusion rate **(ODR)** below $40x\ 10^{-8}$ g per cm^2 per minute.

The growth of root is ceased at $20x10^{-8}$ $g/cm^2/$ minute ODR.

Table : Percentage by Volume

	O_2	CO_2	N_2
Soil Air	20.6	0.25-1.0	79.2
Atmosphere	20.96	0.03	79.2

Field air capacity is the fractional volume of air in a soil at field capacity.It depends upon soil texture.Soil air constitutes in sandy soil,more than or equal to 25%(i.e.> 25%).loamy :15-20% and in clay less than 10% of the total soil volume.Clay soils retain more water hence is the lower air capacity.

Soil temperature :

For optimum root development :10-27^{0}c

For opt. crop root development :15-45^{0}c

Black cotton soils absorb about 86%,alluvial soils 40% and grass covered soils about 60% of solar radiation.On an average,the specific heat of soils ranges from 0.20 to 0.23.

Soil Water :

(A) Physical classification of soil water :

Hygroscopic water :

1. Such water held at hygroscopic coefficient.
2. Tension varies from 31–10,000 atmosphere.
3. held mostly by soil colloids.
4. Mostly non-liquid and therefore biologically inactive.
5. Moves mostly in vapour form.
6. Film thickness not more than 15-20 layers of water molecules and maximum thickness of 4-5 mμ.
7. Related to texture and O.M. content of soil.

Capillary Water :

1. Such water present (or held) in capillaries of soil.
2. Held between field capacity and hygroscopic coefficient in micropores.
3. Soil - Water tension or tension of film varies $\frac{1}{3}$ atm (or 0.1) – 31 atm.
4. Function as soil solution.
5 Moves by film adjustment from thick to thin film.

6 Related to O.M. content and texture of soil.

Gravitational Water :

1. Such water held by a negative tension of 0.3 atm.or less,loosely held between less than 0.1 to 0.3 atm.
2. Free water which drains out,also called drainage water.
3. Usually present in macropores.
4. Undesirable and nutrient leaching takes place.
5. Downward movement of water due to gravitational force.

(B) Biological classification of soil water :

Super fluous water :

1. Moisture tension $\leq \frac{1}{3}$ atm.(less than or equal to).
2. Such water held at field capacity but not beneficial for higher plants.
3. Harmful for crops if too much.
4. Also called gravitational water.

Available water :

1. Portion of capillary water lying between **field capacity($\frac{1}{3}$ atm) and wilting coefficient (15 atm).**
2. Availability of moisture depends upon crops when wilting coefficient is measured at 15 atm pressure it is a measure of soil condition and not that of plant because ability to absorb soil moisture. differs according to plant types.

Unavailable water :

Such water held in soil at the permanent wilting point.It includes hygroscopic water which is removed too slowly by plants to prevent wilting.

Optimum growth of plants occurs when soil moisture content near field capacity with a moisture tension of 1 bar or less.

Field capacity :When macropores drain their water and are filled by air but micropores or capillary pores are still filled with water then soil is called at field capacity.The matric tension is around 0.1-0.3 atm.

The p^F range of capillary moisture is 6.0-4.2 .The p^F range of available moisture is 2.54-4.2 .Moisture with about P^F 2.54 is very favourable for plant growth.

Permanent wilting Percentage : Concept by Briggs and Shantz.Although plants not dead,but they are now in a permanentally wilted condition and will die if water is not added.The tension is 15 atm. For most of the crops the water remaining in the soil is found in the smallest of the micropores and around individual soil particles.The p^F at 15 atm. is about 4.2 .

Hygroscopic Coefficient :

After the removal of liquid water completely from micropores, the remaining water is associated with the surface of soil particles. The tension is about 31 atm. The p^F is about 6.0. The moisture help at this point is called hygroscopic Coefficient.

Moisture Equivalent :

The weight percentage of water retained by a previously saturated sample of soil 1 cm in thickness after it has been subjected to a centrifugal force of 1000 times gravity for 30 minute.

Moisture equivalent was introduced by Briggs and Mc. Lane (1907) . Briggs and Shantz (1912) used the moisture equivalent as an indirect measure of wilting point (W.P).

$$W.P. = \frac{\text{Moisture Equivalent}}{1.84}$$

However this factor i.e. 1.84 can not be used for all soils . Moisture equivalent is assumed to represent moisture held under field conditions or the field capacity. The p^F at Moisture equivalent ($\frac{1}{3}$ atm) is about 2.54.

- Maximum capillary capacity (MCC)

MCC = Water holding capacity – Hygroscopic Coeff.

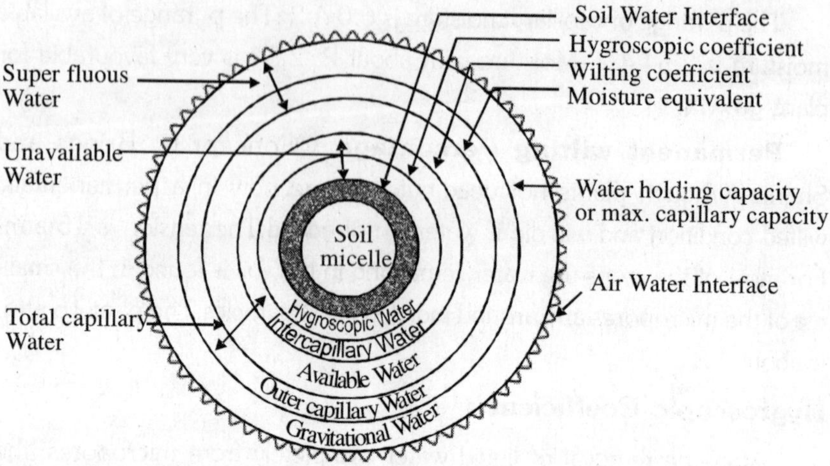

Fig. 16.8 : Different layers of soil water

❒❒❒

17 Soil Colloids & Silicate Clay Minerals

Colloidal particles are generally smaller than 1 micro meter (um) in diameter. Since clay fraction of soil is less than 2 μm, therefore all clay is not strictly colloidal, but even the larger clay particles have colloidal properties.

Colloids do not form true solution like the crystalloids which break into the constituent molecules and get dispersed in a liquid medium like water. The solution is homogenous and the molecular solute particles pass along with the solvent through porous parchment membranes, while the colloids do not do so. Colloids are not soluble but remain suspended in water and do not setlle down like precipitates. 'Sol' state, remains suspended in water. 'Gel' gets back to the solid state.

Minute particles are in a state of constant motion, as all the particles carry like electric charges, repel one another and recoil on contacting the other particles in the system- called Brownian movement.

Colloidal solution : Size of particles in colloidal solution :

1μ– 1mμ (i.e.10^{-6} to 10^{-9}m)

True solution : Size of particles in true solution : less than 1mμ in diameter.

Clay minerals :

A mineral is a naturally occurring inorganic surface with a definite chemical composition and distinct physical characters.

Primary mineral is the original component of rock , generally anhydrous and originally formed by cooling and solidification of molten mass e.g. Feldspar, mica, hornblende etc. Whereas secondary minerals are formed as a result of subsequent changes in rocks due to weathering or metamorphosis of primary minerals and hydrous in nature e.g. Kaolinite, Montmorillonite, illite , limonite etc.

Clay is not amorphous but crystalline and lathe shaped or made up of irregular hexagonal plates (Humus is amorphous).

Composition of clay :

Common clays are mainly made up of silicates of various types and are called silicate clays which is the characteristics of temperate regions. The other type is formed of aluminium and Iron hydroxides, called sesquioxide clay, is not common as it is mostly formed under humid tropical conditions (well weathered soils of tropics and semitropics).

- **Micelles (Micro cells) :** The minute silicate clay colloid particles, ordinarily carrying negative charges is called micelles .

Many positively charged ions or cations are attracted to micelles and formed ionic double layer, also called Holmontz double layer.

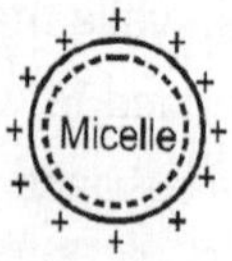

Fig. 1. : Holmontz layer

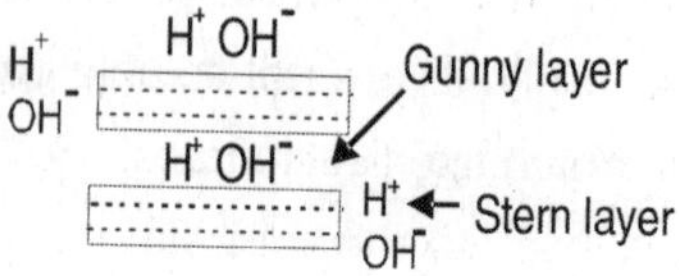

Fig. 2. : Stern & Gunny layer

A large no. of water molecules are carried by the adsorbed cations. Since most of them are definitely hydrated is called stern layer. Some silicate clays, in addition, hold numerous water molecules as well as cations packed between the plates (internal surface area) that make up the clay micelle: is called gunny layer.

Acid Nature of Clay :

The clay being negatively charged is attracted by and shifts to the positive anode like acid radicles in an electrolytic medium. Therfore clay is some times referred to as clay – acid.

In humid regions, H^+ and Al^{3+} are predominant among cations and when it dissociates in the soil moisture, it imparts an acidic reaction to the soil solution. When H^+ and Al^{3+} predominate in the clay complex as in humid regions, the clay is called Aluminium- Hydrogen clay or acid clay.

In arid regions, Ca^{2+} and Mg^{2+} predominate in clay complex and the solution is more or less neutral and the clay is then called calcium clay. When percolation is considerably limited in arid regions, sodium salts accumulate on the soil surface and sodium becomes the predominant cation than calcium in the clay and is called sodium - calcium clay. A sodium calcium clay soil exhibits an alkaline reaction

Region	Predominant cations
Humid	Ca^{2+}, Al^{3+} and H^+
Arid	Ca^{2+}, Mg^{2+}, K^+ and Na^+

- Order of strength of Adsorption or ability to Flocculate soil colloids in decreasing order = $Al^{3+} > H^+ > Ca^{2+} > Mg^{2+} > K^+ > Na^+$.

  Maximum adsorption strength is exhibited by $A1^{3+}$ and the least by Na^+. Na^+ results in a dispersed condition of soil colloids. Ba^{++} flocculates solid colloids.

Sesquioxide clays :

Under the heavy rainfall conditions, soils are gradually leached out of bases and their place is taken up by H^+ in the clay complex. When the H^+ exceeds by a certain limit, silicate material in the soil is acted upon by them and free silicic acid is produced. It is soluble and is leached in drainage , and colloidal aluminium and Iron hydroxides are left behind . The silica content is reduced and the sesquioxides (Iron and aluminium oxides) become predominant in the residual clay . Such clay is called sesquioxide clay (sesquioxide= Fe_2O_3. Al_2O_3. TiO_2) .

Sesquioxide clay does not possess the properties of plasticity and cohesion, has low base exchange capacity and low in fertility. The phosphorus in soil is tied up as Iron and alumiuium Phosphate.

Silica- Sesquioxide ratio :

The proportion of silica to sesquioxides in the clay fraction of soil expressed as a molar ratio, It may be about 6.0 in the original rock and range from 1.25–3.6 in most soils. Red or brown soils have 2.0–2.5 indicating the presence of free iron oxide and alumina, by the leaching of silicic acid during weathering.

Soil Types	Silica-sesquioxide-ratio
1. Black soils	3.0 - 4.3
2. Ashy grey and Black soils	> 2.5
3. Red or Brown soils	2.0 - 2.5
4. Laterite soils	< 1.33 (less than)

The ratio is high in clayey soils and low in coarse light soils . A high ratio means high base exchange, hygroscopic and field moisture capacity and fertility in general.

Gibbsite : $Al_2O_3 . 3H_2O$ or $Al(OH)_3$ }
Goethite : $Fe_2O_3 . H_2O$ or FeOOH } Common in soils

- Allophane : The most significant amorphous silicate mineral, poorly defined aluminium silicate ($Al_2O_3 . 2SiO_2 . H_2O$), most prevalent in soils developed from volcanic ash and high capacity to adsorb cations.

Silicate Clay Minerals :

The clay particles are composed of individual layers or crystal units .The most important silicate clay is known as phyllosilicate (Greek phullon means leaf), suggesting a leaf- like or platelike structure. The silicate clay unit consists of alternate sheets comprised of planes of mineral cations surrounded and linked together by planes of oxygen and OH ions (hydroxyl). One type of sheet is dominated by silicon (Si) and other by Al and /or Mg. **Silica- dominated sheet** is called **Tetra hedral** because of its four sided configuration and **Al and / or Mg sheet** is called **Octahedral** because of eight sided building block . Si^{4+} is surrounded by four oxygen ions and Al/Mg by six hydroxyl or Oxygen ions .

Al- dominated sheet is called Di- octahedral sheet while Mg – dominated sheet is called **Trioctahedral.** The reason- Two Al^{3+} ions have the same charge as three $\mathbf{Mg^{2+}}$ ions. The tetrahedral and Octahedral sheets are the fundamental structural units of silicate clays.

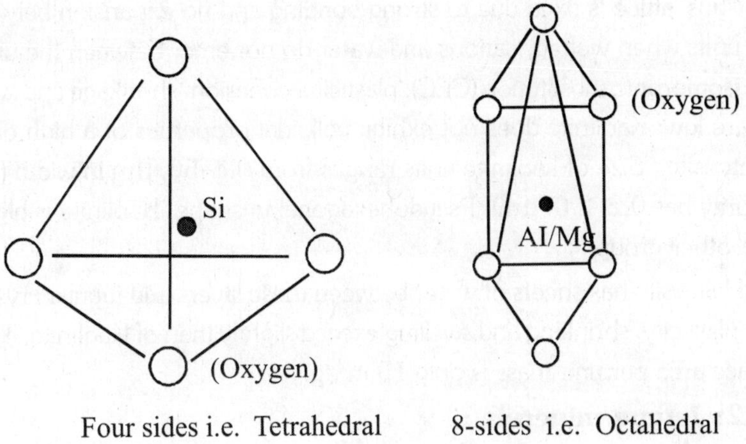

Fig. 17.2

Classification of Silicate clays :

On the basis of number and arrangement of tetrahedral and octahedral sheets, silicate clays are classified into 3-different groups :

(a). 1: 1 type minerals means One tetrahedral to One octahedral (A1) sheet.

(b). 2: 1 type minerals : Two tetrahedral sheet to one octahedral sheet.

(c). 2: 1:1 type minerals .

(a) 1:1 type clay minerals :

Such silicate clay is made up of one silica sheet and one alumina sheet combined. In soils, **Kaolinite** is the prominent member of 1:1 type group. Others are Halloysite, Nacrite & Dickite.

The two sheets are held together by oxygen anions (O^{2-}) mutually shared by Si^{4+} and Al^{3+} in their respective sheets . These units are in turn held together rigidly by H- bonding.

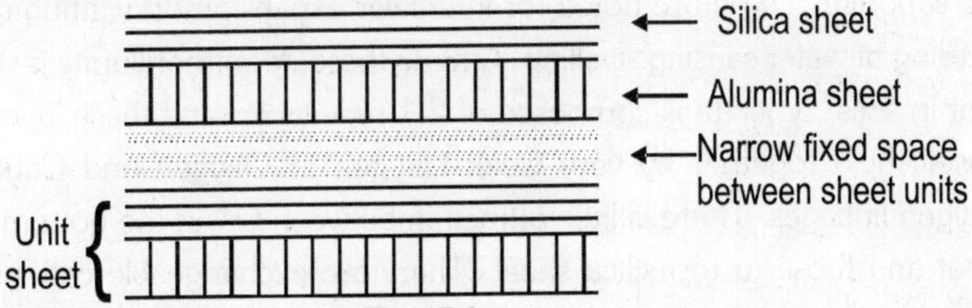

Fig. 17.3

Thus lattice is fixed due to strong bonding and no expansion between two units when wetted; cations and water do not enter between the units; little isomorphic substitution (CEC); plasticity, cohesion, shrinkage and swelling are low. Kaolinite does not exhibit colloidal properties of a high order of intensity . Size of kaolinite units ranges from 0.1–50. µm in width (but majority has 0.2–2.0. µm); Pseudohexagonal in shape. Kaolinite is bigger than other group.

Halloysite has sheets of water between these layers and tubular crystals thus plasticity, shrinking and swelling exceed slightly than of Kaolinite. Total surface area per unit mass is only 15 m^2/g.

(b) 2: 1 type minerals :

Octahedral sheet (alumina) is sandwiched between two tetrahedral (silica) sheet.

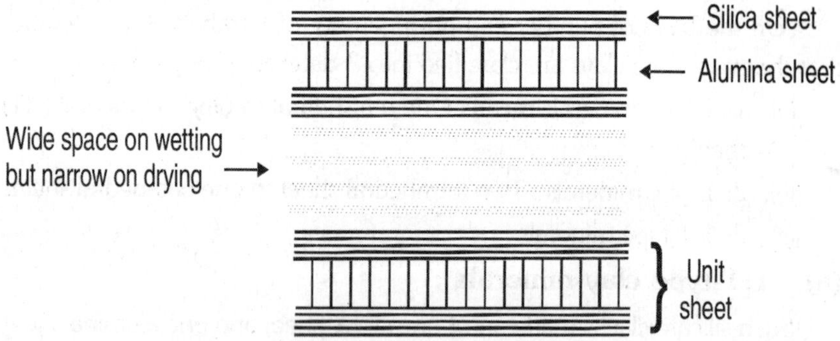

Fig. 17.4

(b1) Expanding Minerals :

Includes Smectite and Vermiculite groups.

Smectite groups :Includes montmorillonite, Beidellite, Nontrornite and saponite. These are noted for interlayer expansion on wetting due to entering of water causing swelling. Among these, Montmorillonite is prominent in soils. Smectte is composed of 2:1 type layers and these layers are loosely held together by very week Oxygen to Oxygen and Cation to Oxygen linkages. There is little attraction between O^{2-} in the bottom silica sheet and those in top silica sheet. Therefore exchangeable cations and associated water molecules are attracted between the layers causing expansion. **Mg^{2+} replaces Al^{3+}** in some sites of alumina sheet and **Al^{3+} replaces Si^{4+}** in some sites of silica sheet. These substitutions give rise to negative charges which account for high Cation Exchange Capacity (CEC). CEC is defined as the amount of a cation species bound at p^H 7 (neutral p^H) and is expressed as cmol (P+) kg^{-1} . cmol means centimol. Previously it was expressed as me/100g soil i.e. mili equivalent. These negative charges are satisfied by a swarm of cations. The specific surface or total surface area per unit mass is 700-800m^2/g.

Montmorillonite is most common smectite in soils where **Mg^{2+}** is substituted for **Al^{3+}** in alumina sheet (Octahedral).

Beidellite : Substitution of Al^{3+} for Si^{4+} in silica sheet (Tetrahedral).

Nontronite : **Fe^{3+}** (trivalent Iron) dominates the alumina sheet (octahedral) and some Al^{3+} replace Si^{4+} in silica sheet.

Vermiculite : has similar structural characteristics to smectite. Most vermiculites are dioctahedral (Al- dominated sheet) and have same isomorphic substitution to smectite in the tetrahedral sheets, considerable substitution of **Al^{3+} for Si^{4+}** accounts for **very high negative charges**. The water molecules, Mg^{2+} and other cations are strongly adsorbed in the interlayer which act as bridge holding the unites together . So degree of swelling is considerably less than of smectite, and therefore called **limited expansion clay mineral.** Expansion is more than of kaolinite but much less than of montmorillonite (i.e.Kaolinite- Vermiculite- Montmouillonite).

The CEC of Vermiculites exceeds that of all other silicate clays due to very high negative charges. (The CEC in ascending order: Kaolinite- Illite – Montmouillonite- Vermicute- Humus) . The Vermiculite crystals are larger than that of montmorillonite but much smaller than of kaolinite.

Table : Clay minerals & Their CEC

Clay Minerals	**Cation Exchange capacity (CEC) expressed as cmol (P+)kg^{-1}**
Kaolinite	3 - 10
Illite	10 -40
Chlorite	10 - 40
Montrorillonite	80 - 150
Vermiculite	100 - 150
Organic Collorids	> 200

(b2) Non- Expanding Minerals :

Example : Micas — Muscovite / Biotite } Often found in Sand and silt.

In clay, fine grained micas (or Illite) are found, **Illite** has 2:1 type crystal. About 20% Si^{4+} of silica sheet is replaced by Al^{3+} which results in high net negative charge in tetrahedral (silica)sheet. To satisfy this charge **K^+** are stongly attracted in the interlayer space. Thus **K^+** acts as **binding agents** preventing expansion. Therefore hydration, CEC, swelling shrinking & plasticity are much less than of montmorillonite.

Table : Comparative study of K.I.M.

S.No.	Property	Mantmorillonite	Illite	Kaolinite
1.	Size (um)	0.01-1.0	0.1-2.0	0.1-5.0
2.	Shape	Irregular flakes	Irregular	Hexagonal crystals
3.	Specific Surface Area (m^2/g)	700-800	100-120	5-20
4.	External Surface	High	Medium	Low
5.	Internal Surface	Very High	Low	None
6.	Cohesion, Plasticity, Swell ing.	High	Medium	Low
7.	Cation Exchange capacity CEC	80-120 me/100g soil	15-40	3-10 me/100g of soil

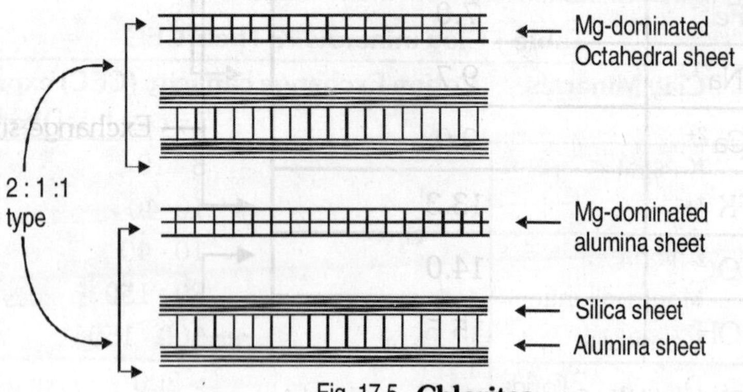

Fig. 17.5 **Chlorites**

(c) 2:1:1 type (or 2:2) type mineral :

e.g. chlorites.

Chlorites are basically ferro- magnesium silicates with some aluminium present.

Chlorites have an extra layer of Mg-dominated alumina sheet say **Brucite** [$Mg(OH)_2$]. But Mg^{2+} also dominates the alumina (trioctahedral) sheet of 2:1 type minerals. Thus crystal unit has two silica sheets (tetrahedral) and two magnesium dominated alumina sheet. That's why sometimes it is also called 2:2 type clay mineral. In other words chlorites are basically silicates of Mg with some iron and aluminimum. The CEC is about same as of Illite, non – expanding nature.

Sources of Negative charge on silicate clays :

1. Isomorphic Substitution :

Table : Ionic radii of common elements of silicate clays

Ion	Radius(nm) (10-9m)	found in
Si^{4+}	4.2	Silica tetrahedra
Al^{3+}	5.1	
Fe^{3+}	6.4	
Mg^{2+}	6.6	Alumina octahedra
Zn^{2+}	7.4	
Fe^{2+}	7.0	
Na^{+}	9.7	
Ca^{2+}	9.9	Exchange sites
K^{+}	13.3	
O^{2-}	14.0	Both sides
OH^{-}	15.5	

Note : Al^{3+}, O^{2-} and OH^{-} can fit in either.

Al^{3+} is slightly larger than si^{4+} hence Al^{3+} can fit into the centre of tetrahedron in the place of Si^{4+} without changing the basic structure of crystal.

In octahedron, Fe and Zn can fit into the position of Al or Mg. Positioning of Mg^{2+} in place of Al^{3+} or Al^{3+} in place of Si^{4+} leaves unsatisfied negative charges from Oxygen anions in the sheets which account for overall negative charges from Oxygen anions in the sheets which account for overall netative charge of clay.

Isomorphic substitution is of great significance in 2:1 type. The resultant negative charge is far in excess of that resulting from broken crystal edges of these minerals . Unlike the charge associated with the exposed crystal edges, those resulting from ionic substitution are not dependent on p^H.

The isomorphous substitution of any cation having lower charge results in an incresase in positive charge. It is commonly occured in trioctahedral layers (Mg – dominated sheets) when Mg^{2+} is replaced by Fe^{3+}/Al^{3+}

$Mg_3\ (OH)_6$ →	$Mg_2Al\ (OH)_6^+$
No substitution	Al^3+ substituted for Mg^{2+}

Isomorphic Subsitution results in Constant (Permanent) Charges

2. **Exposed crystal Edges :**
   - → Ionisation of hydroxyl groups
   - → Ionisation of carboxyl and phenolic groups.

O^{2-} and OH^- groups are exposed at the broken edges and flat external surface as in Kaolinite : At pH> 7, the hydrogen of these hydroxyls dissociates slightly and the colloidal surface is left with a negative charge carried by oxygen.

The loosely help H^+ is readily exchangeable hence called pH-dependent charge of inorganic colloids. This phenomenon apparently accounts for most of the CEC of 1:1 type colloidal clays and for organic colloids.

Ionisation of carboxyl (-COOH) or phenolic (C_6H_5OH) groups is the chief source of negative charges on humus micelles . With the increase in pH, extent of negative charge is increased, therefore is called pH-dependent negative charge or variable charge. As the soil pH increases, more OH^- ions are available to force the reactions to the right ; and negative charge increases.

$$>Al-OH \quad + \quad OH^- \rightleftharpoons Al\text{-}O^- \quad + \quad H_2O$$

$$-\ CO\text{-}OH \quad + \quad OH^- \rightleftharpoons CO\text{-}O^- \quad + \quad H_2O$$

No charge On soil colloids — Soil solution — Charged — Soil solution

Second possible point is that at high p^H, complex aluminium hydroxy ions e.g. $Al(OH)_2^+$ is removed because these ions react with OH^- to form insoluble $Al(OH)_3$, therby releasing negatively charged sites. But at low pH it blocks the negative sites and make them unavailable for cation exchange (the process of exchange of cations between solid and liquid phases).

- In the soils of temperate climates where 2:1 type clays are common , the permanent negative charges are usually dominant . In highly weathered soils of tropics where 1:1 type silicate clays, Iron and Aluminium oxides dominate and in soils high in O.M. the variable negative charges are more common.

❐❐❐

18 Soil Taxonomy and soils of India

Modern classification system :

is US comprehensive soil classification system based on 7th approximation . This system maintains the natural body concept and two major features adopted were (1975)-

(i). Primary basis for identifying different classes are properties of soils rather than genesis of soils.

(ii). Nomenclature employed i.e.Latin or Greek root words are the basis for nomenclature (Soil taxonomy).

Six categories were adopted for this system: order, suborder, great group, sub-group, family and series. There are 10 soil orders in soil taxonomy and the basis of 'order' category is soil forming process associated with presence or absence of major diagnostic horizons.

1. Inceptisols (L.) Incepti means beginning. Embryonic soils with few diagnostic features, found in humid regions.
2. Entisols : Recently formed, little profile development, Enti has no special meaning.
3. Histosols : Histo(Gr.) means tissues, called Organic soils, peat or bog soils have more than 20% O.M.
4. Spodosols: Spodo(Gr.) Means wood ash;such soils are with subsoil accumulation of sesquioxide and humus.
5. Oxisols : Oxi (Fr.) means Oxide; sesquioxide rich soils, highly weathered soils of inter tropical regions.
6. **Vertisols :** Verti (L) means turn; High in dark swelling clays, deep cracks develop when dry; shrinking and swelling properties.

7. Aridisols : Aridi(L) means dry,dry soils of arid regions.
8. Ultisols : Ulti (L) means Last; low base status, forest soils.
9. **Mollisols** : Molli (L)means Soft; Grassland dark soils of steppes and prairies, high base status.
10. Alfisols (non- sense): High to Medium base saturation, forest soils.

- **Cat Clays** : Wet clay soils high in reduced forms of Sulfur that upon being drained become extremely acid due to oxidation of 'S' compounds and formation of sulphuric acid. Cat clay is clay-sulphate which shines like cat's eye.
- **Catena** : A sequence of soils of about the same age, derived from similar parent material, and occurring under similar climatic conditions, but having different characteristics due to variation in relief and in drainage.
- **Caliche** : A Layer near that surface, more or less cemented by secondary carbonates of Ca and Mg precipitated from the soil solution.

Soils of India :

Broadly in five groups -

1. **Alluvial soils** — Entisol
   - Inceptisol
   - Alfisol

Such soils include deltaic alluvium,calcareous alluvium, coastal alluvial soils and coastal sands, formed by transportation in streams and rivers and are deposited in flood plains or along coastal belts; generally deep soils; geologically it is of two types-

(a). Khadar or newar alluvium which is sandy, light coloured and less kankary.

(b). Bhangar or older alluvium, more clayey, dark and full of kankar.

2. **Black soils** — Vertisol
   - Inceptisol
   - Entisol

The typical soil derived from deccan trap is called regur or black cotton soils;the sedentary soil derived from granite and gneiss mostly,clayey in nature, practically uniform throughout profile,parent rocks are generally hornblendic type,rich in lime and magnesia and limesoda-type feldspar,varying quantities of gypsum deposited in the layer,layers in 45% of the black soils.It is highly argillaceous,very fine grained and dark and contains a high proportion of Ca and Mg-carbonates.It is rich in lime,magnesia and alumina,moderate potash and poor in P_2O_5,N and O.M. rich in montmorillonite and beidelitic group of clay minerals, high CEC40-60 me/100 g of soil.Occurence :Maharashtra,west M.P. ,parts of Andhra Pradesh,Gujarat and some parts of T.N.

3. Red soils — Alfisol
Ultisol
Inceptisol

Redsoil in T.N. occupy the largest area.Occurence vast areas of T.N. ,Karnataka, Goa, Daman and Diu, South East Maharashtra, East Andhra , M.P., Orissa and Chhotanagpur, Santhal Parganas, Mirzapur, Jhansi, Hamirpur. Such soils formed from ancient crystalline and metamorphic rocks, **colour due to wide diffusion of iron** rather than to a high content, poor in N, P_2O_5 and humus, potash; Some red soils are of lateritic origin and of a quite different nature, rich in kaolinitic type of minerals. Light soils,silica-sesquioxide ratio is 2.0, Soils generally shallow and low is soluble salts. p^H of soil 7.0-7.5, the soils under lift irrigation are commonly called **garden lands** and hungry and thirsty, opposite of black soils in almost all physical and chemical properties.

In the light and frequent rains of south west monsoon,red soils permit sowing being done earlier than other soils hence called **early soils**.The black soils are not much benefited by light showers in the early season due to its lower permeability.Thus such black soils are moistened sufficiently in the later period of season-called **'late soil'** Aeration and warming up of soil are hampered in heavy soils.

4. Laterite soils — Alfisols
— Ultisol
— Oxisol

Occurence : Hills of Karnataka,Kerala,M.P. Eastern ghat of Orissa, Maharastra, W.B., T.N.,and Assam;poor in lime and magnesia and N,occasionally P_2O_5 but K_2O deficient,occasionally higher humus.

Formed under heavy rainfall and high temp alternating.The heavy rains leach the salts and left it a porous matrix with a characteristics vesicular structure,facilitating easy drainage .Mixture of hydrated oxides of Al and Fe with small amount of manganese oxide,titania and others are present.

5. Desert soils — Entisol
— Aridisol

Occurrence : West Rajasthan,Haryana, Punjab, lying between Indus river and Aravali range. The desert sand is composed of quartz but feldspar and homblende grains also occur with a fair proportion of calcareous grains.

❒❒❒

19 Soil Organic Matter

- Soil Weight (i.e. of furrow slice) = $2x10^6$kg/ha
- Nitrogen status of Soil is 0.03-0.05%N means 1000kg of N/ ha . Total nitrogen of soil means available N + Fixed N.

Nature of soil Organic Matter :

Soil O.M. consists of a whole series of products which range from undecayed plants and animal tissues to fairly amorphous brown to black material bearing no terrace of the anatomical structure of the material that is normally defined as soil humus. In addition to organic constituents present in undecayed plants and animal tissues, soil O.M. contains living and dead microbial cells, microbially synthesized compound and derivatives of these materials produced as a result of microbial decay. The composition of Humus contains 50% C, 35% O,5% N and 5% H.

Soil O.M. — Humic Substance
— Non – humic substances

Humic Substances :

Such materials produced in the soil are either yellow or brown to black, acidic , polydisperse substance of high molecular weight . On the basis of solubility, Humic substances are divided into 3- classes-

(a) Fulvic acid : Lowest molecular weight and both acid and alkali soluble.

(b) Humic acid : Medium mol.wt. and alkali soluble but acid insoluble.

(c) Humin : High mol-wt .and both acid and alkali insoluble except under the most drastic conditions.

- Fulvic acid is most susceptible to microbial attack whereas humin is most resistant.

Non- Humic Substances :

Include all those classes of compounds occurring in plants and mirco-organisms that appear to have relatively definite characteristics e.g.carbohydrates, proteins, fats, waxes, resins, pigments and low mol. Wt. compounds. Most of these could be relatively easily attacked by soil micro-organism and has a rapid turn over in the soil.

Humus :

Humus is a complex and rather resistant mixture of brown or dark brown amorphous and colloidal substances modified from the original tissues or synthesized by the various soil organisms. Fulvic acid, humic acid and humin all come under humus. Humus is in dynamic condition.

In humus, 40-45% lignin and 30-33% protiens and rests are fats, waxes and residual materials. Lignin and protiens constitute about 70-80% hence humus is also called **Ligno- Protien** complex.

In humus, the C:N ratio is 10:1. In most of the Indian soil, C:N ratio is average 14:1. Humus is capable of adsorbing PO_4^{3-} anions from soil solution but not other anions.

Nature and Characteristics of Humus :

(i) Tiny colloidal humus particles (micelles) are composed of C, H and O.

(ii) Surface area of humus colloids is very high, generally exceeding that of silicate clays.

(iii) Negatively charged, the sources of charge being carboxylic (-COOH) or phenolic (C_6H_5OH) groups. The extent of the negative charge is p^{H-} dependent (i.e.high at high p^H).

(iv) At high P^H, CEC : 150-300C mol/kg soil.

(v) Water holding capacity : 4-5 times that of silicate clays.

(vi) Low plasticity and cohesion, thus favourable effect on aggregate formation and stability.

(vii) Black colour.

(viii) Cation exchange reactions with humus are qualitatively similar to those occuring with silicate clays.

- Cellulose is not readily available for the use of bacteria but this is first acted upon by fungi and changed into similar substances and made available for the use of bacteria.
- Wood is decomposed by Actinomycetes.
- O.M. content of Indian soils is low because of the high rate of decomposition under tropical and sub- tropical climate.

Factors affecting O.M. Decomposition :

(i) **Moisture** : when large quantities of O.M. are applied as manure in arid regions, the slender moisture in the soil may be largely used up for the decomposition of O.M. and the crop following may well suffer from lack of moisture.

(ii) **Temperature :** The soil organisms are most active at 24-35^{0}c. O.M. of Indian soils is low because of the high rate of decomposition under tropical and subtropical climate. Except in a few local ised areas in the hilly and high altitutes regions, the O.M. in most of the cultivated soils rarely exceeds 1% .O.M. content in Indian soil is generally 0.5%.

(iii) **Aeration :** In clayey soils, decomposition is less rapid.

(iv) **C: N ratio** : Nitrogenous amendments increase CO_2 evolution and greater loss of cellulose, hemicellulose and other plant polysaccharides. Low N- content or wide C:N ratio slows down decaying process. Therfore C:N ratio is used for predicting the rate of decomposition .It is important in controlling the available N, total O.M. rate of organic decay and in developing sound soil management schemes.The C:N ratio of arable soil is 8: 1–15:1. A little variation is due to climatic conditions viz temperature and rainfall. The C:N ratio is lower in arid soil and subsoil. C:N ratio in plant :20-30 : 1 in legumes and farm manures and as high as 400:1 in sawdust. In microbes it is 4: 1 to

9:1. The C:N value for soil is in between plant and microbes. It is lower in arid soils than that of humid soils when annual temperature is about the same . It is also lower in warmer regions than in cooler ones if the rainfalls are about equal.

Significance of C: N ratio :

Two major significance –

(a) Keen competition among micro-organisms for available N results when residues having a high C:N ratio are added to soils.

(b) Because this ratio is relatively constant in soil, the maintenance of 'C' and hence O.M. in soil depends largely on the soil Nitrogen level.

When decay occurs. the C/N ratio of the remaining plant material decreases since 'C' is being lost as CO_2 and 'N' is conserved .

The older the plants the larger will be C:N ratio and the longer will be the period of Nitrate suppression . Leguminous tissues have a distinct advantage over non – legumes since it promotes a rapid organic turn over in soils.

Legume stubbles decompose rapidly due to high N-content in the stuble.

The practical significance of this relatively constant ratio is that a soil's O.M. content can not be increased without simultaneously increasing its Organic Nitrogen content and vice-versa.

$\therefore$ 58g 'C' present in 100g O.M.

$\therefore$ 1g 'C' present in $\frac{100}{58} = 1.72$g O.M.

Therefore C: O.M. = 1:1.72

1.724 is called Bemlen Factor.

Muck and Peat soils :

Muck soils are having **highly decomposed** O.M. i.e. well mixed O.M. in the soil . Whereas **in peat soils,** mostly O.M. are **partially decomposed** and found under excessive moisture conditions. Peat soils are acidic, p^H 3.9 and below, 10-40% O.M., suitable for paddy when water recedes.

Role of O.M. :

Organic Matters are sources of plant nutrients which are liberated in available forms during mineralisation. Humus can be considered to be a store- house of various nutrients essential to plant growth. During the slow **microbial** decomposition of the soil humus, there is a gradual release, with subsequent mineralisation of C,N,S,P and other elements. It improves soil structure, its drainage and aeration, water holding capacity, buffer and exchange capacities; influences the solubility of minerals and serves as a source of energy for the development of micro – organisms. 95% N and 33%P of soil are obtained from O.M. It has a capacity to control soil temperature.

Mineralisation and Immobilisation :

Mineralisation is the conversion of an element from an immobilised form to an available form as a result of microbial decomposition. Immobilisation is the reverse of mineralisation where available form of an element is fixed as immobile form.

N : 20-40% of total surface soil N is in the form of bound amino acids; 5-10% as combined hexose amines. Usually only 1-3% of the total amount of N present is mineralised during growing season.

When C:N ratio exceeds 30, Immobilisation occurs and when C:N ratio is below 20, mineralisation takes place.

C : N	>	30	Immobilization
C : N	⇒	15-30 20-30	Immobilization and Mineralisation both
C : N	<	20	Mineralisation (i.e. release of NH_3 & NO_3)

Bartholomew & Kirkhan : When immobilisation is faster than mineralisation, no available N is present in soil and vice-versa. As a general rule when C: N ratio of O.M. added to soil is greater than 30, there is Immobilization of soil Nitrogen during initial decomposition process.

Bartholomew (1971) suggested about the natural supply of Nitrogen.

(i) Soil organic N : 20-30 kg N/ha

(ii) Rainfall : 6-8 kg N/ha

(iii) Non- symbiotic N – Fixation : 2-4 kg N/ha

(iv) Dust and organic particles through rainfall : 12-16kg N /ha

Subbhya and Asija (1956): Used alkaline potassium permagnate method and given the following values for available Nitrogen.

Low status : When < 250 kg N/ha

Medium : 250- 500 kg N /ha

High : > 500kg N/ha

P Brammer (1951) : 25- 80 % 'P' present in soil is in the form of **Organic form**. Major portion of **'P'** in humus is in the form of **phytin** or its diratives (40-80%) ; nucleic acid (0-10%). Humus increases the solubility of 'P' due to formation of phospho – humic complexes (which are easily assimilated by plants); anion replacement of PO_4^{3-} by humate ion; coating of sesquioxide particles by humus to form a protective cover which reduces the phosphate fixing capacity of soil and arising of certain organic anion from decompositions to form stable complex with Fe and Al, thus preventing their reaction with 'P'.

C:N : P = 100 : 10 : 1 and if C: P ratio is more than 100: 1, immobilisation of 'P' occurs.

S Barrow (1960) : C : S ratio of O.M. when below 200; SO_4^{2-} usually accumulated in soil, and when C:S > 400; SO_4^{2-} tie up in soil O.M. When the C:S ratio : 200-400; SO_4^{2-} either released or tied up .

Mineralisation of Oragnic Nitrogen :

There are three Steps –

1. **Aminization :** Hydrolytic decomposition of protein and release of amines and amino acids by heterotrops (bacteria like *Bacillus*, *Pseudomonas*) in absence of O_2

$$\text{O.M.} \xrightarrow{\text{hydrolysis}} \text{Polypeptides}$$

$$\text{Polypeptides/Proteins} \longrightarrow \text{R-NH}_2 + \text{CO}_2 + \text{Energy} + \text{other compounds.}$$

2. **Ammonification :** The amines and amino acids so released are further utilised by other hetrotrophs by release of ammonical compounds in absence of O_2

$$R\text{-}NH_2 + HOH \longrightarrow NH_3 + R\text{-}OH + Energy$$

$$2NH_3 + H_2CO_3 \longrightarrow (NH_4)_2\ CO_3 \rightleftharpoons 2NH^+_4 + CO_3^{2-}$$

NH_3 so released are subjected to several fates—

(a) may be converted to Nitrites and Nitrates by nitrification.

(b) may be absorbed directly by higher plants.

(c) may be utilized by heterotrophs organisms in the further decom position of O.M.

(d) may be fixed in a biologically unavailable form (in subsoil about 40-50% and in top soils 6%) in the lattice of certain expanding type of clay-minerals like momtmorillonite,illite and vermiculite.

3. **Nitrification :** By autotrophs in presence of O_2 i.e. biological oxidation. According to Winogradsky (1889), NH_3 or NH^+_4 salt in the soil is oxidised into NO^-_3 in two steps.

(i) NH_3 (or NH_4^+) $\xrightarrow[\text{Nitrosomonas}]{[O]}$ NO^-_2 (Nitrite)

Nitrosomonas is obligate autotrophs.

(ii) NO_2^- $\xrightarrow[\text{Nitrobactor}]{[O]}$ NO^-_3 (Nitrate)

Nitrobactor

Opt. temperature for Nitrifying bacteria is 30-35^0 C and Opt. p^H 6.5-7.5. The activity of bacteria ceases below p^H 5.0.

Nitrobacteria = Nitrosomonas + Nitrobactor both.

Reactions :

$$2\,NH_4^+ + 3O_2 \rightarrow 2NO^-_2 + 2H_2O + 4H^+ + Energy\ .$$

$$2\,NO^-_2 + O_2 \rightarrow 2\,NO^-_3 + energy.$$

When NH_4^+ - Fertilizers are added to soil,three important inferences are drawn by these nitrification equations -

(i) Reactions require molecular oxygen means nitrification takes place most readily in well aerated soil.

(ii) Reactions release H^+ which results in acidification of soil.

(iii) Microbial activity hence their rapidity and extent of transformation will be greatly influenced by soil environmental conditions such as moisture supply,temp etc.

Factors affecting Nitrification :

(i) Supply of NH^+_4

(ii) Population of nitrifing organisms

(iii) Soil reactions

(iv) Soil aeration

(v) Soil moisture

(vi) Soil temp. ($30-35^0$ C)

(vii) C : N ratio

(vIii) Fertility.

Nitrogen losses :

(A) Leaching loss/drainage loss of NO^-_3

(B) Runoff loss like of NH_3

(C) Gaseous losses.

Loss of N_2 may take place in 3 ways—

1. **Voltalization** or non-biological loss of NH_3 : when p^H is above 8, N_2 is lost in the form of NH_3 in alkaline medium.

$$NH_4^+OH^- + H_2O \rightarrow NH_3\uparrow + 2H_2O$$

60% N-loss in India is due to voltalization. Hence in alkali soil, the N-application is raised at least by 25% .

2. Chemical decomposition of NO_2^- under acid conditions to yield nitrogen and nitrogen oxide. This loss takes place when p^H is 5.5 or low under aerobic condition. There are 3-mechanisms-

(i) Decomposition of NH_4NO_2

$$NH_4NO_2 \rightarrow 2H_2O + N_2\uparrow$$

(ii) Van slyke reaction-

$$R\text{-}NH_2 + HNO_2 \rightarrow R\text{-}OH + H_2O + N_2\uparrow$$

(iii) Spontaneous decomposition of Nitrous acid

$$3\ HNO_2 \rightarrow H_2O + 2NO\uparrow + HNO_3$$

$$NO + O \rightarrow NO_2$$

$$2HNO_2 \longrightarrow NO + NO_2 + H_2O$$

NO is unstable under aerobic condition and immediately converted to NO_2.

3. Microbial Denitrification : Leading to liberation of N_2 and N_2O.

Formation and loss of gaseous form of N by biological reduction of NO^-_3 and NO^-_2 is known as Denitrification. Oxidation of NH_3 to NO^-_3 and NO^-_2 and oxidation of carbon compounds run simultaneously as long as supply of elemental O_2 is present. In the absence of elemental O_2, NH^+_4 oxidation ceases. Certain organisms like *Pseudomonas denitrificans* (gram negative), *Achromobactor, Bacillus, Micrococcus*; are capable of using oxygen derived from nitrite and nitrate or both in place of elemental oxygen. Removal of oxygen from NO^-_3 and NO^-_2 reduces the substances chemically . The products of the reductions are mostly gaseous form of nitrogen including nitrous oxide (N_2O), nitric oxide (NO) and elemental Nitrogen (N_2).

$$2HNO_3 \overset{-2[O]}{\rightarrow} 2HNO_2 \overset{-2[O]}{\rightarrow} 2NO \overset{-[O]}{\rightarrow} N_2O \overset{-[O]}{\rightarrow} N_2$$

Waterlogging induces denitrification. Denitrification losses can be reduced by addition of phosphate residues, by providing adequate drainage and with availability of excess accumulation of active nitrogen compounds in the soil.

- After ratoon crop wheat yield is more due to availability of phosphatic fertilizer which is left as residue in the ratoon sugarcane and organic matter content is also high due to sugarcane ratoon.

❑❑❑

20 Manures, Fertilizers

Manures :

The term 'manure' was used originally for denoting materials like cattle manure and other bulky natural substances that were applied to land , with the object of increasing the production of crops. Later chemical substances like $(NH_4)_2 SO_4$ and super phosphate containing plant nutrients in abundance were used for increasing crop production and were also called manures. So in a general way, manure is a substance containing plant nutrients. Later, the organics were specifically called 'manures' and 'inorganics' were called fertilizer. Therefore manures are defined as the plant and animal wastes which are used as sources of plant nutrients.

Advantages of Manuring :

(i) Manures supply plant nutrients including micro-nutrients.

(ii) They improve soil physical properties like structure, waterholding capacity etc.

(iii) increase nutrient availability.

(iv) CO_2 released during decomposition acts as a CO_2 – fertilizer.

(v) Provide food for soil micro organisms.

(vi) Plant parasitic nematodes and fungi are controlled to some extent by altering the balance of micro organisms in soil.

(vii) Provide buffering action in soil reaction .

(viii) Prevent loss of nutrients by leaching or erosion.

On the basis of concentration of nutrients, Manures can be grouped into two-

(i) Bulky organic manures and (ii) Concentrated organic manures.

(A) Bulky Organic Manures :

Contain small percentage of nutrients and they are applied in large quantities. FYM, Compost and Green manure are the most important and widely used bulky organic manures.

1. Farm Yard Manure (FYM)

Decomposed mixture of dung and urine of farm animals along with litter and left over material from roughages or fodder fed to the cattle. FYM contains 0.5%N, 0.2% P_2O_5 and 0.5% K_2O. Urine contains 1%N & 1.35%K_2O.N present in urine is mostly in the form of urea which is subjected to volatilization losses. Size of trench : 6-7.5m long, 1.5-2.0m wide and 1m deep. Heaped upto height of 45-60 cm above the ground level, dome shaped and plastered with cowdung earth slurry. The manure becomes ready for use in about 4 - 5 months after plastering. Chemical preservatives are used to reduce losses and enrich FYM.e.g.gypsum, kainite, and superphosphate. Gypsum absorbs urine and prevents volatilization loss of urea and also adds Ca and S. Superphosphate also acts similarly and increases 'P' content.Bacteria and Actinomycetes play active role in decomposition. 60-70% moisture in the initial stage and 30-40% moisture in decomposed manure (ready to use) as well as 50-60^0 C temperature under the heap are favourable for the activities of these micro- organisms. Generally 10-20 tonne well decomposed FYM per ha is applied but > 20 tonne FYM /ha in case of fodder grasses and vegetables, **at least 15 days** before the sowing to avoid immobilisation of N. Entire amount of nutrient is not available immediately.

About 30% N, 60-70% P_2O_5 and 70% K_2O are available to the first crop.

FYM is the most commonly used organic manure in India . One tonne of cattle dung can supply only 2.95kg. N,1.59kg P_2O_5 and 2.95kg of potash.

Table : Nutrient Contents of Manures & other organic raw materials

Materials	Contents N	P_2O_5	K_2O
Animal refuge			
Cattle dung & urine mixed	0.60	0.15	0.45
Horse dung	0.70	0.25	0.55
Sheep dung	0.95	0.35	1.00
Night soil	1.2-1.3	0.8-1.0	0.4-0.5
Poultry manure (Fresh)	1.0-1.8	1.4-1.8	0.8-0.9
Raw sewage (fresh)	2.0-3.0	--	--
Sewage sludge dry	2.0-2.5	1.0-1.2	0.4-0.5
Sewage sludge activated dry	5.0-6.5	3.0-3.5	0.5-0.7
Cattle urine	0.9-1.2	--	0.5-1.0
Horse urine	1.2-1.5	--	1.3-1.5
Human urine	1.1-1.2	0.1-0.2	0.2-0.3
Sheep urine	1.5-1.7	--	1.8-2.0
Manure compost etc			
Rural compost (dry)	0.5-1.0	0.4-0.8	0.8-1.2
Urban compost (dry)	1.0-2.0	0.9-3.0	1.0-2.0
FYM (dry)	0.5-1.5	0.4-0.8	0.5-1.9
Fitter press cake	1.0-1.5	1.4-5.0	2.0-7.0
Green manure fresh			
cowpea	0.70	0.10-0.2	0.60
Dhaincha	0.60	--	--
Mothbean	0.80	--	--
Greengram	0.72	0.18	0.53
Sunhemp	0.80	0.10	0.50
Blackgram	0.85	0.18	0.53
Plant residues			
Rice hulls	0.3-0.5	0.2-0.5	0.3-0.5
Groundnut husks	1.6-1.8	0.3-0.5	1.0-1.7
Straw & Stalky			
Bajra	0.65	0.75	2.50
Jowar	0.40	0.23	2.17
Maize	0.42	1.57	1.67
Paddy	0.38	0.08	0.71
Arhar	1.10	0.58	1.28
Wheat	0.53	0.10	1.10
Sugarcane trash	0.35	0.10	0.60
cotton	0.44	0.10	0.66

Courtesy : Handbook of Agriculture (ICAR)

- Hot and Cold manure : The manure obtained from the excreta of horses and sheep is called hot manure in temperate countries where– as pig and cattle manure is called cold manure due to comparatively less vigorous decomposition and less rise in temp in manure.
- Short and long manure : Decomposed manure that has lost the structure of the original materials is called short manure while the fresh manure having pieces of straw and other materials is called long manure.
- Fire fanging : There is profuse fungal growth on the surface of moist manure, giving it an ashy grey appearance. It is referred to as fire fanging.

2. Compost :

Composting is the process of reducing vegetable and animal refuse (rural or urban) except dung to a quickly utilisable condition for improving and maintaining soil fertility and decomposed material is called compost. Compost is like well- decomposed cattle manure in general appearance, more powdery and lighter in colour. Ordinary compost can be enriched with N and P through Azotobacter and superphosphate respectively. When superphosphate is used during compost making it is called **supercompost** and compost preapared by using N-fixing bacteria is called **Azo compost.** Azo compost is the cheapest source of N among all organic manures ($N \rightarrow 1.5\%$).

Superphosphate or rock- Phosphate @ 10-15kg/ tonne of raw material is applied at the initial stage of filling the compost pit. Size of pit is about the same of FYM.

The compost made from farm waste like sugarcane trash, paddy straw, weeds and other plants and other waste is called **Farm-Compost** (0.5% N, 0.15% P_2O_5, 0.5% K_2O). Farm compost prepared by Banglore method contains 0.80 -1.24% N, 0.40-0.59% P_2O_5 and 2.0-3.3% K_2O.

The compost made from town refuses like night soil, street sweepings and dustbin refuse is called **Town compost** (1.4%N, 1% P_2O_5, 1.4% K_2O).

3. Night Soil (or poudrette) :

Night soil is human excreta, both solid and liquid . It contains 5.5% N, 4.0% P_2O_5, 2.0%. K_2O)

The dehydration of night soil, as such or after admixture with absorbing materials e.g.soil, ash, charcoal & sawdust produces a poudrette that can be used easily as manure. Poudrette contains 1.32% N,2.8% P_2O_5 &4.1% K_2O.

4. Sewage and sludge :

The solid portion in the sewage is called sludge and liquid portion is sewage water. Both the components are separated and are given a preliminary fermentation and oxidation treatments to reduce bacterial contamination and offensive smell, otherwise soil quickly becomes '**sewage sick**' owing to the mechanical clogging by colloidal matter in the sewage and the development of anaerobic organisms which not only reduce the nitrates already present in the soil but also produce alkalinity. These defects can be removed by thoroughly aerating the sewage in the settling – tank by blowing air through it. The sludge that settles at the bottom in this process is called '**activated sludge**' (**3-6% N**, 2% P_2O_5 & 1% K_2O).

5. Sheep and goat manure :

It contains **3% N,** 1% P_2O_5 & 2% K_2O. it is applied to the field in two ways-

(i) Sweeping of sheep and goat sheds are placed in pits for decomposition and it is applied later to the field.

(ii) Sheep penning, wherein sheep and goats are allowed to stay over night in the field and urine and feacal matter is added to soil.

6. Poultry manure :

3.03% N, 2.63% P_2O_5 & 1.4% K_2O, Litter is the straw, peat, sawdust, dry leaves etc. are used as bedding material for farm animals and birds. It absorbs urine and faeces voided by animals and birds.

7. Green Manure :

Green undecomposed plant material used as manure is called **green manure**. It is obtained in two ways-

(i) By growing green manure crops in the field and incorporating it in its green stage in the same field . It is called **green manuring.**

(ii) Green leaf manuring is the application of green leaves and twigs of trees, shrubs and herbs collected from elsewhere especially waste land fields, bunds & forests to the field. Forest tree leaves are the main sources for green leaf manure e.g. neem, mahua, Glyricidia, Karanji (Pongamia) sesbania, subabul, Indigo etc.

Lentil is used in Kashmir for green manuring in paddy. Sunnhemp is most outstanding green- manure, well suited to almost all parts of country. Dhaincha does well in alkaline and waterlogged soils. Berseem for orchards and irrigated crops of cotton and sugarcane.

Plants at flowering stage contain the greatest bulk of succulent organic matter with low C:N ratio. The incorporation of the green manure crop at this stage allows a quick liberation of N in available form. The fertilizer value of the legume crop can be increased by manuring it with superphosphate (@ 100kg/ha). It increases 'p' content of green manure crops and thus converting an inorganic fertilizer into an organic manures. It has residual affects also.

Green manuring gives 60-80 kg N/ha on an average. Green manuring crops having **stem nodulation** are -

(i) ***Sesbania rostrata*** **:** It is fastest N_2- fixing plant @ 100-285 kg N/ha in 45-55 days.

(ii) ***Aeschynomene afraspera*** **:**

(B) Concentrated Organic Manures :

Concentrated organic manures have higher nutrients than bulky organic manures.

1. Bird guano: The excreta and dead remains of the birds. 7-8% N, **11-14% P_2O_5** & 2-3% K_2O.
2. Fish guano : The refuse left over after the extraction of oil from the fish in factories, is dried in paved yards and used as manure 7% N & 8% P_2O_5.
3. Fish manure : 4-10% N, 3-9% P_2O_5 & 0.3-1.5% K_2O
4. Raw bonemeal : 3-4% N,20-25% P_2O_5
5. Steamed bone meal : 1-2%N, **25-30% P_2O_5**
6. Blood meal : **13-20%** N, rich in **Iron** and its application gives a deep rich colour to the foliage, much appreciated by ornamental gardeners.

7. Meat meal/ meat guano/ tankage : 8-9% N and 7% P_2O_5.
8. Calcined bone : 37% P_2O_5
9. Oilcakes :
   Sunflower cake : **7.8%** N
   Groundnut cake : 7.2% N,Cotton cake : 6.5% N.

Non- edible oilcakes are used as manures especially for horticultural crops(edible fed to cattle). Nutrients present in oilcakes, after mineralisation, are made available to crops 7-10 days after application.

Oilseed cakes need to be well powdered before application for even distribution and quicker decomposition.**Neem cake** reduces nitrification means acts as **nitrification Inhibitor.**

Table : NPK % Present in Oilcakes & Others

Materials	N	P_2O_5	K_2O
Castor cake	5.5-5.8	1.8-1.9	1.0-1.1
Mahua cake	2.5-2.6	1.8-1.9	1.8-1.9
Karanj cake	3.9-4.0	0.9-1.0	1.3-1.4
Neem cake	5.2-5.3	1.0-1.1	1.4-1.5
Safflower cake (undecorticated)	4.8-4.9	1.4-1.5	1.2-1.3
Cotton Seed (decorticated)	6.4-6.5	2.8-2.9	2.1-2.2
Cotton Seed (undecorticated)	3.9-4.0	1.8-1.9	1.6-1.7
Groundnut cake	7.0-7.2	1.5-1.6	1.3-1.4
Linseed cake	5.5-5.6	1.1-1.5	1.2-1.3
Niger cake	4.7-4.8	1.8-1.9	1.1-1.3
Rapseed cake	5.1-5.2	1.8-1.9	1.1-1.3
Seasame/Till cake	6.2-6.3	2.0-2.1	1.2-1.3
Coconut cake	3.0-3.2	1.8-1.9	1.7-1.8
Dried blood	10.0-12.0	1.0-1.5	0.6-0.8
Fish manure	4.0-10.0	3.0-9.0	0.3-1.5
Bird guard	7.0-8.0	11.0-14.0	2.0-3.0
Poultry manure	2.9	2.9	2.4
Hoof and horn meal	14.0	1.0	-
activated sludge (dry)	5.0-6.5	3.0-3.5	0.5-0.7
Settled sludge (dry)	2.0-2.5	1.0-1.2	0.4-0.5
Rawbone meal	3.0-4.0	20.0-25.0	
streamed bone meal	1.0-2.0	25.0-30.0	

Courtesy : Handbook of Agriculture (ICAR)

Fertilizers :

Liebig in Germany and Lawes in England developed independently the idea of treating phosphate with H_2SO_4 and producing a water soluble phosphate by about 1840. It is known as superphosphate or simply super. Lawes set up a factory at Deptford in 1843 for the production of superphosphate. Possibly this was the first factory to be set up for the production of artificial fertizers on a commercial scale.

Fertilizers are the organic or inorganic materials of natural or synthetic origin which are added to the soil to supply certain elements essential to the growth of plants. The term 'Fertilizer' is now commonly restricted to commericial products.

Amendments are the substances other than manures and fertilizers which are added to soils for the improvement of their condition.

Amendments are also termed as ameliorants, improvers or soil conditioners e.g. gypsum and lime though they supply nutrients but the main objective of applying them is for correcting the soil condition.

Classification of fertilizers :

(a) Straight fertilizers : such fertilizers has declarable content of only one major nutrient e.g. urea, ammonium sulphate.

(b) Binary fertilizer : contains two major nutrients e.g. Potassium Nitrate.

(c) Ternary fert: contains three major nutrients e.g. Ammonium potassium phosphate.

(d) Compound/ Complex fertilizer : Such fertilizer has a declarable content of at least two of the major nutrients obtained **chemically** and generally granular in form.e.g. Nitro- phosphate, Amm- phosphate and Diammonium phosphate (DAP).

(e) Mixed fert : Individual or straight fert . materials are blended together **physically** to permit application in the field in one operation . Such fertilizers supply two or three major nutrients in a definite proportion or grade.e.g. nitrophosphate with potash 15: 15: 15 of NPK.

(f) Complete fertilizer : having all the three primary major nutrients viz.N, P, & K.

(g) Incomplete fertilizer : containing any two primary nutrients.

(h) Low- analysis fertilizers ; having less than 25% of the primary nutrients e.g. SSP (16% P_2O_5), sodium nirate(16%N).

(i) High- analysis fert: having more than 25%of the total primary nutrient content e.g. urea, anhydrous ammonia (82.2%N), Ammonium phosphate (20%N + 20%P_2O_5), DAP (18% N + 46% P_2O_5).

- Fertilizer grade: refers to the guaranteed analysis of its plant nutrients. It is the minimum guarantee of the plant nutrient contents in the terms of N, available P_2O_5 and K_2O e.g. 6: 24: 24 .
- Fertilizer Ratio : refers to the relative percentage of N, P_2O_5 and K_2O i.e. 1:4:4 if fertilizer grade is 6:24:24.

Materials used in manufacturing of fertilizers (Mixed) :

1. Suppliers of plant nutrient : Straight fertilizers are used for this purpose.
2. Conditioners: To check absorbing moisture and making one, conditioners like straw, groundnut husk, paddy husk, peat soil etc. are used they just reduce caking and applied in drilling conditions and these conditioners are of low organic materials.
3. Neutraliser of acidity or basicity : Dolomitic limestone is used to reduce residual acidity. Most fertilizers leave residual acidity or basicity.
4. Filler material or make weight material : Sand,soil, earth, coal ash, charcoal such waste materials are added to make up the difference between the weight of the added fertilizers required to supply the plant nutrients and the desired quantity of the fertilizer mixture.

Precautions in Mixing fertilizers :

1. Hygroscopic fertilizers should not be mixed because they form cakes after mixing. **Hygroscopic fertilizers** are in descending order-
   (a) NH_4NO_3 (Amm. Nitrate)
   (b) Urea : $CO(NH_2)_2$
   (c) Ammonium sulphate : $(NH_4)_2SO_4$
   (d) Amm.sulphate nitrate: $(NH_4)_2SO_4\ NH_4\ NO_3$
   (e) Calcium ammonium nitrate(CAN)

The most hygroscopic fert.is Amm.nitrate and urea stands second. All these fertilizers have NH^+_4.

2. Fertilizers containing NH_4^+ should not be mixed with basically reactive materials like lime, basic slag and rock phosphate because mixing results in the loss of N through escape of NH_3 gas.
3. All water soluble phosphatic fertilizers like super phosphate should not be mixed with those fertilizer that contain free lime because it will convert the soluble phosphate into insoluble form.
4. Slightly acidic fertilizer containing chloride may damage the gunny bags and drilling equipments.

Acidity and basicity of fertilizer :

Equivalent acidity : The amount of $CaCO_3$ required to neutralise the acid residue caused by application of acidic fertilizers in the soil e.g. 100kg $(NH_4)_2SO_4$ produces acidity which needs 110 kg of $CaCO_3$ to neutralise it.Therefore equivalent acidity of $(NH_4)_2SO_4$ is 110, Anhydrus NH_3 has more equivalent acidity than other fertilizer.

Equivalent basicity : Residual basicity caused by application of basic fertilizers, expressed in terms of $CaCO_3$ equivalent of basic residue left by a fertilizer material (in kg/100kg of fertilizer salt).

S.No.	Fertilizer	Equivalent acidity	Equivalent Basicity
1	CaCN		63
2	$NaNO_3$		29
3	KNO_3		29
4	$Ca(NO_3)_2$		21
5	Anhydrous Ammonia	148	
6	Nitrate of soda Potash		26
7	NH_4CI	128	
8	$(NH_4)_2SO_4$	110	
9	$(NH_4)_2SO_4$. NH_4 NO_3	93	
10	Urea	80-84	
11	DAP	77	
12	AMM nitrate	60	
13	Mono-ammonium phosphate	55	

Commercial Nitrogenous Fertilizers :

- Except paddy and potato, the ammoniacal form is not directly absorbed by crops. Ammonia always produces acidity. Except NO^-_3, CaCN and CAN; all other nitrogenous fertilizers are acidic in nature. NO^-_3 and CaCN are basic in reaction, whereas CAN is neutral.

Characteristics of :

Nitrate Fertilizer : Highly mobile in soil,suitable for top dressing, highly soluble and subjected to leaching and denitrification in water logged soils, increase alkalinity, more suitable for dry soils.

Ammonical Fert : Readily soluble in water, less leaching losses as NH^+_4 are adsorbed on clay complex, suitable for water logged fields, acidic in nature; paddy in early stage is capble of utilising NH^+_4 form.

Amide Fertilizer : Organic fertilizer, first converted into ammonical and then to nitrate form. Urea is acidic but CaCN is basic. Highly soluble in water. Amide $\longrightarrow (NH_4)_2CO_3 \longrightarrow NO^-_3$

Nitrate-ammonical group : NO^-_3 nitrogen readily available to plants for rapid growth and NH_4^+ nitrogen at the later stage.

1. Sodium nitrate: 16% N, Chilean nitrate, pioneer nitrogenous fertilizer; useful for acidic soils .Its continuous and abundant use causes deflocculation and develop a bad physical condition in low rainfall regions.

2. Ammonium sulphate: 20.6%N and24%S. 20times less used than urea. During rainy season, it sometimes forms lumps, suitable for paddy and jute .Maximum moisture by wt.=1%, Arsenic oxide: max.0.01%, Ammonical Nitrogen: 20.6% by wt.

During fixation, NH_4^+ releases an equivalent quantity of Ca^{2+} from the soil and $CaSO_4$, that results, is leached away under humid conditions.Application of $(NH_4)_2SO_4$ reduces the quantity of reserved calcium in soil.If there is no reserve of Ca in the soil , the H_2SO_4 is formed as a residual product which will increase the acidity in soil.

3. Ammonium Nitrate : 33-35%N, highly hygroscopic and not fit for storage, liable to explode.

Nitro- chalk = Amm. Nitrate + limestone/ dolomite
20.5% N 60% 40% proportion

Nitrochalk is less hygroscopic and formed by mixing amm.. nitrate with about 40% limestone or dolomite,useful for acid soils.

4. Amm. Sulphate Nitrate: 26% N, 15%S, It is a mixture of Amm. sullphate + Amm. nitrate. Three- fourth of N is in ammoniacal form.

5. Calcium Ammonium Nitrate (CAN) : Kisan Khad, 25-28% N (mostly 26%) almost neutral in nature, most suitable for vegetables, commercially prepared from amm. nitrate and ground limestone, half of N is in ammoniacal form. According to standard, moisture content by wt. < 1%, calium nitrate < o.5% by weight.

6. Ammonium chloride : 26% N, extensively used in paddy in Japan. It is usually not recommended for tomatoes, tobacco etc. Cereal crops are not affected by Cl as most of it is retained by straw and is not translocated to the grain. In some crops like potato and sweet potato high quantity of Cl is avoided andinstead K_2SO_4 or KNO_3 is used (KNO_3 : 13.85% N and 46-47% K_2O).

7. Urea : mostly used in India, 46% N, fairly hygroscopic therefore produced in granular or pellet form and is coated with a nonhygroscopic inert material. Max.moisture : 1% by wt. and **Biuret :< 1.5%** according to fertilizer control order.

Biuret is formed during manufacturing of urea when temp. goes high above 150^0C and is toxic to plants when its concentration is more than 2%.`

$$NH_2CONH_2 + NH_2CONH_2 \xrightarrow{\text{above } 150^0c} \underbrace{NH_2CONH - CONH_2}_{\text{Biuret}} + NH_3$$

The concentration of foliar spray of urea varies between 2–6% but general concentration is 2%.

when urea is applied to the soil, following reactions takes place-

$$NH_2\ CONH_2 + 2H_2O \quad (NH_4)_2CO_3$$

$$(NH_4)_2\ CO_{3-} \rightarrow NH_4^+ + CO_3^-$$

$$NH_4^+ \rightarrow NO_2^- + NO_3^-$$

The whole conversion takes about 4-7 days. Therfore it is advisable to apply Urea 3-4 days before sowing.

8. Cyanamide/ Nitrolim/Cal.Cyanamide : 20.6%N, certain immediary compounds formed during its decomposition injure tender and germinating seedlings and therfore it is advisable to apply it at least a week before sowing. Lime is one of the products of decomposition and a certain amount of lime is also present in the original material and these are of high value in acid soils.

9. Aqueous ammonia : 80%N, used as fertigation i.e. in irrigation water.

Phosphatic fertilizers : 3 types -

(A) Water soluble : $Ca(H_2PO_4)_2$ Such water soluble phosphatic fertilizers are used for quick start and short duration like wheat, for neutral and alkaline soils e.g.

1. Super phosphate is of three types-
   - (i) Single super phosphate : 16-20% P_2O_5
   - (ii) Double super phosphate : 32% P_2O_5 (25-30%)or Enriched super phosphate)
   - (iii) Triple super phosphate : 46-48% P_2O_5 (44-52%)
2. Monoammonium Phosphate (MAP) : 12%N and 48% P_2O_5
3. Diammonium phosphate (DAP): 16-48-0; 18-46-0, $(N \rightarrow P_2O_5 \rightarrow K_2O)$

(B) Citrate soluble (but water insoluble) : $Ca_2H_2(PO_4)_2$ or $Ca_2(HPO_4)_2$ used in acid soils, for long duration crops like sugar cane, low land rice, tapioca & tea. e.g.

1. Dicalcium Phophate : 33-40% P_2O_5
2. Thomas/Basic slag : 14-18% P_2O_5
3. Rhemania Phosphate : 23-26% P_2O_5

(C) Citrate and water Insoluble : $Ca_3(PO_4)_2$. Tricalcium phosphate. used in strongly acid soils and suitable for plantation crops e.g.

1. Rock Phosphate : 20-30% P_2O_5
2. Raw Bone meal : 3-4%N + 20-25% P_2O_5
3. Steamed bonemeal : 1-2%N + 20-30% P_2O_5

Single superphosphate (SSP) : SSP contains 12% S + 16% P_2O_5 + 18-21%Ca. Among phosphatic fertilizers, SSP is widely used in India . It is prepared by treating Rock Phosphate with equal amount of Sulphuric acid. After treatment dusty coloured mixture of calcium phosphate and calcium sulphate in equal amount is obtained. But SSP is generally formed by the mixing of Mono calcium Phosphate and Calcium Sulphate (Gypsum) in both forms i.e.powder and granular.In alkaline or acidic soils SSP forms Calcium Phosphate, Iron phosphate or Aluminium phosphate, thereby preventing the fertilizer loss with water and thus is solublised slowly in soil.That's why total amount of this fertilizer is applied as a single dose either at sowing/ planting time or before the time of sowing/planting. In general, SSP is suitable for all the crops and different soils.

According to standard; SSP must contain

Moisture content by wt. : 12%(maximum)

Free Phosphoric acid by wt. : 4%(maximum) (H_3PO_4)

Water soluble Phosphate by wt. : 16%(minimum)

Dicalcium Phosphate :

By neutralising the phosphoric acid by lime, Dicalcium phosphate is prepared and phosphoric acid is obtained by reacting HCl, H_2SO_4 or HNO_3 with rock phosphate.

This fertiliser is equally suitable for acidic neutral and alkaline soils but it is especially useful for acidic soils because it does not form compounds in such soil.

Potassic fertilizers :

K- containing minerals are primary minerals like feldspar/ orthoclase, Muscovite and Biotite.

1. Muriate of potash (MOP):KCl; 60%K_2O, Na as NaCl by wt. less than 3.5%,min 58% K_2O. It is not used in tabacco, rice, tomato, sugarcane, beets and starchy crops .In sugar crops accumulation of sugar is affected by Cl^- Higher content of Cl^- in tobacco leaf reduc-es its burning quality. Although MOP is highly soluble in water but is not lost by water because it is adsorbed in the soil lattices.

2. Potassium Sulphate or Sulphate of Poatash (SOP): 48-52%K_2O and 17-18% S. suitable for light soil and for those crops for which MOP is unsuitable.
3. Nitrate of Potash (NOP) /Potassium nitrate (KNO_3) : 13%N, 44% K_2O, excellent source of K and N; mainly used for fruit trees and crops such as tobacco and vegetables.

Secondary Fertilizers :

S. No	Fertilizer	Ca%	Mg%	S%	Others%
1	Gypsum **($CaSo_4.2H_2O$)**	**29.2%**	-	**18.6**	-
2	Rock phosphate	33.1	-	-	25.2 P_2O_5
3	SSP	19.5	-	**12.5**	16% P_2O_5
4	$MgSO_4$ $7H_2O$ (Epsom salt)	-	9.6	13.0	-
5	Pot Sulphate	-	-	17.5	48(K_2O)
6	Amm Sulphate	-	-	**24.2**	21(N)
7	Basic slag	-	-	**3**	15.6 (P_2O_5)
8	$CuSO_4$	-	-	**11.4**	21(Cu)
9	Fe SO_4	-	-	**18.8**	32.8 (Fe)
10	$ZnSO_4$	-	-	**17.8 (or 15)**	36.4 (Zn)
11	Fe S_2 Iron pyrite	-	-	**22–24**	
12	Borax ($Na_2B_4O_7.10\ H_2O$	-	-	-	**10.6%B**

Micronutrient Fertilizers :

Some of the metallic micronutrient salts when applied to the soil, are transformed into non-available forms due to their reactivity. Organic compounds like EDTA (Ethylenediamine tetra acetic acid), DTPA (Diethylene triamine penta acetic acid) CDTS (Cyclohexane diamine tetracetic acid) have the ability to chelate or loosely hold metallic ions in their cyclic structure and these metal-organic complexes are called metal chelates. Metal chelates are soluble in water but they do not ionise in soil solutions. Metal ions therefore , don't react with soil constituents . Chelate forms of nutrient are more available. Fe,Cu,Zn and Mn are available in chelate form.

50% of Indian soil is deficient in Zinc and 25% of Indian soil in sulphur.

Table : Range of micronutrient conc. for normal plant growth.

Trace elements	Conc. in ppm
Fe	O.5-5 ppm
Mn	0. 1-0.5
B	0. 1-1.0
Zn	0 .02-0.2
Cu	0. 01-0.05
Mo	0. 01-0.05

Nitrification Inhibitors and Slow release Fertilizers:

Among the major nutrients, P.&K are less mobile as compared to N. Nitrogenous fertilizers are easily lost by leaching ,volatilisation and denitrification . To offset these losses to the minimum, N is applied in split doses at the critical phases of crop growth. However split application increases the cost of fertilizer application .In order to reduce the leaching,volatilisation and denitrification losses to the minimum, use of some nitrification inhibitors or slow release fertilizers are useful to regulate the nutrient availability to the crop.

Advantages of Nitrification Inhibitors and slow release fertilizer-

(i) They improve the uptake of N by releasing the nutrient slowly and uniformly.

(ii) Labour saving device since no need for split application.

(iii) Scope for reducing the fertilizer dose due to higher efficiency of fertilizer and uptake of nutrient.

(iv) Low pollution of water and air since leaching losses are minimum.

(v) Application of fertilizer can be done either basal or top dressing thereby greater flexibility in timing.

Nitrification Inhibitors :

Inhibitors are used with good results in case of non-nitrate fertilisers such as ammonia, ammonium salt, urea etc. Inhibitors decrease the activity of nitrifying bacteria e.g.

For lowland :

(a) Oxamide ($NH_2CO\text{-}CONH_2$): 31%N, not hygroscopic, solubility is 0.4g/lit.water.

(b) dicyandiamide (DD): NH_2C (=NH) NHCN: 42%N

(c) Thiourea (TU) : 36.8%N

(d) Urea pyrolyzate : 48%N.

For upland :

(e) AM(2-amino-4-chloro-6-methyl pyrimidine)

(f) N-Serve(2-chloro-6-trichloromethyl pyridine).

Others :

(g) ASU (Guanyl thiourea).

(h) **Nitrapyrin**

(j) ST (Sulphathiazole).

(i) DCS

(K) ATC

(l) **Neem Cake.**

Slow Release fertilizers :

To overcome the problem of leaching, the solubility of nitrogen fertilizers are reduced by-

(A) Synthesizing compounds which are inherently less soluble e.g.

1. Isobutylidene diurea (IBDU) : 31-32%N.
2. Crotonylidene diurea (CDU) : 32.5%N.
3. Guanyl urea sulphate(GUS).
4. **Urea Formaldehyde** (UF/Urea Form) : 38-42% N, less hygroscopic than urea.

5. **Oxamide.**

(B) Coating barriers to the presently available fertilizers e.g. Sulphur coated urea, Neem coated urea, Lac coated or shellac coated urea(34.2%N).

(C) Formation of super Granules/Modified form: Big granules of urea 1-4 g each are made. When these granules are placed in the reduced zone of soil, the losses are substantially reduced.To facilitate deeper placement , Urea is manufactured as super granules, briquetts or mixed with mud and made into balls.

- GROMOR/Gromor: Trade name of urea ammonium phosphate grade 29:29:0.
- Ammophos-B : Earlier name of Ammonium phosphate sulphate. grade: 20:20:0.

Bio fertilisers/Microbial Inoculants :

'Microbial Inoculants' is the more appropriate name of Biofertilizer. It is defined as preparation containing live or latent cells of efficient strains of N-fixing, PO_4^{3-}solubilising or cellulolytic micro-organisms used for application of seed, soil or composting areas with the objective of increasing the number of such micro- organisms and accelerate certain microbial process to augment the extent of the availability of nutrients in a form which can be assimilated by plants.

Application :

(a) Seed Inoculant : 20g of **Rhizobium** culture is required to treat 1 kg seed. One packet of Rhizobium culture contains 200g of Rhizo bium culture.

(i) For small seeded pulses like moong , Arhar, lentil, berseem, lucerne, Kulthi, 500g of Rhizobium culture is sufficient for seeds required to be sown in 1 ha i.e. 2.5packets/ha (since 1packet=200g.

(ii) For Groundnut, 1.5kg of Rhizobium culture for80-100 kg seeds for 1ha i..e.7.5packets/ha.

(iii) For soybean and Bengal gram: 1kg/ha i.e. 5packets/ha.

(b) Soil inoculant : 10 packets (2kg/ha)of carrier based **Azotobacter** and **Azospirillum** culture is mixed with 25kg FYM and 25kg soil and is broadcasted in the field uniformly before transplanting.

Benefits :

1. Rhizobium Biofertilizer can fix 50-200kg N/ha per year.
2. Increases yield by 25-30% and 40-80kg N is left over in the field useful for subsequent crop.
3. BGA can add up to about 20-25 kgN/ha to rice field.
4. BF like BGA, Azotobacter and Azospirillum also supply growth egulators such as IAA,IBA,NAA,GA_1,to GA_{33} and vitamins.
5. Azotobacter and Azopirillum secrete antibiotics which act as pesticides so biofertilizers also act as 'Bio-pesticides.'
6. Azolla not only supplies N but also increases O.M. inform of biomass and increases soil fertility.
7. It increases soil's physical properties such as soil structure, texture, chemical properties such as water holding capacity,cation exchange capacity of soil,buffer capacity of soil etc.
8. It proliferates useful soil micro-organism i.e.biological properties of soil.
9. It is ecological friendly, technologically feasible and socially acceptable input to the farmers.

Biological Nitrogen Fixation

Average fixation of N_2 by legumes -

	Crops	**N (kg/ha)**
1.	Alfalfa (Lucerne)	194
2.	Ladino clover	179
3.	Sweet clover	119
4.	Red clover	114
5.	Kudzu	107
6.	White clover	103
7.	Cowpeas	90
8.	Vetch	80
9.	Peas	72
10.	Soybeen	58
11.	Peanuts	42
12.	Beans	40

Classification

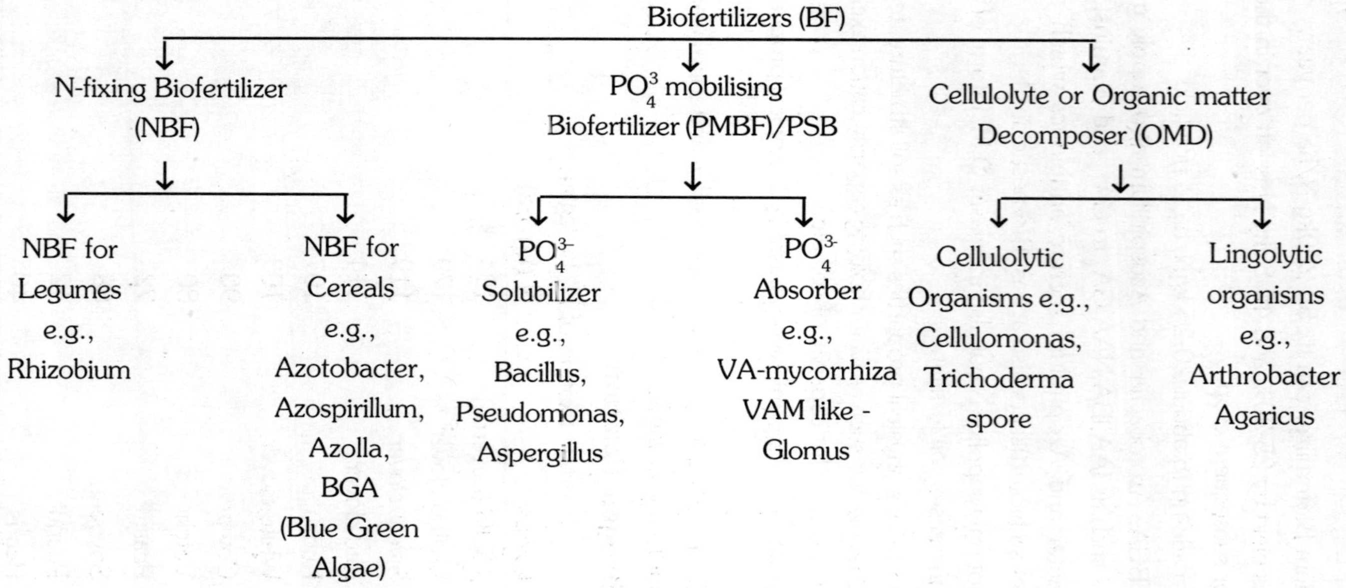

* VAM : Vesicular arbuscular mycorrhiza. The association of a fungus with the roots of a higher plant for mutual benefit; probably symbiotic but may be week form of parasitism.

* PSB : Phosphate Solubilising Bacteria.

Table : A list of Common legumes and the Rhizobium strains by which they are inoculated.

Rhizobium spp.	Legumes innoculated
1. ***Rhizobium meliloti***	(a) **Melilotus** (Sweet clover) (b) **Medicago** (Alfalfa) (c) Trigonella (Fenugreek)
2. ***R. trifoli***	**Trifolium (clover)**
3. ***R. leguminosarum***	(a) **Pisum** (pea) (b) Vicia (vetch) (c) Lathyrus (sweet peas) (d) **Lens** (lentil)
4. ***R. phaseoli***	Phaseolus (beans)
5. ***R. lupini***	(a) Lupinus(Lupine) (b) Ornithopus (serradella)
6. ***R. Japonicum***	(a) **Glycine (soybean)** (b) Vigna (cowpea) (c) **Arachis (Groundnut)** (d) Crotolaria

Fixation of atmospheric Nitrogen :

→ Physio- chemical e.g. discharge of electricity in atmosphere.

→ Bio- chemical by

- → activity of certain saprophytic bacteria
- → activity of symbiotic bacteria
- → activity of Blue green algae.

Bio-chemical or Biological N-fixation is the natural process of fixation of atmospheric N_2 through the acivity of soil micro- organisms. It can be 190 tonnes N/year, where as industrial fixation (by reduction process) is around 50 tonnes.

(A) Symbiotic N-fixation :

(a) Nodule forming :

(i) **With legumes :** On the experimental basis Boussingault first proved that legumes use free N_2 of air for their normal growth. The bacterial symbiont is member of genus ***Rhizobium.*** **Leghaemoglobin** is the **Oxygen -carrier** in the N-Fixation in nodules and it also protects N-fixing enzyme i.e. Nitrogenase from oxygen. Leghaemoglobin has high affinity to O_2 and it relases O_2 to bacteroids slowly and in just sufficient quantity . Rhizobium bacteria is aerobic and heterotrophs. Greengram fixes 20kgN/ha whereas Lucerne, even more than 250kg N/ha .

(ii) **With non- legumes** : About 160spp.from 13 genera of nonlegumes .The important group is angiosperms. The roots of these plants inoculated by soil actinomycetes of the genus **Frankia, and** ***Casuarina alder*** lives in association with casuarina (a xero-phytic tree).

- Actinomycetes(or thread bacteria) are in between true fungi and bacteria. It develop best in moist and aerated soil. The growth of Bacteria and Actinomycetes ceases below p^H 5:5 but no effect on Fungi.

Actinomycetes (Myrica, Casuarina,Elaeagnus)	50-150 kg N/ha year.
Legumes	150-250 kg Nper ha per year
Beans	30-50 kg N/ha/year
Algae of the genus Nostoc	10-20 kg N/ha
Pulses	30 kg N/ha/year

Factors affecting Nodule formation & N-fixation-

1. P^H : Legumes fail to develop nodules when p^H less than 5.0.
2. N & P- status of soil.
3. Excess Moisture in soil : Since Rhizobia are aerobic, there is detrimental effect of excess moisture in soil.

(b) Without Nodule :

The bluegreen algae ***Anabaena azollae*** is having a symbiotic association with water fern, ***Azolla pinnata.*** Anabaena inhabits in the cavities in the leaves of floating fem Azolla. It fix 30-40kg N/ha. Unlike Blue green algae it thrives well at low temperature. Opt. temp:20-30^{0}C, p^H : 5.5-7.0 Nursery for Azolla culture : 4m x2m x 30-40cm(raised), under shade; Azolla application rate @ 0.1-0.5kg/m^2. For faster growth SSP@ 2.5g/square metre and Carbofuran granules@ 1.2g/m^2 to control leaf eating caterpillars, are applied.

Azolla is applied in the main field as a green manure and dual crop. Azolla is allowed to grow on the flooded fields for 2-3 weeks before transplanting.Later water is drained and Azolla is incorporated by ploughing in. As a dual crop 1000-5000kg/ha of Azolla is applied to soil one week after transplanting for better growth, 25-50kg/ha SSP and 5-10 cm standing water continuously is maintained, When a thick mat forms, it is incorporated by trampling. N-fixation is around 25-35kg/ha/year.

(B) Non-Symbiotic N-fixation :

By soil and water free living micro organisms.

(a) Fixation by Heterotrophs : Several groups of bacteria and **BGA.**

(i) Aerobic bacteria :

1. ***Azotobacter chroccum* and *Azotobacter vinefandi* :-** act in temperate zone soils; opt.p^{H}: 6.5-8.0
2. **Beijernckia** : In tropical soils, tolerant to wider p^H (opt p^H 5.0-9.0).

(ii) Anaerobic bacteria : **Clostridium**: tolerant to wider p^H (opt.p^H 5.0-9.0).

The above bacteria fix the nitrogen -@ 5-20kg N/ha/year. These bacteria are applied in the field by mixing 3-5kg its inoculum with 5 tonne FYM per hectare.

(b) Fixation by Autrotrophs: certain photosynthetic bacteria viz. Rhodospirillum (anaerobic) and Blue green algae,

- **Azotobacter:** is used for rice, cotton and **sugarcane** whereas **Azospirillum** for **sorghum.**
- **Saprophytes:** Such microorganisms are capable of decomposing O.M.at a faster rate; hence can be used as a fertilizer for quick release of nutrients e.g. Aspergillus, Penicillium, Trichoderma are cellulolytic fungi.

❑❑❑

21 Soil Fertility and Productivity

Available Nutrient analysis of soil :

For analysing available nutrients,first available nutrients are extracted by using reagent known as extracting reagent e.g. for extracting available N,0.25%$KMnO_4$(Potassium permagnate) is used.For phosphorus 0.5 M sodium bicarbonate ($NaHCO_3$) p^H8.5 is used.For Potash .1N Neutral Ammonium acetate (CH_3COONH_4) is used.

After extracting the nutrients, actual amount of available nutrient is determined by colourimetric method.On the basis of analysis of available nutrient, soil fertility is classified broadly into 3 groups.Viz low,medium and high.

N Low status of N in soil when less than 250 kg /ha.
Medium :250 -500 kg/ ha
High :>500 kg/ha.

P Low < 20 kg/ha
medium :20-50kg/ha;
High : >50kg/ha.

K Low < 125 kg/ha,
medium : 125 - 300 kg/ha,
high : >300kg/ha.

Zn & Cu Low :< 0.5 ppm,
medium :0.5-1.0 ppm,
High >1.0 ppm.

Mn Low < 1 ppm,Medium : 1-3 ppm,High > 3.0 ppm.

B Low < 0.33 ppm,

Medium :0.33-0.67 ppm,

High >0.67 ppm.

Organic carbon (organic matter) :

Low < 0.5%, Medium 0.5- 0.75%, High >0.75%

S. No.	Nutrient Determination	Analysing Process
1.	Total Nitrogen in soil	Kzeldahl Process
2.	Available N	Alkaline permanganate method
3.	Phosphorus in acid condition	Bray No. 1
4.	Phosphorus in neutral and alkaline condition	Olsen's method
5.	K^+ & Na^+	Flame photometer
6.	Sulphate sulphure (SO^{2-}_4)	Terbimatric Method
7.	Organic carbon	(a)Walke & Black method (b)Rapid oxidation process ($KMnO_4$) (c) Morgan's method
8.	Available K (Potash)	Chromic acid digestion method

Crop Logging : defined by **H.F.Clement :**

"The crop log is the graphic record of the progress of the crop, contains a series of chemical and physical measurements. These measurements indicate the general condition of the plant and suggest the changes in management that are necessary to produce maximum yield".

Crop logging is used for sugarcane first time in Hawai. Crop logging is a record of composition of crop,with respect to N.P.K., moisture, sugar and weight of young sheath tissue, and other management at regular intervals during crop growth in the field. From a knowledge of nutrient status, additional nutrient requirements of the sugarcane crop are assessed and necessary application of fertilizer is made. So is the case with moisture status and irrigation .

Soil Fertility and Productivity :

Soil fertility is the inherent capacity of the soil to supply nutrients to plants in adequate amounts and in suitable proportions. Soil productivity is the capacity of the soil to produce crops with specific systems of management and is expressed in terms of yields. All productive soils are fertile, but all fertile soils need not be productive due to some problems like waterlogging, alkali, saline, adverse climate etc. According to modern usage, soil fertility is the capacity of soil to produce crops of economic value and to maintain health of the soil without deterioration.

Soil Fertility	Soil Productivity
1. It is considered as an index of available nutrient to plants.	1. It is a broader term used to indicate crop yields .
2. One of the factors for crop production; the others are water supply etc.	2. It is the interaction of all the factors.
3. Can be analysed in lab.	3. Can be assessed in the field under particular climate con ditions.
4. It is the potential status of the soil to produce crops.	4. Resultant of various factors influencing soil management.
5. Depends upon physical chemical and biological factors of soil.	5. Depends upon location, ferti lity, physical conditions etc.
6. Soil fertility is the function of avai lable nutrients of soil Soil fertility = f (Nutrient status of Soil)	6. Soil productivily is the function of soil fertility, management and climate Soil Productivity = f (Soil fertility + Management + Climate)
7. It is an inherent property of soil	7. It is not an inherent property of soil.
8. The fertility of a certain soil is same in all the climates	8. Soil productivity differs according to the variations in climate and location

9. All fertile soils are not productive	9. All productive soils are cer tainly fertile

- De. Datta and Gomez (1975) observed: The N-efficiency was highest in the first rice crop and declined progressively with succeessive rice crops.

  K- efficiency however, increased sharply over the years, especially during the dry season.
- FAO recommended the following practices for adoption in India :

  (a) In irrigated rice - wheat system,N should be applied to both the crops, 'P' only to wheat and K and Zn to rice.

  (b) In rice-rice greengram/ soyabeen system; N to both the rice crops; 'P' to dry season rice and ; K, S & Zn to the second crop.

  (c) In Maize + Pulse intercropping system. N to maize, P to both crops and K, S & Zn to maize.
- The overall foodgrain production to fertilizer applied ratio is about 10 :1.
- The Nitrate form of N is preferable to saline soils because NH^+_4 will be lost as volatilization.
- Ammonical fertilizers should be applied in the reduced zone and Nitrate fertilizers in the Oxidised zone.
- In crops where vegetative part is the economic yield e.g. sugarcane, beet, potato, tobacco, N- supply must be stopped much before maturity otherwise the quality and maturity may be affected.
- In seed crops, last application of N during seed development phase improves the quality and germinability of seeds. The application of last doze of N should be foliar.
- In determinate grain crops like rice, wheat, maize; N-supply at the begining of the reproductive phase (e.g. panicle initiation) as the last doze increases the number and weight of grains.
- In indeterminate plants e.g. rape, sesame, cotton, N-application at the flowering stage and another at the late flowering stage increases the yield & quality.

- If there is luxuriant consumption of N, a growth modifier say cycocel may be applied to enhance reproductive growth and yield.
- Plants require a greater amount of 'P' at their early stages (for root growth) but plant absorbs K upto harvesting and K- fertilizer becomes available slowly.
- Rabbing means heat treatment applied to soil by burning refuse placed over it.
- Teast : Soil containing high proportion of Mo.
- Mar : Raw humus, a type of forest humus layer of unincorporated organic material.
- Marling is the application of clay to sandy soils.

❑❑❑

22 Mineral Nutrition

The term mineral nutrient is generally used to refer to an inorganic ion obtained from the soil and required for plant growth. The process of absorption, translocation and assimilation of nutrients by the plants is known as mineral nutrition. The elements C,H, and O are not minerals. The rest of the elements are absorbed from the soil and these are called mineral elements since they are derived from minerals. These mineral elements are mainly absorbed in ionic form and to some extent in non-ionic form.

Criteria of Essentiality :

Plant body contains about 30 elements and in some cases as many as 60 elements. Amon and Stout (1939) proposed criteria of essentiality which was refined by **Arnon (1954).**

(a) The deficiency of the element makes it impossible for the plants to complete the vegetative or reproductive stages of its life cycle.

(b) The deficiency is specific to element in question and as such can be prevented or corrected only by supplying that particular nutrient element to the plant.

(c) The elements must have a direct influence on the plant and must be directly involved in nutrition and metabolism of the plant.

Recent investigations show that point (b) can not be accepted absolutely as Mo may be substituted by vanadium (V); Similarly Cl by Br; K by Rb (Rubidium); Ca by strontium (Sr).

Na is known to increase the yield of several crops like beets, turnip, carrot etc. and hence it is found to be essential for the farmers. At Arnon's time **Only 16** elements were considered as **essential.**

Nicholas (1961) from Long Ashton Institute (USA) proposed the term 'Fuctional Nutrient' for any mineral element that functions in plant metabolism whether or not its action is specific . With this criteria, Na, Co, V and Si are also included in functional nutrients in addition to 16 essential elements. The 16 essential elements are :

C	H	O	P	K	i	N	S	Ca	Fe	Mighty good
1	2	3	4	5		6	7	8	9	(Mg)
C	H	O	P	K		N	S	Ca	Fe	(10)
Many			Cu		rB		Zones		Monthly .	Closed
(Mn)			Cu		B		(Zn)		Mo	Cl
11			12		13		14		15	16

Out of these 16 essential elements C,H,O,N,P,K,Ca,Mg,S are called Macro- elements or Macro- nutrients and the remaining elements (or nutrients) are called Micro- nutrients.

N, P,K macronutrients are called Primary nutrients and Ca, Mg, S are called Secondary nutrients.

Functional Nutrients = 16 essential elements + Co, Na,V, Si

Thus Functional nutrients are 20 in number for the plants. Presently Ni (Nickel) is considerd as essential hence total essential nutrients are 17

Table : Forms of elements absorbed by plants

Element	Ionic Form	Non- ionic form	Source
1. C	CO^{2-}_3, HCO^-_3	CO_2 mostly	Air (mostly), water
2. H	Molecular	form	Air & water
		H_2O	
3. O	Molecular	form	Air & water
4. N	**NO^-_3** (mostly), NH^+_4	$CO(NH_2)_2$ amide	Parts of N from air but mostly from soil
5. P	**$H_2PO^-_4$,** HPO^{2-}_4	Nucleic acid **Phytin**	Soil.

6. K	K^+	-	
7. Ca	Ca^{2+}		
8. Mg	Mg_2^{2+}		Soil.
9. S	$\mathbf{SO_4^{2-}}$	SO_2 from air	Soil and air.
10. Fe	$\mathbf{Fe^{3+}}$, Fe^{2+} (Ferric)	$FeSO_4$ with EDTA	Soil
11. Mn	Mn^{2+}	$MnSO_4$ with EDTA	Soil
12. Zn	Zn^{2+}	$ZnSO_4$ with EDTA	↓
13. Cu	Cu^{2+}	$CuSO_4$ with EDTA	
14. B	$\mathbf{B_4O_7^{2-}, H_2BO_3}$ $\mathbf{HBO_3^{2-}, BO_3^{3-}}$		
15. Mo	$\mathbf{MoO_4^{2-}}$ (Molybdate ions) HMO_4^-		
16. Cl	Cl^-		
17. Ni	Ni^{2+}		
18. Na	Na^+		
19. Co	Co^{++}		
20. Si	$Si(OH)_4^-$ mono silic acid		

- EDTA means Ethylene Diamine Tetracetic Acid.

Table : Essential elements and their discoverers

S.No.	Elements	Scientist who discovered	Year
1.	O	De Saussure	1804
2.	N	Ruther Ford	1872
3.	Fe	Gris	1844
4.	Ca	Solm Harstmar	1856
5.	C	Sachs	1882
6.	H	Sachs	1882
7.	K	Schimper	1890
8.	Mn	G. Bertrand	1897

9.	P	Posternack	1903
10.	Mg	Willstatter	1906
11.	B	Agulhon	1910
12.	S	Peterson	1911
13.	Cu	Brenchley	1914
14.	Zn	Maze	1915
15.	Mo	Arnon and Stout	1939
16.	Cl	T.C.Broyer	1954

Table : Elements and its major role

Elements	Major role
C,H,O	Provide Basic Structure
C,H,O,N,P,S	Tissue Building elements
K,Mg,Ca,Cl	Electrolytic Balance
Ca,Mg,P	Skeletal elements
H,O	Energy exchange elements
N,P,S	Energy Storage, Transfer & Bonding
C,N,S,P	Plastic and storage (of energy)elements
Fe,Co, Mn, Cu, Zn	Oxidation Reduction (Redox) regulators.
Cu,Fe,Mn,S	Catalytic elements

Mobility of Nutrients :

Mobility in Soil :

(a) Mobile nutrients : Such elements are highly soluble and are not adsorbed on clay complex e.g.
NO^{-3}, SO_4^{2-}, BO_3^{3-}, Mn^{2+}, Cl^-

(b) Less Mobile: Such elements are also soluble but adsorbed on clay complex and thus their mobility is reduced e.g. NH_4^+, K^+, Ca^{2+}, Mg^{2+}, Cu^{2+}

(c) Immobile : highly reactive and get fixed in the soil e.g. $H_2PO_4^-$, $HPO^{2-}{}_4$,Zn^{2+}

Mobility in Plants :

(a) Highly Mobile : N,P & K

(b) Moderately mobile : Zn

(c) Less mobile : S,Fe, Cu, Mn, Mo &Cl

(d) Immobile : Ca & B.

Classification of essential Nutrients

Basis : Quantity of nutrients present in plants.

1. **Basic Nutrients :** C, H, O constitute about 96% of the total dry matter of plants. Among them C & O constitute 45% each.

2. **Macro/Major nutrients :** Such elements are required in larger amount i.e. more than 1 ppm. NPK are Primary / Major nutrients. Ca. Mg.S are Secondary nutrients because they are indirectly applied to the soils when NPK fertilizers which contain these nutrients are used.

3. **Micronutrients :** required in small quantities (0.01-1000 ppm) and generally less than 1 ppm (Parts per million) Micronutrients are also called **trace elements, Oligoelements,** or **Spurne elements.** These are very efficient, and minute quantities produce optimum effects. On the other hand even a slight deficiency or excess is harmful to the plants.

   Micronutrients are all the essential nutrients except NPK, Ca Mg S.e.g.Mn,Cu, Mo, B, Cl, Co, Fe and Zn. Fe and Zn behave as macronutrients but for historical reasons, included in micronutrents.

4. Ultra micronutrients : Concentration required 1ppb(Parts per billion). Nicholas (1963) put forward this concept giving the example of Mo and Co but it is faulty classification and is not acceptable today because the quantity of nutrients absorbed depends upon the types of plant.

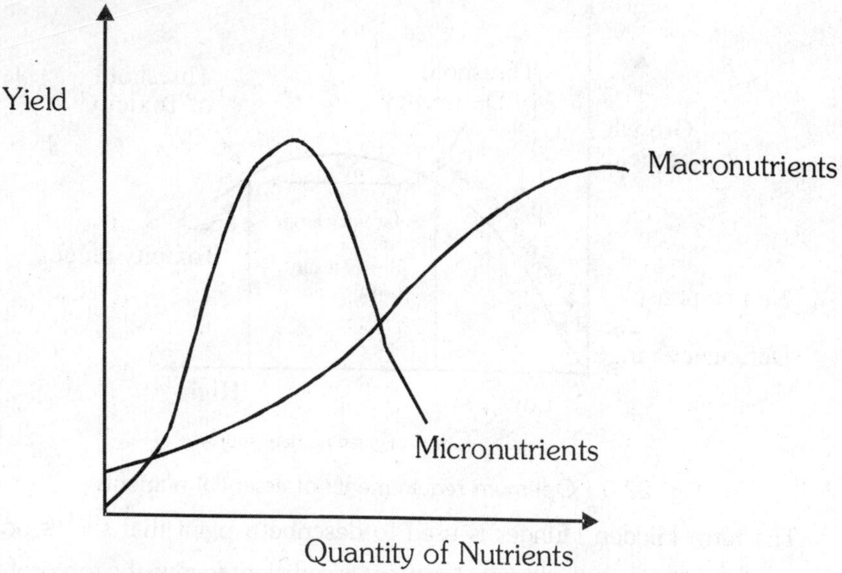

Fig. 22.1 : Yield curves for macro and micronutrients

- **Beneficial nutrients** or Potential micronutrients : Such elements at very low concentrations and often under specific conditions have been shown to stimulate the growth of certain plants or to have Sure rate other beneficial effects e.g. Selenium (Se), Al, Rb, Sr, Ni, Cromium (Cr) and Arsenic (As).

Hidden Hunger :

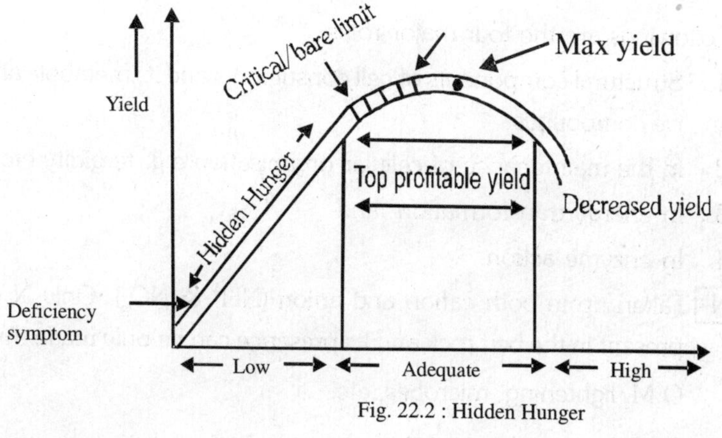

Fig. 22.2 : Hidden Hunger

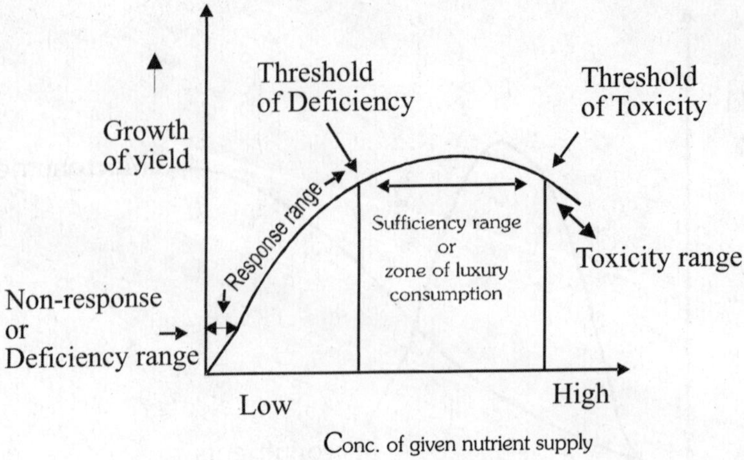

Fig. 22.3 : Optimum requirements of essential nutrients

The term Hidden Hunger is used to describe a plant that shows no obvious symptoms yet the nutrient content is sufficient to give the top profitable yield. Fertilization with the sure rate rather than bare economic optimum for on average yield helps to detain the top profitable yield. Sure rate is a rate a little above critical limit to be sure that there is no reduction in yield.

The Hidden Hunger is an expression in soil science to designate a condition in plants brought about through the lack of certain essential elements in the soil and is not easily detected by outward appearance.

Role of Essential Plant Nutrients :

Followings are the four major roles-

1. Structural components of cell constituents and its metabolically active compounds.
2. In the maintenance of cellular organisation e.g. tergidity etc.
3. In energy transformation ;and
4. In enzyme action.

N - Taken up in both cation and anion (NH_4^+ & NO_3^-). Only N is not present in the bed rock and its presence can be only made through O.M. lightening, microbes, etc.

$$NO_3^- \xrightarrow[\text{reductase}]{\text{Nitrate}} NO_2 \xrightarrow[\text{reductase}]{\text{Nitrate}} HNO_2 \longrightarrow \text{Hydroxyl amine} \rightarrow NH_3$$

NH_3 is first used in plants in the synthesis of amino acids. N is the essential constituent of protein, nucleic acid, nucleotide, amino acid, chlorophyll, phospholipids, alkaloids,enzymes, hormones, vitamins etc. Imparts dark green colour to plants,improves quality and succulence of leafy vegetables and fodder crops. Deficiency of N provokes the synthesis of Anthocynin which gives different types of colouration. (Compost is mainly used for horticultural and ornamental flowers because it is economical and helps in their physiology like proper colour, size etc).The content of N in healthy plants ranges between 1 - 5% upon the species or variety.

P Available form of P in acid soil: $H_2PO_4^-$; In less alkali or neutral : HPO_4^{2-}; and when alkalinity is more i.e. high pH : PO_4^{3-}.

p^H-scale :

1	acid	7	alkali	14

$$H_2PO_4^- \rightleftharpoons HPO_4^{2-} \rightleftharpoons PO_4^{3-}$$

In high acidic condition, Fe and Al antagonises the uptake of P and in alkaline condition, Ca antagonises the uptake of P. Its content in healthy plants ranges between 0.1 - 0.4%

Due to deficiency of single element phosphorus, plants can not complete their life cycle hence 'p' is called key to life. Nitrogen governs the above earth growth whereas 'P' govems the root growth i.e.below earth growth. constituent of nucleic acid, phytin, phospholipids, ATP; and essential of reproductive functions; stimulates early root growth; counteracts the excess N; enhances the activity of Rhizobia and root nodules; increases the ratio of grain/bhusa. It gives rapid and vigorous start to plant and strengthens straw and decreases lodging tendency. Excess of P may cause in some cases the deficiency of trace elements particularly Fe and Zn.

K Unlike the other elements, K does not enter into the composition of any of the important plant constituents.It is present in a plant in a state of solution in cell sap. Imparts disease resistance;produces strong stiff straw especially in paddy and wheat; regulates water balance, better utilisation of available water;essential in the formation and transfer of starches and sugars thus required in large quantity for potato, sweet potato, turnip, banana, suran, counteracts injurious effects of excess N in plants; improves the quality of tobacco leaf, quality of fibre, size and keeping quality of fruits. In citrus fruits, however an excess of K has a bad effect on quality. Its content in healthy plants ranges between 0.1 - 0.5%

K regulates osmo- regulation, and stomatal movement. It acts as a traffic policeman, root booster, stalk strengthener, food former, sugar and starch transporter, protein builder, breathing regulator, water stretcher and as a disease retarder but it is not effective without its co- nutrients such as N & P.

Ca Constituent of middle lamella of cell-wall (calcium pectate) which gives turgidity of cell,highly required in Telophase for cell plate formation. Its content in healthy plants ranges between 0.2 - 0.1%. Structural component, neutralises organic acids, essential in activated the growing point especially root tips; Its hydration effect is opposit to that of K which promotes hydration whereas Ca depresses it.

Mg Consitituent of chlorophyll,chromosomes, polyribosomes;carrier of P in plants particularly in connection with the formation of seeds of high oil content; promotes formation of oils & fats, translocation of starches, catalytic action. Its content varies from 0.1 - 0.4% in healthy plants.

S Constituent of amino acids like Methionine,cystein, cystine and vitamins (like glutothion, biotine, Thiamine) lipoic acid and Acetyl Co-A. 'S'is associated with aromatic compounds and creates a type of fragrance, aroma and smell. Animal protein is rich in 'S' than plant protien; helps in chlorophyll formation, involved in forming and stabilising the tertiary structure of enzymes and other proteins;promotes nodule formation. Its content varies from 0.1 - 0.4% in healthy plants.

B Essential for translocation of sugar; involved in the reproduction of plants and germination of pollen grains; primary role concerned with Ca- metabolism; keeps Ca in soluble form within the cell and acts as a regulator of K/Ca ratio; constituent of cell membrame and essential for cell division, primarily needed to maintain the apical growing point; concerned with buffer action ;helps in active salt absorption, hormone movements, N and carbohydrate metabolism. Its content is 10 - 200 mg B /kg dry matter.

Fe absorbed mainly in Fe^{3+} (soil availability) but physiologically active in Fe^{2+} (root sites).Fe^{3+} (ferric) is reduced in the plant into Fe^{2+}(ferous) in presence of Mn. Component of Ferredoxin (Fe & S- protein). cytochrome, Flavoproteins, enzymes like catalase, peroxidase, cytochrome oxidase, acts as catalyst in Nitrate reductase, activator in the synthesis of chlorophyll; acts as oxygen carrier. Its content is 100 - 500 mg /kg dry matter.

Mn Associated with Fe- metabolism, role in respiration and structure of chloroplast, protects the structure of chloroplast. role in the production of Super oxide dismutase (SOD). SOD protects the chloroplast membrance from free radicles of O_2 which has deleterious effect on membrane; essential for hill reaction; A good Mn-supply helps in counteracting the bad effects of poor aeration. Its content is 25 - 500 mg /kg.

Zn Helps in production of SOD; constituent of carbonic anhydrase and alcoholic anhydrase; RNA polymerase enzymes; important in the synthesis of IAA; essential for water uptake (At high P^H, Zn^+ is converted into negatively charged zincate complex whose availability is reduced in alkaline soil. Calcium Zincate is formed at high P^H and thus fixed). Its content is 20 - 150 mg /kg. of dry matter.

Cu Its presence in soil is less than 1ppm; more found in bed rock in the form of Copper Iron pyrite ($CuFeS_2$); takes part in electron transport i.e.plastocynin, peroxidase and certain oxidases, involved in production of SOD. Location of SOD production is stroma and protects from disintegration of thalacoides: involed in production of flowers . Cu is more required in the flowering and development of grain. Its content is 5 - 20 mg /kg of dry matter.

Mo Two components of Nitrogenase (1) Fe & (2) Fe +Mo. When 1000 atoms of Fe is present then 4 atoms of Mo is required. In bacteria, Mo can be replaced by vanadium, obvious role in N fixation. Its content is 0.2 - 2 mg /kg of dry matter.

Cl Plants like tobacco, tomato, potato, sugerbeet ,wheat, peas etc. generally require about 1 kg Cl^- for each 4000kg of drymatter (weight). In tobacco, it increases the water content of tissues and affects the carbohydrate metabolism. According to D.I. Arnon, Cl^- is required for primary photosynthetic reactions Isolated chloroplast when is given Cl^-, hill reaction is better. Due to Cl^-, tobacco leaf becomes thick, and brittle and don't burn uniformly whereas proper 'K' results shining, lustre and proper burning. Cl^- interferes in starch formation hence is avoided in rice. Its content is 100 - 500 mg c /kg dry matter.

Ni Absorbed by plants in Ni^{2+} plants contain 1.0 - 10 mg Ni/kg ; essential for hydrogenase, methyl reductase and urease activities that regulates N - metabolism. It is needed for grain filling & seed vitality.

Co Structural component of vit B_{12} which in turn essential to the formation of leghaemoglobin needed for N_2- fixation; essential for growth and metabolism of micro organisms; required by Rhizobia, activator of certain enzymes like orginase, Lecithinase etc.

Na Essential for sugarbeets and such crops; influences water relations in sugarbeet and increases their resistance to drought. In higher plants, Na is essential and to some extent it can supplement for K.

V In symbiotic N_2^- fixation ; reported in BGA e.g. nostoc, anabaena etc. required in 0.05 to 0.1ppm and 0.1-1ppm.

Si Essential to certain diatoms. Silicious cellwall (made of Si) protects food material which is oils in diatoms; gives resistance to rice and perhaps other crops like cucumber,barley etc. Si is reported to increase the top length, no. of stems, fresh and dry weight of rice plant grown in nutrient cultures.

selenium Can be replaced with 'S' in the production of Methionine, cystein and cystine because it is analogous to 'S' (i.e.same function) but origin different. (Homologous means same origin).

Al Activator of glandular system but more amount has toxic effect.

Deficiency Symptoms :

The deficiency symptoms can be distinguished on the basis of the followings--

(a) Region of occurrence.

(b) Presence or absence of dead spots and

(c) Chlorosis of entire leaf or interveinal chlorosis.

Deficiency Symptoms on old Leaves :

(a) Without deadspots :

N Uniform chlorosis of leaves including veins.The leaves become stiff and erect especially in cereals.Cereals show characteristic 'V' shaped yellowing at the tip of lower leaves.

P Small,erect, usually dark green leaves with a greenish red,greenish brown or purplish tinge.The rearside develops bronzy appearance ripening of crop.

Mg Yellowing (chlorosis) in between the veins and veins remain green.Leaf is not erect,necrosis in extreme case only in the margins; upward curling of leaves along the leaf margin,most prevalent towards the end of growing crop season. Cereals fairly resistant but oat is more susceptible.

(b) With dead spots :

K Yellowing starts from tips or margins of leaves extending to the centre of leaf base;Necrosis very soon,sharp difference between green and yellow and necrotic portions;curling of leaves towards undersurface,**Tip burn,margin scorching,** Brown spots near margin rolling of laminae;Barley most

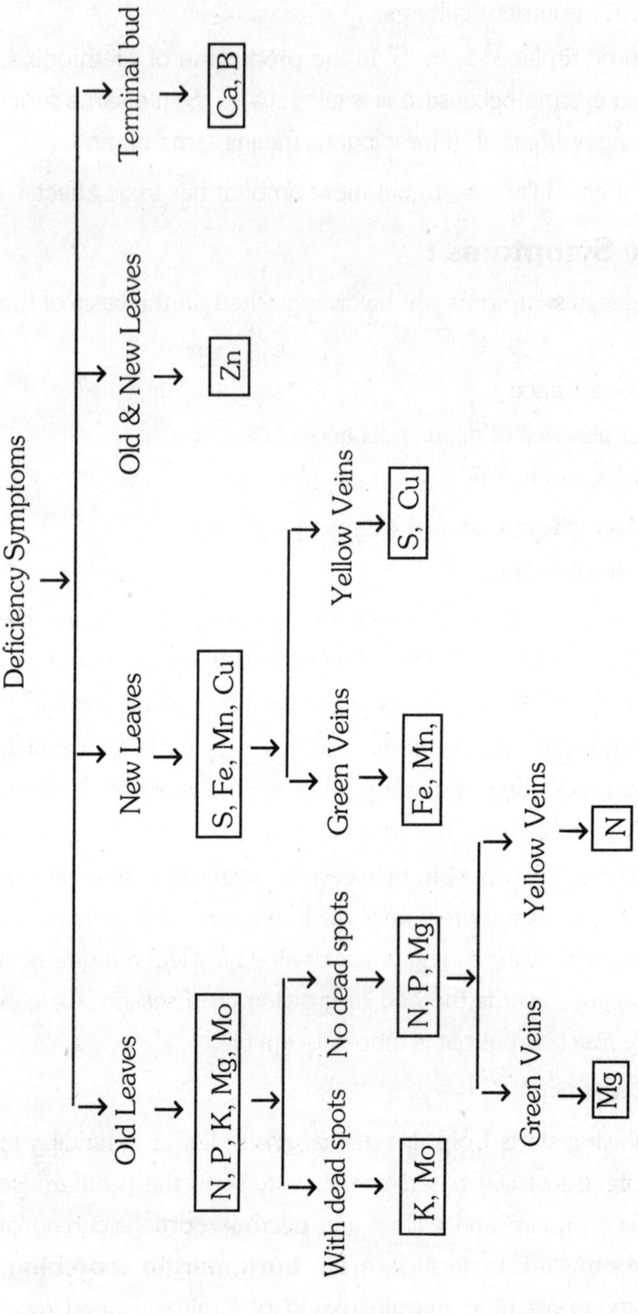
Deficiency Symptoms
Old Leaves
New Leaves
Old & New Leaves
Terminal bud
N, P, K, Mg, Mo
S, Fe, Mn, Cu
Zn
Ca, B
With dead spots
No dead spots
Green Veins
Yellow Veins
K, Mo
N, P, Mg
Fe, Mn,
S, Cu
Green Veins
Yellow Veins
Mg
N

susceptible, Potato shows abnormal dark green colour of foliage followed by browning,cotton bolls are either small, knotty or poorly opened.

Mo Translucent spots of irregular shape in between the veins of leaves;these spots are light green, yellow or brown;chlorotic motling between veins of middle leaves.The affected spots are impregnated with resinous gum which exudes from rear side of the leaf from the reddish brown spots;Brassica sp. most susceptible, **yellow spot of citrus,whiptail of cauliflower** (rat diseases).

Symptoms on new Leaves :

(a) Veins remaining green : Interveinal chlorosis:

Fe The principal veins remain conspicuously green and other portions of the leaf turn yellow tending towards whiteness; chlorosis starts from base of leaf.

Mn The principal veins as well as the smaller veins are green . Interveinal portion is yellowish not tending towards whiteness; **Complete Interveinal chlorosis;** chequered appearance to leaf; oat and Soybean most affected; The minimum level in healthy oat at flowering is 14 ppm; leaf top remains green and base dies; **Grey speck of oat; Pahala blight of sugarcane.**

(b) Veins not remaining green :

S Leaves small and veins are paler than interveinal portion, No dead spots; plants not loose the lower leaves as in N-deficiency. Vegetable leaves develop yellowish green colour and become thick and firm.

Cu Leaf yellowish tending towards whiteness; In extreme deficiency chlorosis of veins occurs and leaf loses lustre. Chlorosis spreads alongside the margins towards the base of leaves; development of several auxillary buds when shoot apex dies; wheat very susceptible; multiple bud formationin leaf axils, **rosetting and excess gumming.**

Symptoms on Terminal buds :

Ca bud leaf becomes chlorotic white with the base remaining green. About one-third chlorotic portion of the tip hooks downward and becomes brittle, root systems poorly developed or lack fibre.

B Yellowish or chlorosis which starts from base to tip . Tip becomes very much elongated into a whip like structure and becomes brownish or blackish brown : leaves become thick and margin upwardly rolled; **Hollow stem of cauliflower, Browning of cauliflower curd, Top sickness of tobacco and Hard fruit of citrus.**

Deficiency symptoms on both old and new leaves :

Zn Lamina becomes chlorotic and veins remain green; dead spots over leaf including veins, tips and margins; Structural aberrances in root tips; dwarfing of vegetative growth: New leaves emerge white in colour **(White bud of maize); Khaira disease of rice;** In cereals deficiency appears in 2-4 leaves from top during vegetative stage. Plants bushy due to reduced internodal elongation **(malformation).**

Toxicity symptoms

N Lodging and abortion of flowers; susceptible to pests and dieases.

P Deficiency of Fe and Zn.

Fe Tiny brown spots on lower leaves of rice starting from tips.

Mn stunted plant and tillering is often limited in lowland rice.

B Chlorosis of tips of older leaves especially along the margins.

❑❑❑

23 Saline, Alkali and Acid Soils

Saline and Alkali soils :

When the Chloride (Cl^-), Sulphate (SO_4^{2-}), Carbonate (CO_3^{2-}) and Bicarbonate (HCO_3^-) salts of Sodium (Na^+), Calcium (Ca^{2+}) and Magnesium (Mg^{2+}) are increased in Soil; the soil becomes saline and alkaline. On the basis of amount of soluble salts, average quantity of exchangeable sodium and P^H; such soils are classified as saline soil, alkali soil and saline-alkali soil.

$$\textbf{ESP} = \frac{\text{Exchangeable Sodium (in milli equivalent per 100g. of soil)}}{\text{Total Cation Exchange Capacity (in m.eq/100g soil)}} \times 100$$

ESP means Exchangeable Sodium Percentage and is also called Soluble Sodium Percentage i.e. SSP.

$$\text{SAR (Sodium Adsorption Ratio)} = \frac{[Na^+]}{\sqrt{\dfrac{[Ca^{2+}] + [Mg^{2+}]}{2}}}$$

Saline soils/ White Alkali :

Characteristics :

Toxic concentration of soluble salts in the root zone. Soluble salts are Cl^- and SO_4^{2-} of Na, Ca and Mg. Electrical Conductivity (EC) of soil solution saturated extract is more than 4.0 milli mhos/cm. [or decisiemens/ metre] at 25^{0}C. The Exchangeable sodium percentage (ESP)is less than 15% and p^H is less than 8.5.

- Two types of saline soils -
  (a) Having substantial quantity of Ca and Mg.
  (b) Having chiefly Na and damages the colloidal complex.

Formation of saline soils : Common in arid and semi-arid regions having annual rainfall less than 55cm. Lots of salts are deposited on soil surface in saline soils and saline soil layer shines white in dry-season.

Favourable Conditions :

(a) High water table with a fairly high salt concentration.

(b) High temperature.

(c) Low rainfall.

The sources of salts are soil itself, ground water , irrigation water, canal and sea.

Reclamation :

Principle : Removal of excess salts to a desired level in the root zone.

Leaching with water of good quality an adequate drainage are two essential component of any permanent solution of the salinity problem.

Ponding of water is the most commonly used for leaching. Intermittent ponding is more efficient than continuous ponding.

Management practices : are water management and cropping system. Rice is grown during initial years of reclamation.

Sodic soils /Black alkali/ Non- saline alkali soils :

Characteristics :

High exchangeable sodium percentage interfere with plant growth.

(a) EC< 4.0m.mhos/cm at 25^0C (less than)

(b) ESP> 15% (more than).

(c) $P^H > 8.5$ but generally 8.5-10

(d) Low infiltration rate.

Formation : Salt accumulation is the first stage in the sequence of processes and it is common to the family of salt affected soils; dominant in mean annual rainfall of 55-90 cm and relatively low lying areas with insufficient drainage.There are three distinct stages in the evolution of alkali soils viz.

(a) Salination

(b) Saline- alkaline soil

(c) Alkalization i.e desalination and intense alkali soil formation .

Occurrence : Indo- Gangetic plains of U.P. Punjab and Haryana.

Reclamation :

Basic Principle : To replace exchangeable Na^+ by Ca^{2+} and thus released Na^+ salt be leached out of root zone .The following amendments and practices are -

(i) Gypsum ($CaSO_4 . 2H_2O$)

(ii) Iron pyrite (FeS_2)

(iii) Several other amendments like calcium salts, acids, acid forming materials. The choice of amendment depends upon the nature and quantity of CO_3^{2} Present in soils

(iv) Bulky Organic manures,green manures, crop residues and others produce weak organic acids . These are applied in conjugation with Gypsum.

(v) Leaching with water of good quality.

Management Practices :

(i) Cropping with green manure

Rice- Dhaincha in U.P.

Dhaincha -Rice -Berseem in Punjab.

(ii) Higher dose of N because of volatilization . Application of zink in initial years of reclamation.

(iii) Frequent irrigation with small quantities of water.

Alkali soils are further divided into three classes viz.

(a) Class I : Soils having alkali earth carbonate.

(b) Class II : $p^H > 7.5$ but practically free from alkali earth carbonate.

(c) Class III : (or Degraded alkali) : $p^H < 7.5$ and contain no alkali earth carbonate. To reclaim it , limestone ($CaCO_3$) is applied .

Saline - alkali Soils :

Characteristics :

(i) EC > 4 mmho/ cm at 25^0 C

(ii) ESP > 15%

(iii) $p^H < 8.5$

Favourable factors are :

(i) aridity

(2) poor external or internal drainage

(3) irrigation by saltish water

(4) Permanent water courses

(5) rise in water table by excess irrigation

(6) erratic use of irrigation water i.e.flooding followed by intense drought.

Saline- alkali soil is difficult to manage since its physical condition is very bad.

Detrimental effect of soil salinity and alkalinity :

1. Soil structure : in alkali condition, soil is dispersed and becomes compact.
2. Less water permeability due to compactness.
3. Low aeration
4. Low microbial activities.
5. Unavailability of nutrients like P,Ca,N.
6. Nutritional disturbances at p^H 8.5 and more .
7. Hinderance in water absorption.
8. Effects of osmotic pressure : Increase in osmotic pressure of soil solution badly affects the plant's growth.
9. Salt toxicity : In alkali soil, Na_2CO_3 is highly toxic .

Table : Difference between saline and alkaline Soils .

Saline	Alkaline
1. Mainly Cl^- and SO_4^{2-} of Na^+ but also Cl^-,SO_4^{2-} and HCO^- of Ca^{2+} and Mg^{2+} in small amount.	1. Mainly CO_3^{2-} of Na^+ but also CO_3^{2-} of K^+, $Ca^{2+}$and Mg^{2+} in small amount.
2. Soluble salt concentration is equal to or more then 0.1%.	2. Soluble salt conc.<0.1%

3. Exchangeable sodium % < 15%.	3. ESP > 15%
4. $p^H < 8.5$	4. $p^H > 8.5$
5. EC of saturated soluble extract at 25^0C is more than or equal to 4 milli mho per cm. i.e. EC ⩾ 4 mmhos/cm.	5. EC < 4 mmho /cm.
6. White/light grey colour hence called white alkali.	6. Black colour hence called black alkali.
7. Flocculated soils therefore soil, aeration and permeability is normal.	7. Dispersed & compact soil, aeration and permeability is low.
8. Easy to manage because physical condition of soil is good.	8. Such soils can be managed because physical condition is not so good.
9. O.M. or humus is always found in soil.	9. Very less amount of O.M. or humus or even absent.
10. Can be reclaimed by mechanical methods upto some extent.	10. Use of amendments is must.
11. in rainy season, some natural vegetation is grown.	11. No any natural vegetation except some grasses.

Reclamation of USAR soils :

Occurence : In U.P., 1.2 million ha usar soils and forestry as a measure is suggested. Other states are Punjab and Haryana.

(a) Mechanical Method :

Includes the following practices .

1. Scrapping of salts.
2. Flooding and leaching salts.
3. Leaching solouble salts.

Two points are necessary for leaching soluble salts-

(a) Water table at least 3 metre deep from surface.

(b) There should not be hard pan in soil. Leaching is possible where ground water table is very low. Generally after leaching, flushing/ washing is used to wash the remaining salts.

4. Ditching / Trenching : Ditch or trench is made and this ditch is refilled up with soils of next trench.

5. Drainage : 1-2 feet deep trenches are made around the field and water is applied to this trench. Then salts come to this trench by seepage and this water can be drained out.

(b) Chemical methods :

1. Application of Gypsum : 12-15 tonne/ha; this practice is common in India.

$$Na_2CO_3 + \mathbf{CaSO_4 + 2H_2O} \rightarrow CaCO_3 + Na_2SO_4 \downarrow + H_2O$$

Alkali Soil Gypsum

2. Application of Sulphur : 2.5–4.0 tonne 'S'/ha

$$2S + 3O_2 \rightarrow 2SO_3$$

$$+ [SO_3 + H_2O \rightarrow H_2SO_4] \times 2$$

$$\mathbf{2S + 3O_2 + 2H_2O} \xrightarrow[\text{Microbial}]{\text{Sulfonification}} \mathbf{2H_2SO_4}$$

$$H_2SO_4 + Na_2CO_3 \rightarrow Na_2SO_4 \downarrow + H_2O + CO_2$$

3. Use of Iron Pyrite (FeS_2) : recently in use.

$$\mathbf{2FeS_2 + 7O_2 + 2H_2O} \rightarrow 2\,FeSO_4 + 2H_2SO_4$$

$$CaCO_3 + H_2SO_4 \rightarrow CaSO_4 + H_2O + CO_2$$

4. Lime Sulphur

$$CaSO_4 + SO_2 + 4H_2O \rightarrow CaSO_4 + 4H_2SO_4$$

$$H_2SO_4 + CaCO_3 \rightarrow CaSO_4 \downarrow + CO_2 + H_2O$$

$$H_2SO_4 + Na_2CO_3 \rightarrow Na_2SO_4 \downarrow + CO_2 + H_2O$$

5. Use of sulphuric acid and Iron sulphate.

$$Na_2CO_3 + H_2SO_4 \rightarrow Na_2SO_4 + CO_2 + H_2O$$

$$FeSO_4 + H_2O \rightarrow H_2SO_4 + FeO.$$

Among amendments Gypsum, Sulphur and Iron Pyrites are generally used.

6. Use of molasses : Microbes get energy for decomposition from molasses and decomposition produces organic acid.

(C) Soil and crop management :

1. Proper drainage of water to remove the salts from the rootzone.
2. Use of proper irrigation -water.
3. Proper time of irrigation : Seedlings should not be irrigated with saline water. Salt tolerant capacity is more for matured tree.
4. Use of salt free irrigation -water.
5. Retardation of water evaporation from soil surface by vegetation.
6. Application of organic manures e.g. FYM, Compost; green manuring increses the permeability and aeration and thus improves physical condition of soil.
7. Selection of suitable crops and varieties.

- Salt tolerant :
- High salt tolerant : _ Rice, sugarcane, Oats.
  Medium salt tolerant : - Cotton , Sorghum, maize, Mustard, wheat.
  Low salt tolerant : - Pulses, sesame, radish, beans.
- **Alkali tolerant :** Paddy, cotton, mustard, barley, oat, wheat, beet, tomato, onion, potato, cauliflower, carrot, brinjal, dates ,ber,guava, grapes, orange and citrus; Berseem.
- **Crops and Varieties :**

  Paddy - Jaya, Padma, IR-8, Pusa 2-21

  Wheat-Sonalika, Kalyansona.

  Mustard- Kanpur tori.

  Barley - R.S-6, NP-13

  Pea - T-163.

  Jowar - CHS-1,

  Bajara - Syn-1, Syn-3.

In usar land, first of all, grow high tolerant crops. And in the first 3 years, only grow grass after excess flooding the field.

Table : Classification on the basis of Soil Salinity Tolerance

Saline Sensitive Crops	Less Salinity Tolerant	High salinity Tolerance
Electrical Conductivity (EC) < 4.0m.mho/cm	EC 4-10 m mho/cm	EC > 10 m mho/cm (more than)
Beans, Radish, Carrot Pear, Apple, Orange, Almond,Citrus	Wheat, Oat, rice, Maize, Linseed, sun-flower, Soybean, Tomato, Cabbage, Cauliflower, Potato Chilli, Onion, Cucum-ber, Bottle gourd, Grape, Mango, Anar, Sudangrass.	Barley, Sarson,Cotton, Beet, Palak, Date

Table : Classification on the basis of Soil Alkalinity tolerance

Sensitive Crops	Low tolerant	Tolerant crops
ESP < 15% (less than)	ESP : 15-40%	ESP > 40% (more than)
Bean, Maize, Carrot Orange, Peach	Carrot, Salad, Oat, Onion, Radish, Rai, Jowar, Palak, Wheat, Pea	Oat, Beet, Cotton, Rhodes grass

8. Other measures -

(i) Forestry

(ii) Soil should be removed after digging the small plots and refilled it with pond soil and dung.

(iii) Grow jungle of Acacia, which reclaims it in 15-20 years because tap root breaks the hardpan.

(iv) Flooding and drainage also improve.

(v) Only use acid forming fertilizer like $(NH_4)_2 SO_4$.

(vi) Use $ZnSO_4$ -@- 40-50 kg/ha because zinc- deficiency at high p^H.

(vii) Frequent light irrigation.

(viii) East- west bunds are made and crops are grown on the slope of north side. Due to high temp, soluble salts accumulate in south side and at the top of bund (through evaporation) and thus crops escape from salt injury.

(d) Biological method :

Grow green algae after flooding and apply sufficient ammount of FYM, which change the soil structure and aeration.

In all these cases, restricted drainage is usually the main problem.

Acid Soils :

From practical standpoint, soils having less than p^H 5.5 are called acid soils. There are two criteria used to distinguish acid from non- acid soils.

(1) The percent base saturation :

Percent Base Saturation (PBS)

$$= \frac{\text{Exchangeable bases (Cmol/kg soil)}}{\text{CEC (Centimol/kg Soil)}} \text{X100}$$

unit of PBS are Cmol/kg = m.eq./100g of soil

Humid region soils, dominated by silicate clays and humus, are acid if their % base saturation (PBS) is much below 80. A good cultivable soil has 60% bases occupied by calcium. It means Ca-dominant soil is fertile soil.

At a given percent base saturation, the p^H is normally highest in hydrous oxide clay minerals, intermediate in Kaolinitic and lowest in 2:1 type minerals.

The greatest buffering occurs at about 50% base saturation . The buffering is uniform over p^H range 4.5-6.5. It means same amount of lime is required to the soil p^H from 5.0 to 5.5 as from 5.5 to 6.0.

(2). p^H : The major source of H^+ in most very acid soils is adsorbed aluminium.

$$Al^{3+} + H_2O \rightarrow Al(OH)^{2+} + H^+$$

Adsorbed Hydrogen is the second but minor source of H^+

Acid soils have relatively high amount of exchangeable H^+ & Al^{3+}. Kaolinite and Illite clay minerals are dominant.

Table : Soil acidity and p^H

Soil acidity	**p^H**
Low acidity	6.1 - 6.5
Medium acidity	5.6 - 6.0
Strong acidity	5.1 - 5.5
Very strong acidity	4.5 - 5.0
Highly strong acidity	less than 4.5

Genesis :

1. Humid climate : Due to continous rains, the bases, obtained by ionic exchange and mineral decomposition , leached away and insoluble acid residues are remained in which Si, Al and Fe oxides are dominant and responsible for acidity.

2. Carbonic acid formation by nitrification and reduction.

3. Soils from acid rocks : Acid rocks are rich in Quartz and silica. Rocks like granite and Rhyolite produce silicic acids mainly . Orthosilicic and trisilicic acids. When these rocks lack bases, produce acidity in soil after decomposition by weathering.

4. Acid soils from non- acid minerals : Some parts of bases may be converted into CO_3^{2-} and HCO_3^- after decomposition by chemical process. These CO_3^{2-} and HCO_3^- are leached away due to soluble in nature and hydrated aluminium silicate causes acidity.

5. Long use of physiologically acid fertilizers e.g. $(NH_4)_2 SO_4$.

6. Farm practices that increase acidity : Leaching of salts in fallow land causes acidity . Vegetation reduces leaching.

7. Absorption of salts by crop like potato and beets : Potato and beets absorb salts from soil thus reduces bases in the soil.

The mean annual temperature, type of vegetation, hydrological conditions also govern the extent of acid soils and the degree of acidity.

Processes involved in acid soil formation :

1. Laterisation of varying degrees.
2. Podzolization with sub-temperate to temperate climate.
3. Intense leaching of light alluvial soils .
4. Marshy conditions with significant amount of partly decomposed O.M.

Occurrence : Acid soils occur in almost all the major groups except black soils (vertisols). Northern-east, eastern and peninsular regions, coastal plains (Coastal alluvium, peaty soils, marshy soils) and Himalyan regions. Large area of acid soils- Assam, Manipur and Tripura.

Adverse effect on plant growth :

(a) Direct influences :

(i) Toxic effects of H^+ ions on root tissues.

(ii) Influence of soil acidity on the permeability of the plant mem branes for cations.

(iii) Desturbance in the balance between basic and acidic constituents through the roots.

(iv) Direct harmful effect by affecting the enzymic changes.

(b) Indirect influences :

(i) Availability of various nutrients e.g. P,Cu, Zn.

(ii) High solubility and availability of elements like Al, Mn and Fe in toxic amounts due to high soil acidity.

(iii) Beneficial activities of soil micro-organisms are adversly affected e.g.Nitrification, Nitrogen fixation.

(iv) Prevalence of Plant diseases.

(v) Due to soil acidity, nutrients such as Ca and K may be deficient.

(vi) Disturbances in the physical condition of soil. Deflocculated conditions of soil is due to high acidity.

Characteristics of acid soils :

[A] Physical :

1. Light texture
2. High permeability
3. Poor water holding capacity.
4. Poor cation exchange capacity.
5. Poor O.M. content.
6. Mainly kaolinite and sometimes illite clay minerals .

[B] Chemical :

1. Base unsaturated soil, more anions than cations.
2. Active and potential soil acidity.
3. Availability of nutrients -

   at Low p^H : Al, Fe, Mn, Zn, Cu, Co (all trace elements except B & Mo).

   High p^H : Mo, N, K, Ca,Mg, S (all major nutrients except 'P')

   Neutral p^H : P and B. Boron available at p^H 5.0-7.0.

Table : Soil p^H and nutrients availability

pH	Available nutrients	Remark
Low pH	Al, Fe, Mn, Zn,Cu, Co	All trace elements except B and Mo.
High pH	Mo,N, K,Ca, Mg,S	All major nutrients except 'P'
Neutral pH	P,B (boron)	Boron is available at p^H 5.0 - 7.0.

4. Toxic effects on plants : Al^{3+} Conc. more.

[C] Biological :

1. Population of Fungi more than that of bacteria . There is no effect of p^H on the population of fungi.
2. Fungi cause diseases .
3. Rate of decomposition of biological materials and rate of mineralisation and nitrification etc. are reduced when acidity is increased.

Management of acid soils : In two ways - (i) Either by growing crops suitable for particular soil p^H or (ii) By ameliorating the soils through the application of amendments.The former is rather risky. Intensive and continuous cropping of arable crops in humid regions will aggravate soil acidity and use of acid producing fertilizers will further accentuate the process. Acid soils are made more suitable for agricultural use by liming which raises the soil p^H .

Liming is generally recommended for soils with p^H less than 5.5 .

Dose : Dose of liming material depends on soil p^H, soil texture, capacity and types of liming material.. Generally at lower p^H higher amount of lime and at higher pH lower amount of lime is required . For normal and 4.8p^H, 35 tonne/ha and for p^H 6-7, 4 tonne/ha lime is required . On an average hydrated lime $Ca(OH)_2$ @ 5 tonne/ha is applied in the field and is throughly mixed in soil. The gap between lime application and crop sowing is at least two to three months. Application of small quantity of lime every year gives good result, although lime can be applied once in two to three years for convenience.

Reactions in soil after application :

$$CaCO_3 + H_2O + CO_2 \rightarrow Ca(HCO_3)_2$$

$$Ca(HCO_3)_2 \rightarrow Ca^{2+} + 2HCO_3^-$$

$$H\text{-}Soil + Ca^{2+} \rightarrow Ca\text{-}soil + 2H^+$$

$$H^+ + HCO_3^- \rightarrow H_2CO_3 \Leftrightarrow H_2O + CO_2$$

Liming Materials :

1. Calcium limestone ($CaCO_3$) . more than 90% use in India .
2. Dolomite (rich in Mg)
3. Quick lime (CaO).
4. Slaked lime $Ca(OH)_2$.
5. Coral shell lime.
6. Chalk $CaCO_3$
7. Blast furnace slag $CaSiO_3$ and Ca_2SiO_4.
8. Miscellaneous sources like wood ashes etc.

Effect of lime in the soil : Relative lime requirement of selected crops.

High lime requirement : Barley, Beans, Cotton, Pea, Soyabeen, Sugarbeet and sunflower.

Medium lime requirement : Corn (maize). Sorghum, tobacco, wheat.

Low lime required : Oat, potato, Rice, Rye.

Very low lime required : Pine apple

- Optimum p^H range for tea and rice is 4 to 6.

❑❑❑

24 Extension Education

Introduction :

1. Extension = Ex + tensis (Latin word)

   ↓ out   ↓ Stretching

2. Education : Production of **desirable changes** in human behaviour through gaining desirable Knowledge, Attitude and Mental and Manual Skill,is called Education.

   Thus Ext. Education is the type of education which is stretched out the villages and field beyond the limits of schools and colleges to which the formal type of education is normally confined to it.

3. The word 'Extension' was used first time in USA.
4. Extension activity was started first time in USA.
5. First extension work was done in 1785 in Philadelphia state of USA when few farmers got together and organised to sell their produce..
6. In 1862, Morril Land Grant Act was passed in USA and signed by the President Lincon. Morril was an educationist. The objective of this act was "All the agricultural institutes would pass on their innovations to rural people"
7. In 1885, the concept of demonstration was given by Dr. Seeman A. Knapp. He generated a new variety and demonstrated on one million acre farm in Lausannia. Therefore he is known as father of method demonstration.
8. On the basis of American pattern i.e. Land grant college pattern, the first agricultural university at Pant Nagar was established in

India. But the study of extension education as a course study first time started at **Sabour** (Bhagalpur) in Bihar in 1956.

Agricultural Extension Education

Definition :

Extension Education is an **applied behavioural Science** to bring about **desirable change** usually through various strategies and programmes of change and by applying the latest scientific and technological innovations.

Features :

1. Participation is purely voluntary.
2. It is an informal education i.e.out of school and college.
3. Teacher or extension worker starts with practical first and there after theoretical aspects are to be touched i.e. First Practical then theory.
4. There is no fixed curriculum or course of study.
5. Farmers study the problems.
6. Authority rests with the farmers.
7. It is Bi-directional learning means teacher also learns from the farmers during the course of teaching to the farmers.
8. Teaching is also through local leaders or lay-leaders.
9. Teaching is mainly horizontal.
10. Large and heterogenous audience.
11. It is flexible.
12. Freedom to develop programmes locally.
13. It is based on the needs and expressed desires of the people.
14. It is more Practical.
15. It is continuous education i.e. throughout the life.

Objectives :

There are two dimentions of movement -

1. Fundamental objective : Development of the people.
2. In India there are three main objectives of agricultural extension-

(a) Dissemination of useful and practical information.

(b) Practical application of useful knowledge. The word '**Learn ing by Doing** ' was coined by John Dewey in 1966.

(c) To improve all aspects of the life of the **rural** people.

Principles :

1. Principle of Interest and Need : means it is based on the needs and interests of the people.
2. Principle of cultural difference : i.e.based on the cultural background of the people.
3. Principle of Participation.
4. Principle of Adaptability i.e.Extension programmes should be flexible.
5. Grassroot principle of Organisation : means programmes fit in with the local conditions so that more and more people would participate .
6. Leadership Principle : Programme is based on full utilisation of local leadership.
7. The whole- family principle : means programme is useful for all the members i.e. male, female and youth of the family.
8. Principle of co-operation : Rural people cooperate with their village, block and state officials to pursue a common cause.
9. Principle of Satisfaction : Satisfaction is the key to success in extension work . A satisfied customer is the best advertisement .
10. Principle of Evaluation : To know the merits and demerits of the programme, analytical study is necessary . Therefore it needs constant evaluation.

Basic Philosophy of Extension :

1. Self help.
2. People are the greatest resource.
3. Programme of change.
4. Cooperative effort.

5. Democratic nature means Development of people by the people and for the people.
6. Two channel of knowledge and experience :

   Problems : are brought to laboratory.

   Solutions : are taken to farmers/villagers.
7. Programme is based on creating interest by seeing and doing.
8. There is voluntarily cooperative participation.
9. Persuation and education of people.
10. Programme is based on attitude and values of the people with whom it works.
11. It is never ending process.

Learning & Teaching in Extension & Communication :

1. Extension is an educational process to bring about desirable changes
2. Essential role of an extension worker is to create effective learning situations.
3. Essential elements of effective learning are five in number.
   (a) Teaching materials and Plan.
   (b) Subject matter.
   (c) Physical facilities and environment.
   (d) Teacher or instructor .
   (e) Learner.

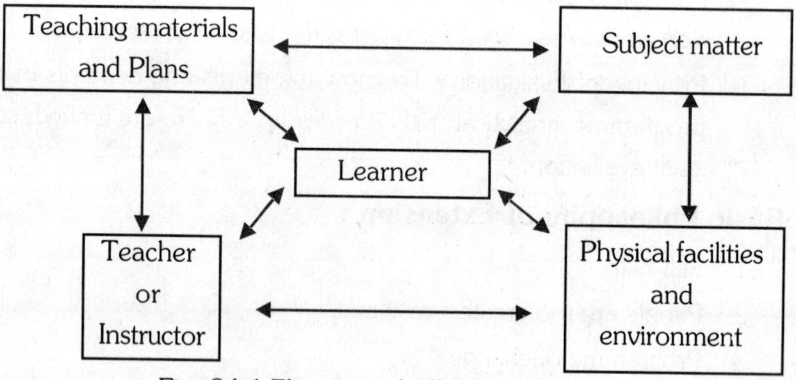

Fig. 24.1 Elements of Effective learning

Teaching Steps :

1. Attenton → 2. Interest → 3. Desire
↓
6. Satisfaction ← 5. Action ← 4. Conviction

Summary to remember : AIDCAS

Adoption Process :

1. Adoption is an Individual matter or phenomenon or behavioural socio-eonomical phenomenon or mental process.
2. Elements of Adoption process:
   (a) Innovation
   (b) Diffusion
   (c) Motivation
   (d) Adoption
3. Diffusion is the process between persons.
4. Stages of Adoption :
   (a) Awareness
   (b) Interest
   (c) Evaluation
   (d) Trial
   (e) Adoption
5. Categories of Adopters : There are five groups of adopters :
   (a) Innovators : Such people are called innovators who adopt im mediately after getting knowledge. In India Innovators constitutes only 2.5% of the total population.
   (b) Early adopters : Such people adopt through local leaders and constitute only 13.5% of the total population.
   (c) Early majority : Such people adopt just before adopting the common people and not through the local leaders and constitute 34% of the total population.
   (d) Late Majority : Such people adopt after seeing their relatives and neighboures and constitute 34% of the total population.

(e) Leggards : Such people adopt in the last and constitute only 16% of the total Indian population.

Communication

1. Teaching and learning is the process of communication.
2. Teacher acts as catalyst.
3. Communication means sharing of ideas and feelings in a mood of mutuality.
4. 'Communication'is derived from latin word 'communis' meaning 'common'. It means to establish commonness between sender and receiver of message is termed as communication.
5. It is a two way or Double way Process.
6. Communication involves :
   (i) Source and his message.
   (ii) Receiver and his response.
7. The chief emphasis in communication on Target Audience (not on material or medium).

8. Communication systems :

K → Knowledge, G → Generating,

S → System, D → Disseminating

C → Consuming.

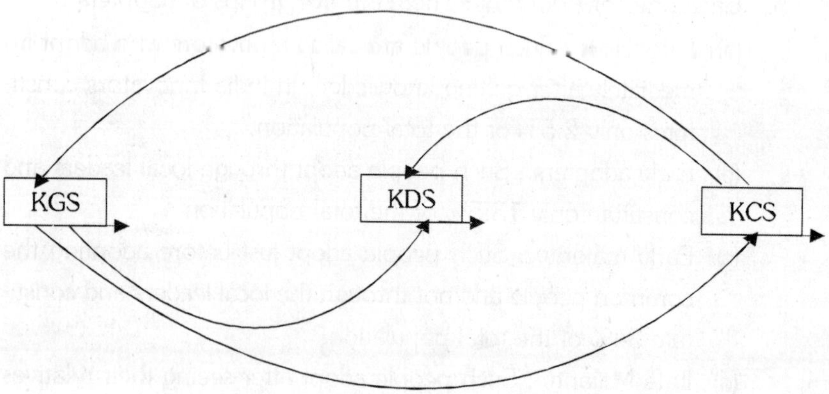

Fig. : 24.2

Agencies →

KGS	KDS	KCS
Universities	Mass media	Farmers
Research Institutes	Universities	(in Agriculture)
Private Organisation	Research Institutes	
Voluntary Organisation	Govt. Agencies	
	Voluntary Organisations	

9. Key elements of communication :
   (a) Sender
   (b) Message
   (c) Message Treatment i.e. encoding
   (d) Communication channel/Medium
   (e) Receiver : It is the most important element because every element is directed towards receiver.
   (f) Receiver's response.

10. Communication Process

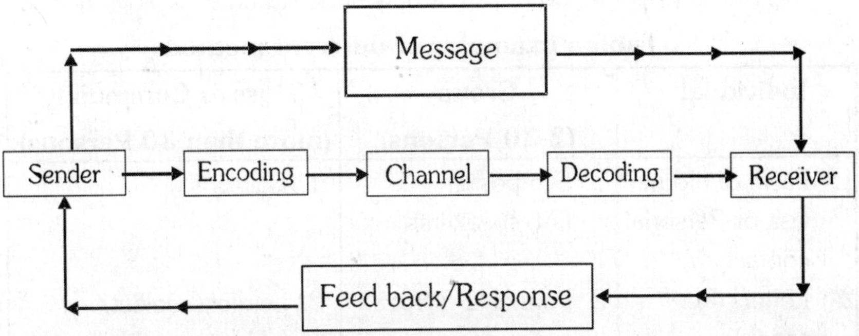

Fig. 24.3

11. Ingredients in communication :

Source ⟶ Massage ⟶ Channel ⟶ Receiver

KGS ↳ KDS ↲ KCS

12. According to Berlo (1960)

Source → Encoder → Message → Channel → Decoder
↓
Receiver.

Classification of Communication
(Ext. Teaching Methods)

A. According to use and nature of contact :

On the basis of use and nature of contact, the extension teaching methods are categorised under three groups-

(i) Individual contact method (or Interpersonal channel of communication) : Here contact is face to face or person to person. It is very effective in teaching ; quantum of feedback available is very high but slow communication.

(ii) Group contact method : Group means a group of 2-30 persons but usually consists of 20-25 persons. Group is usually formed around common interest. Here is face to face contact.

(iii) Mass or community contact method : It is more useful for making people aware of the new agricultural technology quickly.

Table : Examples of different methods

Individual	Group (2-30 Persons)	Mass or Community (more than 30 Persons).
1. Farm & Home visit or Personal contact	1. Symposium (2-5 speakers)	1. Bulletins
2. Letters (Personal)	2. Pannel (2-8 Speakers)	2. Leaflet/booklet/ folder/pamphlet
3. Telephone Call	3. Discussion	3. News paper
4. Office call	4. Lecture	4. Magazine
5. Result Demonstration	5. Tour and field days	5. Journal
	6. Method Demonstration (or Demonstration)	6. Exhibition

	7. Meeting	7. Television
	8. Role playing	8. Radio
	9. Puppet show	9. Circular letter
	10. Slide	10. Fairs
	11. PAE set (public Address equipment)	11. Posters
	12. School	12. Cinema (Film)
	13. Conference	13. Movie
	14. Seminar	14. Drama & songs
	15. Training camp	
	16. Flash cards	

B. According to Equipments i.e. Projected and Non-Projected :

1. **Projected Aids :** Here projection is controlled by motor and electricity e.g.
   - (i) Film projector
   - (ii) Slide projector : Slide is of 35 mm. and direct projection.
   - (iii) Overhead projector : It is indirect projection and transparencies are used.
   - (iv) Epidiascope/Opaque projector : There is no need of transparency.Epidiascope means both transparent and opaque materials can be seen (or focused) on the screen by magnifying The image of a picture is made on the screen directly and indirectly. One additional mirror is required in it .
2. **Non-Projected Aids :** Here no machinery is required e.g. Flash cards, Graphs, Charts, Leaflets, Folders , Blackboard.

C. According to Audio - visual aids : Examples are

(i) Audio : Radio, Taperecorder

(ii) Visual : Flash cards , overhead projector,slides etc.

(iii) Audio -visual :Puppet, Drama, Television, Cinema etc.

Demonstraion

Demonstration means showing by doing. The basic principle of Demonstration is learning by seeing and doing. There are three types of demonstration -

(a) Method Demonstration :

(i) It is short -type demonstration.

(ii) It's main purpose is to provide only skill.

(iii) It does not compare between the old and new technique or skill. It means comparative study can not be done.

(iv) It is the oldest form of teaching.

(v) The father of method demonstration was Dr. Seeman. A. Knapp.

(vi) It is a single practice demonstration i.e.how to operate tractor, how to apply fertilizer in the field. In this demonstration any process is shown and made clear to the people by doing in a sequence starting from the beginning to the end and zealous person is given opportunity for doing.

(b) Result demonstration :

(i) It is long type demonstration.

(ii) It shows the value or worth of the (new) practice.

(iii) Comparison between two practices i.e. old and new is always done.

(iv) Results of both practices are shown.

(v) It is very effective in adoption.

(vi) It is conducted by a farmer under direct supervision of an extension worker.

(vii) It is used to show the superiority of practices .

(c) National or Composite Demonstration :

(i) It is a composite of both method and result demonstration.

(ii) It is first line demonstration or Front line extension project.

(iii) It is conducted by the researchers on the farmer's field.

(iv) The main objective is to show how production can be increased per unit area and per unit time.

(v) Multiple cropping system are usually followed and High yielding varieties (HYV) are used in this demonstration.

(vi) National demonstration was first time started in 1965 by Ministry of Food and Agriculture.

(vii) This programme was given to ICAR in 1967.

(viii) Now it has become part of agricultural production programme.

Training and Visit System

1. T and V system is a new extension service approach. The concept of T and V system was evolved in 1973 by Israeli extension expert Daniel Benor who was consultant on extension programme in World Bank. This system was first tried in seyhan irrigation project in Turkey.
2. T and V system is also called Baster and Benor Scheme and firstly evolved in Turkey by inccidental learning.
3. T & V system has two phages/ stages-

   (i) Training and (ii) Visit.

   (i) Training : Training is imparted for transfer of technical knowhow from subject matter specialist (SMS) to extension worker.

   SMS $\xrightarrow{\text{Technical Knowhow}}$ Extension worker

   (ii) Visit : Extension worker has to visit to the farmers for transfer of knowhow obtained at the training.

   Ext. worker $\xrightarrow{\text{Technical Knowhow}}$ Farmers
4. It is Govt. sponsored rural development programme.
5. It is also called Reformed Extension or Intensive Agricultural Extension.
6. T and V system is introduced in India in 1974 in Rajasthan canal area and chambal command area in M.P. to improve the effectiveness of agricultural extension.

7. According to Benor et. al. (1984) T and V system is an effective management system of known extension principles.
8. Systematic programme of In-service training of village extension workers and planned schedule of visit to the farmer's field is the basic feature of T & V system.
9. In brief ; under T & V programme extension workers are imparted training of the latest agricultural technology by the agriculture experts and the trained worker after visiting the village diffused the obtained knowhow to the selected farmers and instigate them to adopt it . Besides, the problems of the farmers are brought to the agriculture expert during the next training where the solution is to be found out.

Lab to Land Programme

Lab to land programme was launched in India on the occasion of ICAR golden jubilee celebration year in July 1979 for only one year . But later on this programme was extended further year after year. The objective of LLP is the transfer of proved and viable low cost agricultural technologies to small and marginal farmers and landless laboures to improve their overall socio-economic condition.

The benefits of new technology or technical knowhow of agricultural production were being utilised by the big and rich farmers. Therefore it was felt that for holistic developmental approach of agricultural sector, the evolved technology in the laboratory (or at the agriculture farm)would be stretched out to the lands of small and marginal farmers . That's why lab. to land programme was launched. Under lab. to land or land to lab. programme, the problems related to the agriculture were brought to the laboratory where the particular solution was to be found out or searched out and later on were communicated to the farmer. But due to agro- climatic regional differences lab. to land programme has been changed into Farmer's First Programme where the agricultural problems are sorted out and are solved at the farmer's field itself.

Krishi Vigyan Kendra
(Farm Science Centre)

1. Education committee (1964-66) recommended to establish agricultural polytechnique.
2. Finally ICAR mooted the idea to establish Krishi Vigyan Kendra (KVK) and Mohan Singh Mehta Committee was appointed in 1974 by ICAR to formulate KVK scheme.
3. First KVK was started at Pondichery in 1974 under the adminstrative control of Tamilnadu Agricultural University, Coimbtore. With the aim of transfer of technical literacy to increase agricultural production. The teaching design is based on philosophy of teaching by doing and learning by doing.
4. The main objective of KVK is to impart training (or vocational training) at the kendra (centre) or out of the centre i.e. in the village to the people according to their needs about the Agriculture and allied subjects viz. Animal husbandary, Fisheries, Horticulture, Agriculture Engineering, Home Science etc.
5. In the initial five years financial assistance was given to KVK by the Indian Agriculture Research Institute (IARI) and for the next five years it was availed by IARI and State Govt. or State agriculture university in the 50 : 50 ratio.There was a plan to finance full assistance by the state govt. or state agriculture university after the lapse of initial ten years. At present ICAR provides 100% financial assistance.
6. There were 500 KVK in number by the year 2006 and the plan is to establish one KVK in every district of India.
7. The credit for the success of KVK goes to Dr. Chandrika Prasad.

Technology Transfer

The essential and major part of technology transfer is extension.It means technology transfer is broader in meaning than extension i.e.

Technology Transfer $\neq$ Agl. Extension or Extension.

Technology Transfer includes the additional functions of input supplies and agri-services.

Technology Transfer = Agl. Extn. + Input supplies and agri-services = Ext. + Supplies & Services.

Etawah Pilot Project

1. Under the leadership of Albert Mayer,the office (or centre) of Etawah Pilot Project was established by the U.P. Govt.in Oct.1948 at the place called Mahewa in the Etawah district.
2. Initially 64 villages were selected around Mahewa for the development.
3. It was also called 'Average District Plan' because the project was initiated in the normal environment.Everything like climate, soil, fertility etc. was normal.
4. 'Pilot' means doing a thing in a particular area.
5. This project is called pilot project because Community Development Programme has been initiated or launched in India due to success of Etawah Project.Therefore Etawah Pilot Project was the **forerunne**r of CDP.
6. Village Level workers (VLW) were appointed in this project who become very much helpful in the development works.Hence **Multipurpose Concept** of **VLW** is the outcome of Etawah pilot project.

Community Development Programme

1. CDP was a programme/project/method/ or a process means it may be called community development project or CD Method or CD process.
2. In the formulation of CDP .major role was played by Grow More Food Campaign.
3. Initially 15 pilot projects were launched in 1952 with financial assistance from Ford Foundation.

4. **On 2nd Oct,1952;** 55 Community Development Projects were started in different parts of the country with Indo-USA agreement for three years.
5. This project covered nearly 25,260 villages and a population of 6.4 millions.
6. At outset,each project was consisted of 300 villages.400-500 square miles area and a population of 2 lakhs.
7. Later on, the project area was divided into three development blocks and thus the whole country has been divided into Development Blocks.
8. Each development block consists of 100 villages,60,000-70,000 people and 150 square miles area.
9. Again each development block has been divided into groups of 5-10 villages and the charge for development has given to multipurpose VLW.
10. In each block, there are 15 VLWs.VLW is now called VDO i.e. Village Development Officer.
11. The basic idea and concept of Community Development was bor rowed from USA.

National Extension Services

1. Due to success of Community Development Project.People demanded more of CDP.Therefore National Extension Service was inaugrated on 2nd Oct.1953 owing to limited financial and technical manpower resources.
2. Under CDP,intensive development work was required in all fields but scheme of National Extension Service (NES) was to provide essential basic staff and a small fund for people to start development work essentially on the basis of self-help.
3. The operational unit of NES block was 100 villages comprising of 60,000-70,000 people.

4. Later on,NES block was converted into Community Development Block with higher budget to take up more intensive development programmes.
5 Under the chairmanship of Blawantrai Mehta one committe was formed in 1957 to revise CDP and the committee was called Committee on Plan Project.And the pattern of CDP was revised with effect from 1 April 1958.
6. A block is the Unit of Planning & Development.
7. The entire country has been covered with CDP by the year 1963.

Panchayati Raj :

Panchayati Raj was first started at Nagaur (Rajasthan) and Andhra Pradesh in 1959.As a whole state Rajasthan was covered first by the Panchatati Raj system.In the structure of Panchayati Raj,village Panchayat is the lowest unit.The general body of village Panchayat is Gram Sabha.The entire electorates of Panchayat constitute gram sabha.

IRDP : Integrated Rural Development Programme is a non-plan programme which means IRDP comes under five year plan.The main aim of IRDP is the all round development of family IRDP was launched on 2nd Oct 1980 (sixth five year plan) by integrating following programmes--

(i) SFDA :Small Farmers Development Agency.

(ii) DPAP : Drought Prone Area Programme.

(iii) :Operation Flood II :Development of Dairy,Poultry,Piggery,etc. in Gujarat & Haryana.

(iv) NREP : National Rural Employment Programme.

(v) MNP : Minimum Need Programme.

Operation Floods :

(i) Operation Flood I :This programme was experimental includes all the agricultural and health development.

(ii) Operation Flood II :Development of Dairy.Poultry,Piggery,etc in Gujarat & Haryana.

(iii) Operation Flood III :It's objective is only Milk Production and Cattle care.

TRYSEM : Training of Rural Youth for Self Employment (TRYSEM) was launched in 1979 (Sixth five year plan) with the basic principle of knowledge by doing or learning by doing to provide technical skills to the rural youth between age of 18 and 35 years old.

Young Farmers' Association :

4H-club was established in 1900 in America to provide training to youth for agriculture and home development.It was first govt.step.4H means Heart,,Head, Hand and Health as explained by A.B. Graham.On the basis of 4H-club of USA,Young Farmers association of India was started by P.S Deshmukh in 1956.He published a magazine 'Rural Youth'

Swarna Jayanthi Grama Swarozgar Yojana

Govt. of India have launched 'Swarna Jayanthi Grama Swarozgar Yojana'(SGSY) with effect from 1st April 1999 (1999-2000) as the main programme for promoting poverty alleviation through self employment. SGSY is a holistic package and replaces the earlier IRDP, TRYSEM, DWCRA, (Development of Women and Children for Rural areas),SITRA (Supply of Improved Tools Kitts for Rural Artisans).Ganga Welfare Plan and Million Wells Scheme (MWS).Unlike in the earlier programme where the aproach adopted was one of selecting the beneficieries first and then deciding on the activity for which assistance should be given to them, in the SGSY the emphasis is on selecting key activities and activity clusters under which beneficieries are selected keeping in view their suitability for the activities. The requirement of assisting only below poverty line households will continue. Also major part of the programmes is aimed at covering groups of beneficieries including self help groups rather than providing assistance to individuals.

SGSY Philosophy is based on the assumption that carefully selected viable economic activities suitable to each area should be promoted by provision of infrastructure and credit facilities. BPL (Below Poverty Line) households who are assisted to take up such activities are then expected to derive substantial incremental income and cross the poverty line. Thus, the emphasis is on identifying and developing sectors of economic activity in each district/block which can when suitably developed to lift large numbers of people above the poverty line. . Once such sectors are identified the infrastructure required for enabling that sector to provide self employment opportunities can be built up or strengthened using SGSY infrastructure funds as also normal plan allocations of the concemed departments. Further, to equip large number of people to take up self employment in such sectors intensive training programmes for both skill development and development of entrepreneurial skills are also provided for creation of marketing and other support services to ensure the viability of the economic activities selected is given emphasis in SGSY.

The economic activities selected under SGSY may produce either goods or services for the market. The marketing opportunities need not be confined to the local rural or urban markets and may involve distant market centres including export opportunities. To make marketing in distant markets feasible it is necessary to aim at adequate volumes of production which means a large number of small producers equip to produce high volumes of goods and services of a given variety. That is why activity cluster approach is adopted.

Objective of SGSY : Swarozgari should be able to cross the poverty line within 3 years. A minimum net income of Rs. 2000 P.M. is anticipated.

Target : To cover 30% of BPL familes in each block during the next five years (1999- 2004).

Difference between IRDP & SGSY

IRDP	SGSY
1. IRDP was essentially credit-cum-subsidy Programme.	1. It is holistic programme covering all aspects of self-employment .
2. Additional income generation for BPL families.	2. Micro-enterprise prog. for bringing the rural poor above poverty line.
3. No time frame for the assisted BPL families to cross the poverty line.	3. 3-years time frame.

4. No estimate regarding coverage of target groups.	4. Plans to cover 30% of the BPL families in each block (this is subject to availability of funds) in five years.
5. Dispersed activities.	5. Activity cluster identification, according to the nature of each block, 4-5 specific activities.
6. Group given maximum of 25% coverage.	6. Focus on group- approach.
7. Procedure of selection of beneficieries by the gram sabha.	7. Selection of beneficieries by 3-member team by visiting each habitation
8. No specific role for 'Panchayats'._	8. Greater role for Panchayats (a) Lists of swarozgar is families selected by placed before Gram Sabha (b) Panchayat monitors performance. (c) Responsibility for recovery. (d) Panchayat Samiti to activity clusters.
9. Linkages between groups formed under DWCRA and trainees of TRYSEM and IRDP not strong.	9. Holistic programme.
10. No provision of Technology and Marketing.	10. Provision exist for linkages.
11. Training not provided for groups.	11. A sum of Rs. 10,000/- earn marked.
12. Did not envisage specific role for NGOs.	12. NGOs role emphasized.

13. Bank role mostly confined to sanction of loans.	13. Banks's role integral to all aspects.
14. No monitoring of progress of beneficieries.	14. Swarozgari-wise monitoring.
15. Recovery of loan not given due weightage.	15. SGSY lays stress on recovery. Recovery to be monitored. Specific provisions of disincentives.

National Rural Employment Guarantee Act (NREGA) : NAREGA was enacted in September 2005 and came in force with effect from (w.e.f) 2nd February 2006 in most backward districts with the objective of providing 100 days of guaranteed unskilled wage employment to each rural household opting for it. It provides a legal right for guaranteed employment to the rural population through an act of Parlimament. The SGRY and NFWP have been subsumed in it. SGRY means Sampoorna Grameen Rozgar Yojana & NFWP means National Food for Work Programme. NAREGA would cover all districts of the country within five years. The focus of the act is on work, development, flood control/protection, (including drainage in waterlogged areas) and rural connectivity in terms of all weather roads. Panchayats have a key role in planning, implementation, & monitoring of the Act through preparation of perspective plan, approval of shelf of projects, execution of works at least to the extent of 50% in terms of costs. At least 1/3rd of the beneficiaries are to be women.

Extension Activities : Agriculture ext. is a state subject. The Dept. of Agriculture and Co-operation lays down major policy programme in close collaborations with the state Dept. of Agriculture; the state Agriculture Universities. Current ext. services include : Agriculture-Technology Management Agency (ATMA), Encouranging Public Private Partnership (PPP), Augmentation of Mass Media Suppot, operation of Kisan call centres.

Table : Rural Welfare Activities & Other Programmes in India

Name of Activity or Programme	Starting year	Name of the worker	Remarks
1. Co-operative Movement	1904	**F. Nicholson**	Individually financed
2. Sri Niketan (W. Bengal)	1908 (1921)	**Rabindra Nath Tagor**	Concept of **VLW**
3. Gurgaon Project (Haryana)	1920 (1927)	**F. L. Brayne**	Concept of **Village Guide**
4. Sevagram (near Wardha in Maharashtra)	1927	M. K. Gandhi	To spread charkha and Khadi
5. Marthandom Project (Keral)	1928	**Spencer Hatch**	With the help of YMCA (Young men Christian Association)
6. Social Legislation	1929	Govt. of Indian (British)	
7. Rural Development	1935	Rural Development Department of British India.	
8. Grow More Food Compaign	1942 (1947)	Rural Development Department of British India.	
9. Firka Development Scheme	1947 (1943)	Madras Govt.	Firka means a group of 5-20 village, promoting Khadi and village industries.
10. Nilkoheri Project (Haryana)	1943 (1948)	**S. K. Dubey**	also called Refugee Rehabititation Project, Created Mazdoor Manzil a new town ship.
11. Indian Village Service	1948	**W. H. Wisher**	

12. Etawh Pilot Project	1948	**Albert Mayer**	
13. FAO (Food Agriculture Organisation)	1948	International Organisation	
14. UNICEF	1948		
15. Planning Commission	**1950**	Govt. of India.	
16. Sarvodaya (from Bombay)	1950	**Acharya Vinobha Bhave**	
17. Five year plan	**1951**	Govt. of India.	
18. Training and Development Centres	1951	India Govt.	
19. Bhoodan Movement	**1951**	**Vinoba Bhave**	Started in Pochampali (Telangana in Andhra)
20. Gramdan Movement	1952	Vinoba Bhave	at Mangroth Village in U.P.
21. Community Development	**2nd Oct. 1952**	Govt. of India	
22. Gro More Food Enquiry Committee	**1952**	Govt. of India	Chairman–V.T. Krishna machari
23. National Extension Service	2nd Oct. 1953	Govt. of India	
24. UNESCO	1955	International Organisation	
25. Family Planning Programme	1956	Govt. of India	
26. Kharif and Rabi Movement	1958	Govt. of India	
27. **Nalagarh Committe Report** (or Agricultural Adminstrative Committee)	1958	Govt. of India	Recommended for **All India Agricultural Service**

28. Intensive Agriculture District Programme IADP	1960-61	Govt. of India	Commonly known as Package Programme (1961)
29. Multiple Cropping Scheme	1960	Govt. of India	
30. Democratic Decentralisation (Panchayati raj)	1958	Govt. of India	on the recommendation of Mehta committee
31. National Seed Corporation	1963	Govt. of India	
32. Agricultural Refinance Corp. (Later Agriculture Refinance and Development Corporation)	July 1963	Govt. of India	
33. Intensive Agriculture Area Programme (IAAP) except Kerala	1964	Govt. of India	
34. National Demonstration Project	1965	Ministry of Food and Agriculture	
35. High Yielding Variety Programme (HYVP)	1966-67	Govt. of India.	
36. **Administrative Reforms Commission**	1966	Govt. of India	
37. **Agricultural Adminstration Team Report**	1967	Govt. of India	
38. State Farms Corporation of India	1969	Govt. of India	

39. Small Farmers Development Agency (SFDA) and Marginal Farmers and Agriculture Labour (MFAL)	1970-71	Govt. of India.	
40. Drought Prone Area Prog. (DPAP)	1973-74	Govt. of India	To tackle the special problems in drought areas.
41. Command Area Development Programme (CADP)	1974-75	Govt. of India	
42. 20-Point Programme	1975	Indira Gandhi	
43. National Seed Programme	1975-76	Govt. of India	
44. Special Livestock Production Programme	1975-76	Govt. of India	
45. Regional Rural Banks (RRBs)	1976 Act	Rajasthan Govt.	
46. Antyodaya Programme	1977	Govt. of India	On the Philosphy of Gandhiji
47. Desert Development Prog.	1977-78	ICAR	
48. Lab to Land Programme	1979	Govt. of India	
49. TRYSEM (Training of Rural Youth for self Employment)	1979	Govt. of India	Basic Principle; learning by doing.
50. National Rural Employment	1980	Govt. of India	
51. Integrated Rural Development Programme (IRDP)	2nd Oct. 1980	Govt. of India	Non-plan programme aimallround developmnet of **family.**

52. Special Component Plan	2nd Oct. 1980	Govt. of India	For S.C. and S.T.
53. National Bank for Agriculture and Rural Development **(NABARD)**	12 July 1982	Govt. of India	First Canairman Sri B. Sivaraman
54. Drought Prone Area Prog. (IDPAP)	1982-83	Govt. of India	
55. Rural Landless Employment Guarantee Prog. (RLEGP)	1983-84	Govt. of India	RLEGP renamed in 1986-87 as Indira Awas Yojna.
56. Indira Awas Yojna	1986-87	Govt. of india	
57. Council for Advancement of People's Action & Rural Technology (CAPART)	Sept. 1986	Govt. of India	To promote voluntary action through community participation and to propagate appropiate rural technologies for the benefit of rural masses.
58. Integrated wastelands Development Prog. (IWDP)	1989-90	Govt. of India	Projects senctioned to the areas not covered under DPAP or DDP
59. SGSY	April 1999	Govt. of India	To bring the assisted poor families abvoe the proverty line by organising them into self help groups (SHG's)

60. District Rural Development Agency (DRDA) Administration	April 1999	Govt. of India	A centrally sponsered scheme for strengthening DRDA and making then more professional in their functioning.
61. Pradhan Mantri Gram Sadak Yojna (PMGSY)	Dec. 2000	Govt. of India	Connectivity to all the unconnected habitations of more than 500 persons in the rural areas (250 persons in the hilly & desert areas) by good quality all weather roads.
62. Sampoorna Grameen Rozgar Yojna (SGRY)	Sept. 2001	Govt. of India	Additional wage employment and food security alongside creation of durable community assets in rural areas.
63. Swajaldhara	Dec. 2002	Govt. of India	To formulate, implement, operate & maintain drining water projects by the village community.
64. Hariyali	April 2003	Govt. of India	To empower Panchayati Raj institutions both financially and administratively in implementation of Watershed Development Programme.

65. National Food for Work Programme (NFWP)	Nov. 2004	Govt. of India	To provide additional resources apart from the resources available under SGRY to 150 most backward districts for generation of supplementary wage employment and provision of food security.
66. National Rural Employement Gaurantee Act (NREGA)	Sept. 2005	Parliamentary Act.	100 days of guranteed unskilled wage employement to each rural household opting for it. SGRY and NFWP have been subsumed in NREGA.

25 Agricultural Economics

Introduction

Definition

Economics has been defined by various Economists :-

1) **Smith :** Economics is the science of Wealth.
2) **Marshall :** It is defined as the study of **human behaviour** in the daily business of life i.e. economic and non-economic activities.
3) **Robins :** Human wants are unlimited and the study of choice making behaviour. It is Modern definition.
4) **Pigue :** Emphasis given to the human welfare.

Subject Matter of Economics

(1) Consumption (2) Production (3) Exchange (4) Distribution and (5) Public Finance.

Factors of Production :

1) **Land :** Anything above the earth surface which is free given by the nature. It is fixed.
2) **Labour :** Labour is an active factor.
3) **Capital :** It is a passive factor . All capital is wealth but all wealth is not capital . Capital is the part of wealth used for further regenerating wealth.
4) Organisation and
5) Enterprise :

Reward of land → Rent

Reward of Labour → Wages

Reward of Capital → Interest

Reward of Enterprise → Profit

But according to modern Economists, the subject matter of Economics is grouped into two heads i.e. Micro - Economics and Macro-Economics .

a) **Micro - Economics :-** Such activities and services of Consumption, Production, Exchange and Distribution concerned with individual units (viz single industry, single farm, single consumer etc) i.e. at micro-level; are grouped under Micro-Economics.

b) **Macro - Economics :-** It deals with whole economic set -up and related additions or averages e.g. total production , total income, total employment, total expenditure, total savings, price level and economic development of whole economic set-up.

While we study the problems of resource - allocations with the objective to maximise the benefit of any single farm or individual family; it is called micro-economics. But while we study the problems of resource allocations between different areas of different units of the same or different types of farms; is called Macro-economics .

Cooperatives in India :

History :

1) Cooperative movement in India started in 1904 when cooperative credit societies act was passed . Father of cooperative movement in India was **F.Nicholson**.

   Limitations of 1904 act. : No any provision for the formation of non–credit societies.

2) 1912 act : Provision for the formation of non -credit societies with the aim to improve the economic status of its member as well as to meet the daily requirement. There was Provision for the formation of central cooperative society also (no provision in 1904 act).

3) The Govt. appointed a commitee under Sir Macglan in 1914 and

it observed that people were had the impression that cooperatives were mainly the govt. agencies. Therefore he recommended :-

a) for the proper education of the cooperative principles so as that members may participate efficiently.
b) The formation of cooperative banks to help the primary co-operative societies .
c) To form the provincial cooperative society in each state.
d) For careful scrutining of members before granting loans.

4. In 1944, Govt appointed an agricultural finance sub-committee under Prof. D.R. Gadgil to study the cooperative movement and to give suggestion for their improvement.
5. In 1945, Govt. appointed a Cooperative Planning Committee which submitted its report in 1946.
6. In 1951, All India Rural Survey Committee Report was published. The chairman of the committee was A.D. Gorwala . Its recommendations were:-
   (a) Organisation of large size cooperative society
   (b) Organisation of Marketing Cooperatives
   (c) Assistance from govt . in the form of loans and subsidies to the society.
   (d) adoption of system of controlled credit.
   (e) Establishment of SBI as institution which may assist the Cooperative Society through its rural branches.
7. In 1954, the All India Rural Cooperative Society recommended for the formation of large size society.
8. 3rd Indian Cooperative Congress held at New Delhi in April 1958 and recognised the importance of cooperative sector.
9. Indian National Cognress in 1959 at Nagpur session accepted the resolution of Agrarian Economy and this resolution brought a radical change in the policy and programme of cooperative movement.
10. Multipurpose Cooperative societies were first established in 1939 in Orissa.

11. The year 1954 is a landmark in the history of rural credit and Rural Credit policy in India.
12. 'Co-operation' is originated in Europe. Britain is the homeland of cooperative store movement.
13. Primary objective of cooperative marketing is to reduce marketing margin and to assure the farmer a better price.
14. Multi-unit Co-operative Societes Act, (1942) governs the working of Co-operative societies whose objects and area of operation extends to more than one state.
15. National Co-operative Development Corporation (NCDC) Act (1962) by repealing earlier Act. NCDC Act was enacted which replaced the earlier NCDC Board.
16. To provide greater functional autonomy to Co-operative to reduce bureaucratic interference and to professionalize the management of these institution, based on the recommendations of Ch. Brahm Prakash Committes, the Multi-State Co-operative Societies (MSCS) Act, 2002 has been enacted. It came into force with effect from 19 August 2002, replacing the MSCS Act, 1984. The Co-operatives have been empowered to hold their elections, appoint the auditors and also to riase resources by receiving deposits, raising loans & grants.

RBI and SBI :

Reserve Bank of India came into existence on 1st April 1935 with the RBI act 1934 and has been nationalised on Jan 1, 1949. **Mr. Osporne A. Smith** was the first Governer of RBI.

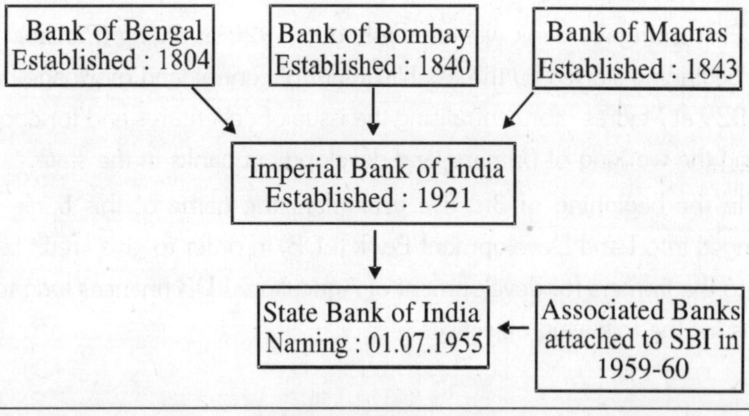

- Finance according to Period :-
    a) Short term loan :- Such loan is granted for one season of crop i.e.for 1to $1\frac{1}{2}$ years.
    b) Medium term loan :- Such loan is granted for the period of one year to 5 years.
    c) Long term loan :- Such loan is for the pericd of 5 to 30 years.

Cooperative Banking Structure

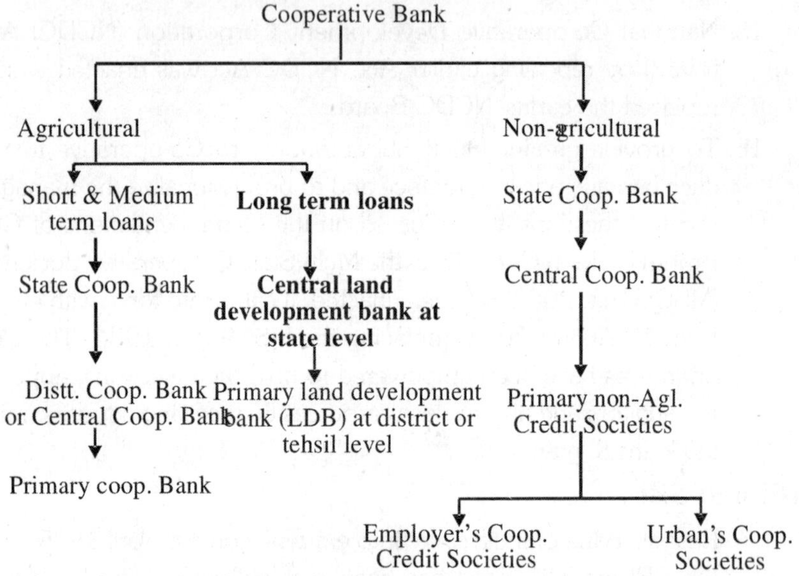

Land Development Bank (LDB)

First mortgage bank was established in 1920 at Jhang (Punjab). But real beginning started by the establishment of central land mortgage bank in 1929 at Madras - for centralising the issue of debentures and for coordinating the working of primary land development banks in the state.

In the beginning of 3rd five year plan, the name of this bank was changed into Land Development Bank (LDB) in order to give credit facilities to the farmers for development of Agriculture.LDB finances long term loans for the following objectives :-

a) The settlement of old debts

b) Improvement of agricultural lands

c) giving credit facilities to the farmers for purchasing irrigation equipments and tractors.

Crop Loan Scheme :

Crop loan scheme was one of the recommendations of All India Rural Credit Survey Committee Report in 1954. Mr. B.L. Mehta also recommended the need for crop loan in 1960 . The crop loan is sanctioned in one-third of the total loan. First crop loan was given to **cotton** at Berar in Maharashtra.

Regional Rural Banks (RRBs)

1) Committee on Rural Banks headed by M. Narasimhan (1975) .

2) RRB came into existence as a result of measures taken under 20 point economic programme in 1975 by Indira Gandhi.

3) First of all 5 RRBs were established ,in U.P. (Muradabad and Gorakhpur), Haryana(Bhiwani), Rajasthan (Jaipur) and in W.B. (Malda District). Syndicate Bank, SBI, Punjab National Bank, United National Bank (UCO) and United Bank of India were sponsored bank respectively . The share capital of RRBs was subscribed by Central Govt, State Govt and sponsored bank; and share of each agencies were 50% , 15% and 35% respectively . Authorised working capital of RRBs was one crore and paid up capital was 25 laksh for each bank.

4) Main objective of RRBs : Provision of credit and other facilities,especially to small and marginal farmers, agricultural labourers, artisans and small entrepreneurs in rural areas.

5) In June 1977, a committee under Prof M.L. Dantwala to evaluate the performance of RRBs; report in 1980 and its recommendation : Establishment of RRBs was very essential for making available credit needs of the farmers and weaker people even in those areas where primary cooperative credit societies were functioning.

- Taccavi : It is the name of Govt. Loan which is given by the govt.in case of famine and distress condition.
- FSS means Farmer Service Society.
- PACS : Primary Agricultural Cooperative Society.
- Employment Guarantee Scheme was started in Maharashtra.

Marketing Services

Marketing Services or Functions are the activities performed by different institutions or agencies which are involved in the process of marketing.

The marketing services are :-

1) Selling
2) Buying or assembling
3) Traffic Management or Transportation
4) Storage
5) Financing : is the provision and management of money and credit to get goods from the hands of producer to the consumer. Financing is presently the most important function.
6) Risk Management : is also the most important function today.
7) Standardization and grading.

Market Agencies or Functionaries

1) Village merchant
2) Intenerant dealer : There is lack of competition for them in the village.
3) Commission agent
4) Brokers
5) Whole Seller
   (a) Kaccha Arhatia : deals with the assembly of produce.
   (b) Pakka Arhatia : deals with the distribution to retailers. Both Arhatias advance loan to the producer with a term that produce would be sold only to them.
6) Retailers : are the last link between producer and consumer.

Characteristics of Indian Agriculture :

These are the following characteristics of Indian agriculture :-

1) Low Land Productivity of Agriculture.
2) Low Labour Productivity in Agriculture
3) Uncertainty in Agriculture
4) Subsistence farming
5) Prevalence of foodgrain crops
6) Traditional methdos of cultivation still practiced
7) Population Pressure on agricultural land
8) Labour dominance
9) Small holdings in India : The average holding size in USA is 158ha, 188ha in Canada and 1933 ha in Australia whereas average size of land holdings in India is only about 1.8 ha.
10) Uneven distribution of land : Only 10% people have ownership on 51.7% of total cultivated land whereas 10% farmers have only 0.18% . About 14%rural people are landless.
11. Mixed farming
12. Regional specialisation of crops
13. Dependence on Monsoon

Problems in Indian Agriculture :

Problems of Indian agriculture are the resultant of natural, social and economic reasons.Following are problems---

1) Uncertainty of rainfall and Problems in irrigation
2) Excess population pressure on agriculture
3) Faulty land ownership
4) Uneven land distribution
5) Smaller land holdings
6) Division and sub-division of land holdings
7) Low soil fertility status
8) Natural hazards

9) Lack of education
10) Lack of organisation in farmers
11) Lack of Capital
12) Indebtness of Indian farmers
13) Lack of suitable and proper seeds
14) Lack of manures and fertilizers
15) Lack of modern agricultural equipments
16) Lack of production technique
17) Lack of agricultural research
18) Lack of proper agriculture planning
19) Faulty agricultural marketing systems
20) Ups and downs in agricultural prices
21) Lack of timely payments to cane-producers
22) Lack of transportation and communication
23) Lack of cold stores and godowns
24) Lack of processing units
25) Lack of electricity,oils & fuels
26) Lack of highly productive animals
27) Irrational views of the farmers
28) Dependence on fate by the farmers

Economic Holding :

The size of holding which provides full employment to the farm family and provides a reasonal standard of living is termed as Economic Holding.

According to Central Land Reforms Committee,the economic holding size for a family (having 5 members) is -

i) 10 acres i.e. 4 ha for irrigated land, or

ii) 27 acres i.e. 10.9 ha for partially irrigated land,or

iii) 54 acres i.e.21.85 ha for non-irrigated land.

On the basis of ownership of agriculture land holding, there are three types of common farmers in India :

1) **Marginal Farmers :** Such farmers who have less than one hectare unirrigated land are called marginal farmers. Marginal farmers constitute 51% of the total farm holdings.

2) **Small Farmers :** Such farmers has one to two (1-2) hectare unirrigated land and constitute 19% of the total farm holdings in India.

3) Agricultural L abour : Such farmers are landless and having no any landed property.

Agriculture Price

In agricuicural sector, conditions of Perfect Competition are more prevalent than that of any other market.

Conditions for Perfect Market :-

1) Large no. of buyers and sellers
2) Product produced by all sellers is homogenous.
3) Both sellers and buyers have perfect knowledge about price.
4) No legal or institutional restrictions on buyers and sellers.

A competition that is not pure is called Imperfect Competition. It exists in two forms :-

(a) **Monopolistic Competition :** Many producers but each produces and sells differentiated products . Each one has a monopoly on its own product.

(b) **Oligopoly :** A few sellers, dealing in homogenous or differentiated product is small, complete interdependence among the seller with regard to their price output policies .

There is extreme case also i.e.

(c) **Monopoly :** means practically one seller or producer.

Practically all these three market conditions do not prevail in agricultural sector.

Marketable Surplus : The portion of the total produce in stock which the farmer is willing to sell.

Marketed Surplus : The portion which is actually brought to the market at a particular time for sale.

- Agriculture Price Commission came into existence in 1965

Price Elasticity :=

$$E_p = \frac{\text{Relative Change in Quantity}}{\text{Relative Change in Price}} = \frac{\frac{\Delta Q}{Q}}{\frac{\Delta P}{P}}$$

1) Completely Inelastic Demand : when demand completely unresponsive to the price change i.e. $E_p = 0$.

2) Relatively Inelastic Demand : When the increase in quantity sold is proportionately smaller than the decrease in price i.e. $0 < E_p < 1$.

3) Unit Elasticity : When proportionate change in the quantity demanded is just equal to the proportionate change in its price $E_p = 1$.

4) Elastic Demand : When proportionate change in quantity demanded is more than proportionate price change i.e. $E_p > 1$.

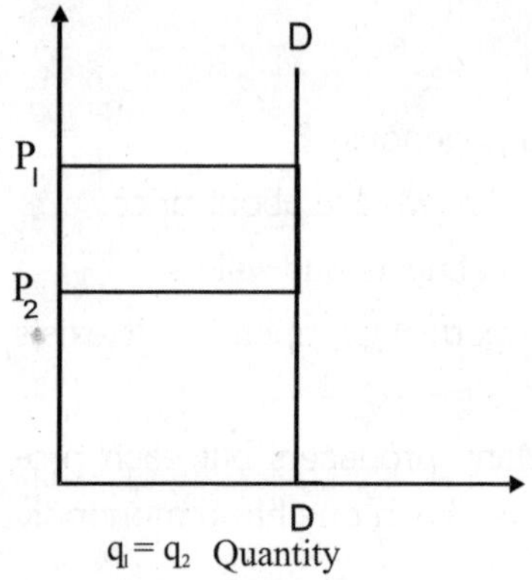

(1) Perfectly Inelastic
E = 0

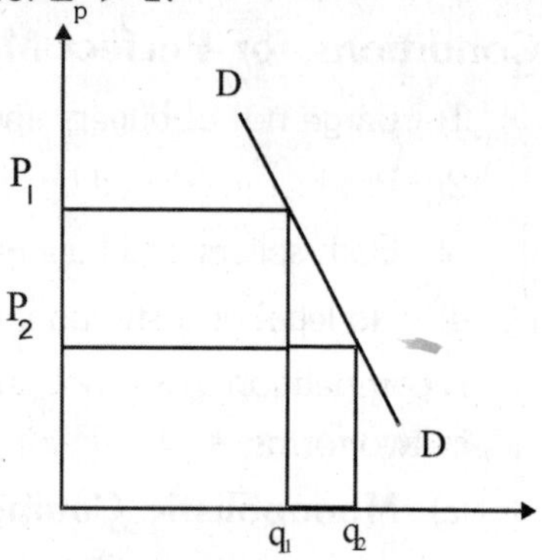

(2) Relative Inelastic demand curve
E >0; E<1
or 1>E>0

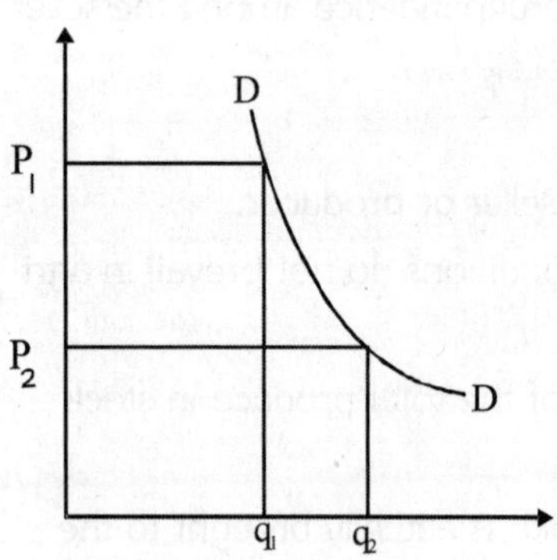

(3) Unit Elasticity
E = 1

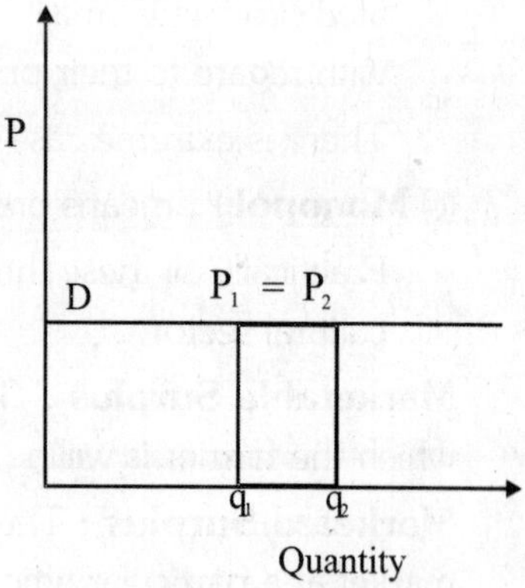

(4) Perfectly Elastic
E = ∞

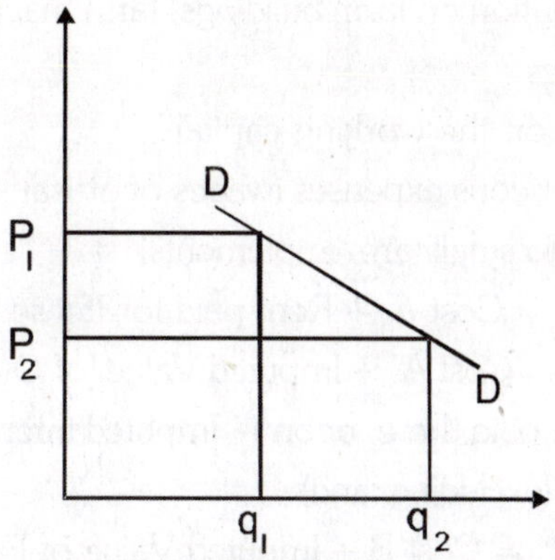

(5) Relative Elastic
E < ∞; E > 1 or 1< E >∞

Cost Concepts

The cost is the outlay of funds for productive services . There are many cost concepts used in agricultural economics and farm management i.e. Cost A_1, : Cost A_2, Cost B, Cost C; working or variable or operational cost and fixed costs; Explicit and Implicit cost; opportunity cost; labour costs etc.

i) Cost A_1 : consists of following 16 items of costs :

1) Value of hired human labour (Permanent & Casual)
2) Value of owned bullock labour
3) Value of hired bullock labour
4) Value of owned machinery
5) Hired machinery charges
6) Value of fertilizers
7) Value of manure (Produced on farm and purchased)
8) Value of seed (both farm-produced and purchased)
9) Value of insecticides & fungicides
10) Irrigation charges (both of the owned & hired tubewells, pump ing sets etc.
11) Canal-water charges
12) Land revenue, cesses and other taxes
13) Depreciation on farm implements (both bullock drawn & worked with human labour)

14) Depreciation on farm buildings, farm machinery and irrigation structures.

15) Interest on the working capital

16) Miscellaneous expenses (wages of artisans, cost of ropes & repairs to small farm implements)

ii) Cost A_2 = Cost A_1 + Rent paid for Leased Land

iii) Cost B = Cost A_2 + Imputed Value of Owned Land (less land revenue paid there upon)+ Imputed interest on Owned Fixed Capital (excluding land)

iv) Cost C = Cost B + Imputed Value of Family Labour where cost C is the cost of cultivation.

It means cost C consists of 20 items viz.

Cost C = Cost A_1 (having 16 items) +

17) Rent Paid for Leased land +

18) Imputed value of owned land (less land revenue paid there upon)+

19) Imputed interest on owned fixed capital (excluding land) +

20) Imputed value of family labour

or Cost C = Cost A_1 + (17th + 18th + 19th + 20th items)

or Cost A_1 = Cost C– [17th +18th + 19th + 20th items]

⇒ Cost A_1 = Cost C– Rent paid for leased land–Imputed value of owned land– Imputed interest on owned fixed capital–Imputed value of family labour.

⇒ Cost B = Cost C– Imputed value of Family labour.

⇒ Cost A_2 = Cost B– Imputed value of owned land – Imputed interest on owned fixed capital.

= Cost B– [Imputed value of owned land + imputed interest on owned fixed capital]

⇒ Cost A_2 = Cost C–[Imputed value of Family labour + Imputed value of owned land (less land revenue paid there upon) + Imputed inter est on owned fixed capital (excluding land)]

Net Income = Gross Return – Cost C

$$\text{Cost of Production} = \frac{\text{Cost C}}{\text{Output}}$$

Cost of Production means how much fund is outlayed to produce unit output for example to produce, 10 quintals the cost of cultivation (Cost C) incurred by a farmer is Rs. 900; then the cost of production will $\frac{900}{10}$

=Rs. 90/q or Rupee Ninety per quintal.

Fixed Cost or Overhead Cost : - Such cost does not change in magnitude as the amount of the production process changes and are incurred even when production is not under taken. These are Sunk Cost and may be cash or non-cash fixed costs. The examples of fixed cash cost are Land taxes, Interest , Insurance premiums, annually hired labour etc. Whereas the non-cash fixed costs are depreciation on capital investment, cost of family labour and costs of management, machinery equipment, interest on capital managememt.

Variable Cost or Prime Cost :- It is the cost of using the variable inputs . Such variable inputs are directly related to the production. Examples of variable cost are cost of seed , feed , fertilizer, water, labour hired occasionally , interest on current investment, current repair replacement, diesel etc. Here farming expenses are the function of farm output.

Farming Expenses = f (farm output)

It means higher the production, higher will be the variable cost. Variable cost is also known as Prime Cost or Special Cost or Direct Cost.

Implicit and Explicit Cost :- Implicit cost is the money value of those inputs which are supplied by the farmer himself.Whereas Explicit cost is the actual money expenses directly incurred in raising a farm commodity and monetary estimates of implicit cost. Explicit cost is also known as accounting cost. Both Explicit cost and Implicit cost together constitute Economic cost.

Economic Cost = Accounting (Explicit) Cost + Implicit Cost

Opportunity Cost or Social/Alternative Cost :- The farm resources have normally a number of alternative uses.For example a farmer raises paddy on his farm instead of maize,it means the farmer utilises the other opportunity by giving up the first alternative.Here the Social cost of raising paddy will be the amount of maize sacrificed in the process.Therefore in modern economics the real cost is presented as opportunity cost or Alternative Cost.

Cost Elements

The cost has different elements but broadly two categories viz- Fixed Cost and Variable Cost . The cost elements are derived from these two costs.

1) Total Cost (TC):-

TC = Total Fixed Cost + Total Variable Cost

$= TFC + TVC \Rightarrow TC = FC + VC$

= Explicit Cost + Implicit Cost

In the beginning total production cost is lower but increases gradually. When total cost is less than the Gross Income at the farm, the profit will be maximum.

2) Average Total Cost /Average Total Unit Cost :-

Average Total Cost (ATC) or Average total unit cost (ATUC) is also known as simply Average Cost (AC)

$$\text{A.C. or ATC/ATUC} = \frac{TC}{\text{Output (Q)}}$$

ATC is called unit cost of production.

3) Total Fixed Cost (TFC or FC)

4) Average Fixed Cost (AFC) :-

$$AFC = \frac{TC}{\text{Output (Q)}} = \frac{FC}{Q}$$

Fixed cost per unit production is called Average Fixed Cost.]

5) Total Variable Cost (TVC or VC) : It is also called Prime Cost, Special Cost or Direct Cost .

6) Average Variable Cost (AVC) :-

$$AVC = \frac{TVC}{\text{Unit of Output (Q)}} = \frac{VC}{Q}$$

The variable cost per unit production is called AVC.

7) Marginal Cost (MC)

$$MC = \frac{\text{Increase in Variable Cost}}{\text{Increase in Output}} = \frac{\Delta VC}{\Delta Q}$$

The additional increase in variable cost with the additional increase in output is called Marginal Cost or Additional Cost.

Marginal Fixed Cost is always zero because fixed cost does not change with the change in output. Therefore Marginal Cost is necessarily marginal Variable cost and the change in fixed cost (FC) will not affect marginal cost (MC). For example, the cost of producing few more vegetables by farming a given amount of land morje intensively, is not affected by the amount of rent paid for the fixed amount of land. Marginal Cost (MC) is independent of the size of fixed cost.

Table of Different Costs

(Cost in Rupees)

Units of Production	FC	VC	TC = FC + VC	AFC $=\frac{FC}{Q}$	AVC $=\frac{VC}{Q}$	ATC $=\frac{TC}{Q}$	MC= $\frac{\Delta VC}{\Delta Q}$
0	25	0	25	∞	-	∞	-
4	25	4	39	6.25	1.00	7.25	1.00
12	25	8	33	2.08	0.66	2.70	0.50
18	25	12	37	1.38	0.66	2.50	0.66
23	25	16	41	1.08	0.69	1.78	0.80
27	25	20	45	0.92	0.74	1.66	1.00
30	25	24	49	0.83	0.80	1.63	1.33
32	25	28	53	0.78	0.87	1.65	2.00
33	25	32	57	0.75	0.96	0.72	4.00

Three points are inferred from the above table :-

1) Fixed Costs are the same at the whole production level.
2) Variable cost changes with the change in production level.
3) Total Cost and Variable Cost increase with the increase in production.

Cost Curves

The relationship between production and cost curve is shown below.

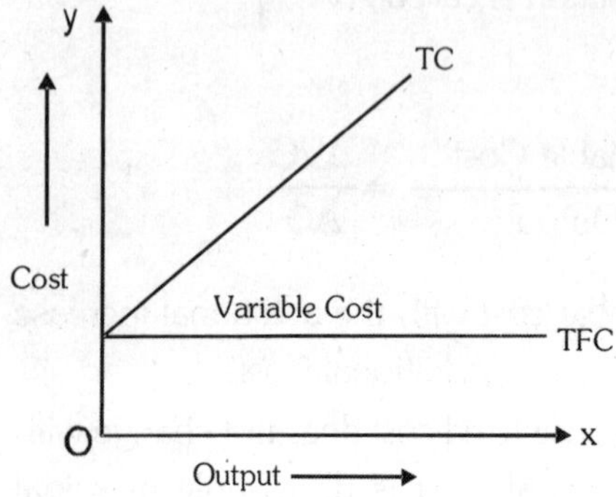

Fig. 1

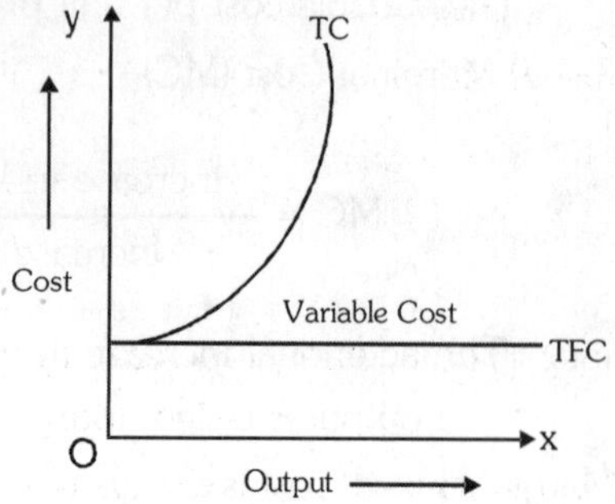

Fig. 2

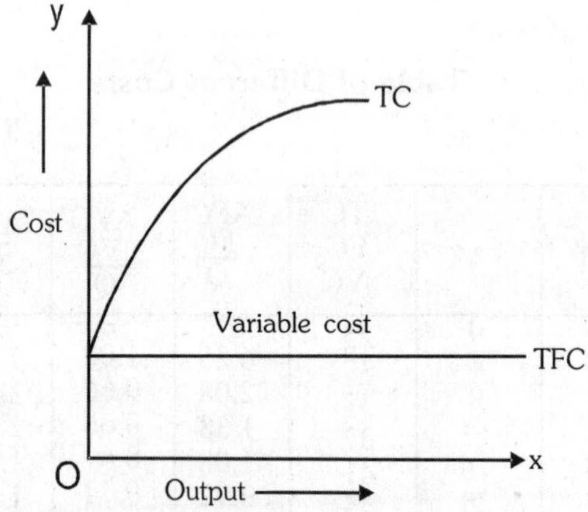

Fig. 3

The exact nature of the total cost curve depends on the nature of the production function.

Fig .1. First figure shows total cost curve for linear Production function.

Fig.2. Second figure shows total cost function where the factor–product (Input-Output) relationship is one of diminishing marginal pro ductivity.

Fig 3. T.C. function reflecting increasing returns throughout.

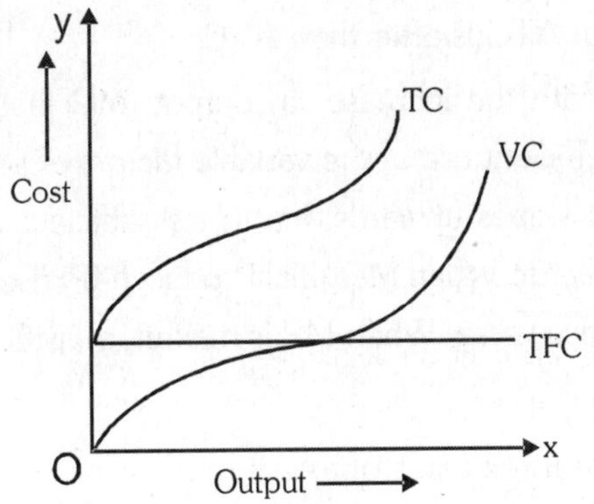

Fig. 4 : Generalised Cost Curves

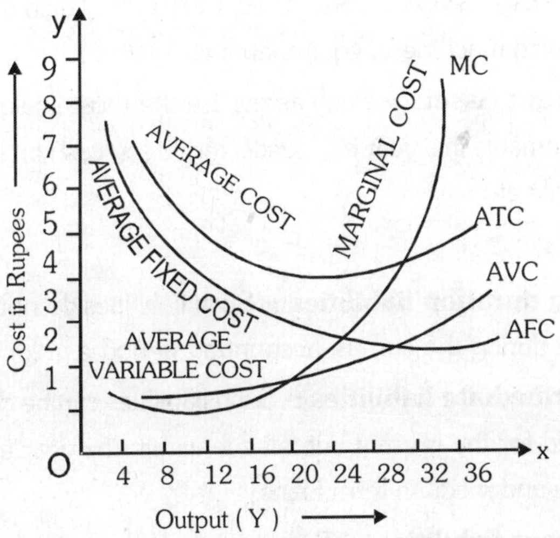

Fig. 5 : Different Cost Curves

The inference may be drawn from the above different cost curve :-

i) AFC : is continued to decline and never shows upward movement because the input investment after the achievement of maxi mum product becomes irrational .

ii) AVC : with the rise in prodcution, AVC firstly declines , reaches the lowest point at highest APP and again increases .

iii) ATC : The trend of ATC is same as of AVC but the lowest point of ATC is after than AVC.

iv) MC : With the increase in output, MC first falls due to more efficient use of the variable factors of production and then it slopes upwards due to less efficient use of the variable factor. When Marginal Product (MP) rises, MC declines and vice-versa. When MP is maximum, MC is the lowest.

Assets

There are three types of assets :

a) **Fixed Assets :-** Such assets are difficult to convert into cash to meet any current obligations e.g. land, buildings .

b) **Working Assets : -** Such assets are more liquid than fixed assets e.g.farm machinery, equipments etc.

c) **Current Assets :-** Such assets are the most liquid assets and are consumable in a year e.g. seeds, fertilizers, cash on hand , livestock for sale etc.

Liabilities :

a) Long duration liabilities :- Such liabilities do not require repayment during the current accounting period e.g. long term loans.

b) Intermediate liabilities :- Such liabilities can be defered or postponed for the present but fall due within the year e.g. promissory notes and medium term loans.

c) Current liabilities :- Repayment of these liabilities may be demanded at once e.g. Short term loans, fertilizers, seeds etc.

Efficiency Measures :

There are two methods to evaluate efficiency measures for volume of business–Ratio methods and Aggregative measures. Here ratio methods are discussed :

A Assets - Liabilities Ratios :

i) Net Capital ratio $= \frac{\text{Total Assets}}{\text{Total Liabilities}}$

ii) Working Capital ratio $= \frac{\text{Working assets and current assets}}{\text{Intermediate \& Current Liabilities}}$

iii) Current ratio $= \frac{\text{Current assets}}{\text{Current Liabilities}}$

iv) Debt-equity ratio $= \frac{\text{Deferred liabilities}}{\text{Net Worth}}$

B Income Ratios

i) Rate of **Turn Over** $= \frac{\text{Gross Income}}{\text{Total Assets}} \times 100$

It is useful in the case of capital-starved situation.

ii) Net Income per ha. $= \frac{\text{Total returns from fixed resources}}{\text{Total area in hectare}}$

or

Return from fixed

farm resources per ha

C Cost Ratios

i) Gross ratios $= \frac{\text{Total Expenses}}{\text{Gross Income}}$

ii) Fixed ratio $= \frac{\text{Fixed Expenses}}{\text{Gross Income}}$

iii) Operating ratios $= \frac{\text{Operating Expenses}}{\text{Gross Income}}$

Monotonic and Non-Monotonic Relationships :

a) Monotonic Relationships in the following graphs :

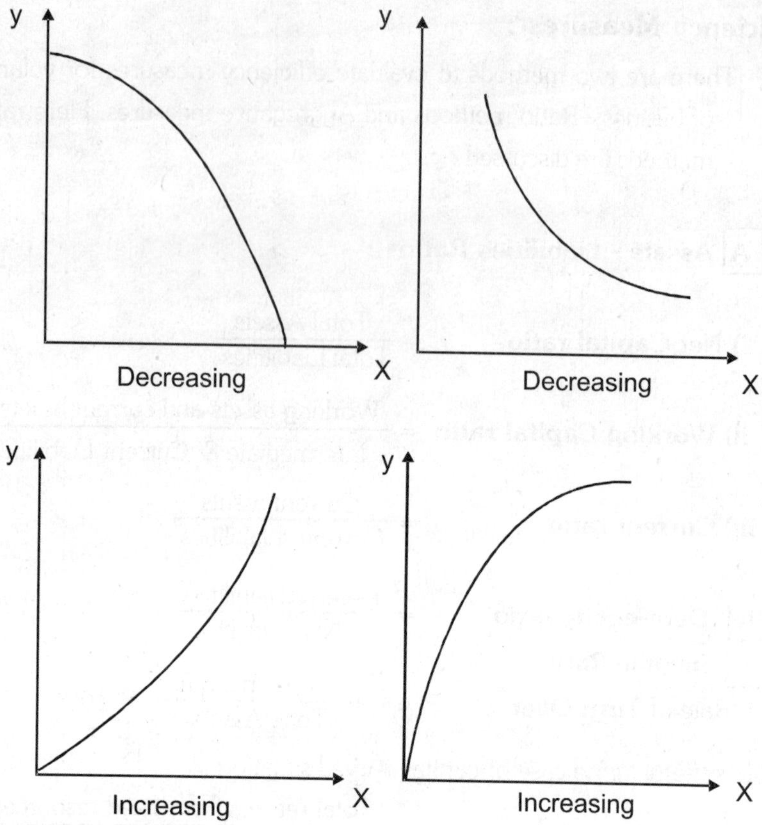

Fig : Monotonic Curves

b) Non-Monotonic Relationships in the following graphs :

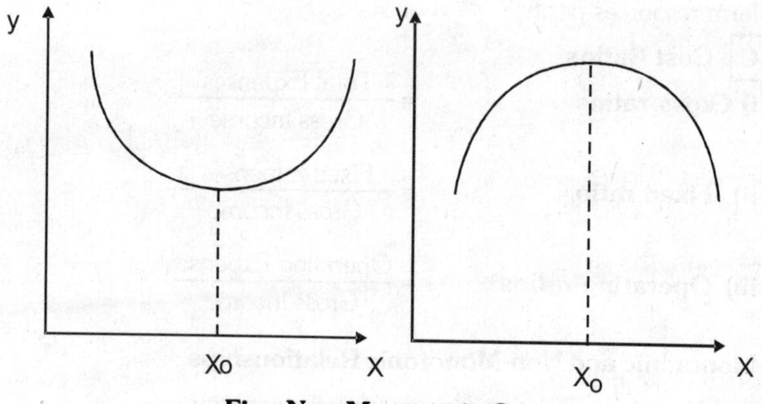

Fig : Non-Monotonic Curves

AGMARK

The Agricultural Produce (Grading Marking) Act, 1937 empowers the Govt. to fix quality standards, known as 'AGMARK', standards and to prescribe terms and conditions for using the real of 'AGMARK'. So far, good standards have been notified for 182 agricultural and allied commodities. The purity standards under the provision of the prevention of Food Adulteration (PFA) Act, 1954 and Bureau of Indian Standards (BIS) Act, 1986 are invariably taken into consideration while farming the grade standards. International standards framed by Codex/International Standards Organisation (ISO) are also considered.

Department of Agriculture & Co-operation has 3 Organisations dealing with Agricultural Marketing :–

a) Directorate of Marketing and Inspection at Faridabad (Central Agmark Laboratory at **Nagpur** and 16 Regional Agmark Lab. functions under it.) for the promotion of standardisation and grading of Agricultural and allied produce, construction of rural godowns, development of Agricultural marketing infrastructure and marketing information network.

b) National Institute of Agricultural Marketing, Jaipur (1988) involved in

   (i) training activities (in 2003-04, Management development programmes were introduced).

   (ii) Research

   (iii) Project formulation and

   (iv) Undertaken postgraduate programme in Agri-business management to assist agl. graduates to acquire the critical know-how to compete in the domestic and global business arena.

c) Small Farmers Agri-Business Consortium, 1994 to support innovative ideas for generating income and employment in rural areas by promoting private investments in agri-business projects.

❑❑❑

26 Farm Management

Meaning:

Farm generally means an area of land under single management system.Farm is defined by D.S.Chauhan as "a piece or pieces of land operated as single unit of agriculture enterprise under one management". Hence farm is an economic unit,used for agriculture and livestock production.The main factor of a farm are land, labour and capital.But all these factors are limited. Therefore, Farm management may be defined as a science which deals with judicious decision on the use of scarce farm resources, having alternative uses to obtain the maximum profit and family satisfaction on a continuous basis from the farm as a whole and under sound farming programmes.It deals with the allocation of resources at the level of an individual farm i.e. micro-economics.

Objects of Farm Management

1) Minimisation of costs, Losses and Damages, Inefficiency and unemployment.
2) Maximisation of profit, productivity, efficiency and Employment.

Minimisation of		Maximisation of
1. Cost		1. Profit
2. Losses and Damages	⇌ F.M.	2. Productivity
3. Inefficiency and		3. Efficiency
4.Unemployment		4. Employment

There are three main points to be considered in Farm Management :-

- 1) How to maximise the production.
  2) How to get maximum profit.
  3) How to minimise the cost of production

Production Decisions

Individual farmer facing the production descisions are of three types :

1) To produce a certain amount of a given product having certain amount of resource to use.

Input (Resource) ⟶ Product (output)

$x \longrightarrow y$

One resource ⟶ One product

What is the most profitable amount of resource to use in the production ?

How are various factors of production converted into final form i.e. products?

What is the optimum level of resource for maximum production ?

Such decisions are studied in Factor (resource) - Product Relationship.

2) Which combination of resources is most profitable?

Certain combinations of resources ⟶ Certain amount of products

$X_1, X_2 \quad \therefore y = f(X_1, X_2) \quad y$

'How to produce', is studied in Factor-Factor or Resource-Resource relationship.

3) What to produce or which enterprise is to be selected is studied in product -product relationship.

$X \longrightarrow y_1, y_2$ (Products)

The above points may be presented in a tabular form :

In order to provide answer to all these questions, farmer will take help of production function analysis.

Production Functions :

S. No.	Management or Production Decision	Explaining	Principle of Economics
1.	How much to produce? (Optimum level of resource use)	Factor-Product Relationship	Principle of Diminishing Returns or increasing cost.
2.	How to Produce (Least cost method)	Factor-Factor-relationship	Principle of substitution or Least cost combination of Resources.
3.	What to produce (Enterprise selection)	Product-Product relationship	Principle of opportunity cost or equi-mar-ginal Returns

According to L.E. Bishop, The production function is a mathematical relationship describing the way in which the quantity of a particular product depends upon the quantities of particular inputs used .

$y = f(x_1, x_2, --x_n)$.

Production function is also called Factor-Product Relationship. It is of two types:

1) Production function for short run period when the quantities of some inputs are kept constant and the quantities of one input are varied i.e. **Law of Variable Proportions.**

2) Production function for long run period when all inputs are varied i.e. **Law of Returns to Scale.**

Cobb Douglas Production Function :

$Q = K. L^{á} C^{â}$

Where Q = Production; L = Labour and C = Capital.

In the production function, solved by Cobb-Douglas, the share of labour was $\frac{3}{4}$ and that of capital $\frac{1}{4}$

$Y = K. L^{3/4} C^{1/4}$

Here á +â=1, it means Law of Constant Return to scale.

After changing the value of power of labour and capital i.e. á and â ; the different laws of production function can be explained :-

(i) when á + â = 1 means Law of Constant Return.

(ii) á + â > 1 Law of Increasing Return to scale.

(iii) á + â < 1 means Law of Diminishing Return to Scale.

When á+ â = 1, it is a Linear Homogeneous Production Function at constant return if function is taken in Logrithm (Log) but not in natural scale.

Cobb- Douglas function has constant Elasticity of substitution.

Types of Production Functions :

(a) Linear Production Function : When production function is homo geneous of the first degree

$y = a_0 + a_1 x_1 + a_2 x_2 + \ldots\ldots\ldots + a_n x_n$

where y = production, a_0 = constant

$a_1, a_2 \ldots\ldots\ldots\ldots a_n$ = coefficient factors

$x_1, x_2, \ldots\ldots\ldots\ldots x_n$ = factors of production

(b) Quadratic production Function

$y = a + bx + cx^2$

where a, b,c = constant.

x = Production factor

(c) Square Root Production Function :

$y = a + b\sqrt{x_1} + cx_2$

(d) Cubic Production Function

$y = a_0 + a_1 x + a_2 x^2 + a_3 x^3$

Forms of Factor - Product Relationship :

When only one factor (or input) of production is variable remaining other factors constant, then the nature of the relationship between single input (factor) and single output (product) is of three forms viz. Increasing marginal return, Constant marginal return and Decreasing marginal return (product) .

Factor- Product relationship is also **Input - Output relationship**.

Each Input-Output relationship is measured by two ways :-

(i) Average Physical Product (AP) = $\frac{\text{Output}}{\text{Input}} = \frac{Y}{X}$
or Average Product (APP)

(ii) Marginal Physical Product (MPP) = $\frac{\text{Additional Output}}{\text{Addional Input}} = \frac{\Delta Y}{\Delta X}$

or Marginal Product (MP)

MP is the addition of output with the addition of one unit of input. APP or AP is also written as APP_x or AP_x means Average Physical Product of x (i.e.input). Similarly MPP or MP is also written as MPP_x or MP_x.

a) **Increasing Marginal Productivity (IMP) or Increasing Marginal rate of Returns (IMR)** : It is clear from the table and graph.

Table : IMR or IMP

x (Kg / ha)	y (q / ha)	Δy	Δx	$\frac{\Delta y}{\Delta x}$
0	5	-	-	-
5	7	2	5	0.4
10	12	5	5	1.0
15	19	7	5	1.4
20	29	10	5	2.0
25	44	15	5	3.0

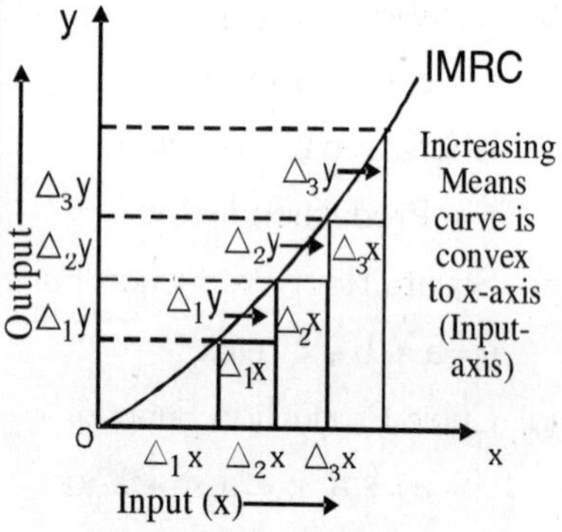

Fig : IMRC (Increasing Marginal return curve)

From the table and graph, generalised formulae for IMR or IMP is

$$\frac{\Delta_1 y}{\Delta_1 x} < \frac{\Delta_2 y}{\Delta_2 x} < \frac{\Delta_3 y}{\Delta_3 x} - - - - - \frac{\Delta_n y}{\Delta_n x}$$

b) **Constant Marginal Productivity (CMP) or Constant Marginal Return (CMR)**

Table : CMR or CMP

x (kg/ha^{-1})	y (q/ha^{-1})	Δy	Δx	$\frac{\Delta y}{\Delta x}$
0	10	-	-	1
5	15	5	5	1
10	20	5	5	1
15	25	5	5	1
20	30	5	5	1
25	35	5	5	1

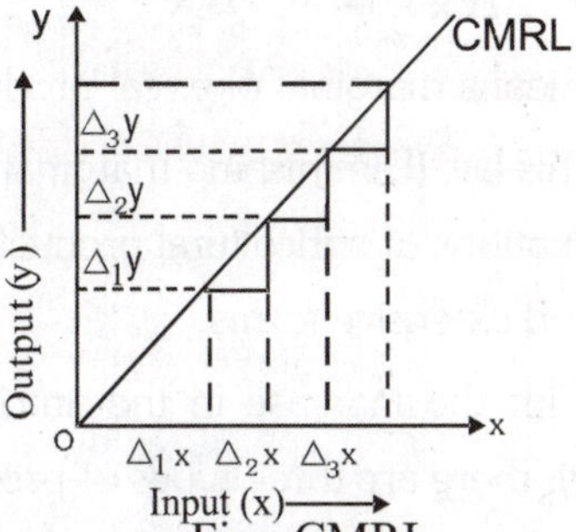

Fig : CMRL
(Constant marginal reaturn Line)

General formulae for CMP or CMR is

$$\frac{\Delta_1 y}{\Delta_1 x} = \frac{\Delta_2 y}{\Delta_2 x} = \frac{\Delta_3 y}{\Delta_3 x} \text{ — — — — } \frac{\Delta_n y}{\Delta_n x}$$

Here Production Function is Linear and is Presented by an Algebraic equation

$y = a + b x$

constant marginal return (CMR or CMP) is also called Equal Marginal return. At the time of constant (equal) marginal productivity in factor - product relationship , the Production function is called Constant Maginal Production (return) function.

c) **Diminishing Marginal rate of Returns (DMR) or Decreasing Marginal Productivity (DMP) :**

Table : DMR or DMP

x (kg/ha^{-1})	y (q/ha^{-1})	Δy	Δx	$\frac{\Delta y}{\Delta x}$
0	10	-	-	-
5	25	15	5	3.0
10	35	10	5	2.0
15	42	7	5	1.4
20	47	5	5	1.0
25	50	3	5	0.6

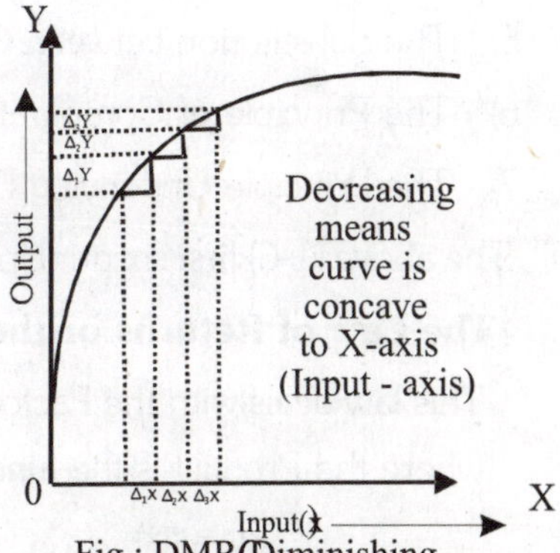

Fig : DMR(Diminishing Marginal return curve)

General formulae for DMR or DMP

$$\frac{\Delta_1 y}{\Delta_1 x} > \frac{\Delta_2 y}{\Delta_2 x} > \frac{\Delta_3 y}{\Delta_3 x} \text{ — — — — } \frac{\Delta_n y}{\Delta_n x}$$

means marginal physical product in decreasing order.

This law (Diminishing marginal return) is applicable in almost all practical situations of agricultural production. To land it applies both in its intensive and extensive forms.

With the increase in the amount of the variable resource (input or factor), there are three types of productional changes seen in the marginal return i.e. Increasing marginal return, constant marginal return and Decreasing marginal return. According to modern Economists the above three proportional changes together is known as the **"Law of Variable Proportions ."**

For obtaining maximum profit and successful farm business, a farmer has essentially to know some different economic principles . Such economic principles involved in making rational farm management dicisions are :-

1) The Law of Returns or the Principle of Variable Proportions .
2) The Cost Principle
3) The Principle of Equi - marginal Return or the Opportunity Cost Principle.
4) The Substitution between Inputs or Factors
5) The Substitution between Outputs or Products
6) The Principle of Comparative Advantage
7) The Principle Underlying Decision under Risk and Uncertainty.

The above (1-6) first six principles are applied where there is certainty.

The Law of Returns or the Law of Variable Proportions

This law deals with the Factor -Product or Input-Output relationship where the Product is the function of the factor.

$$y = f(x)$$

The conditions for obtaining the output levels of the variable input and that of the output, giving positive maximum profits are in the fol lowing two alternative forms :-

Input approach	Output approach
i) MVP = MIC ii) MVP must be diminishing iii) Average value Product must be diminishing	i) MR = MC ii) MC must be rising iii) Average cost must be rising.

Where ,

a) Marginal Value Product (MVP or VMP)

$$\text{MVP or VMP} = \frac{\text{Change in Total Value Product}}{\text{Additional Input}}$$

$$= \frac{\Delta TVP}{\Delta x} = \frac{\Delta y}{\Delta x} Py$$

b) Marginal Cost (MC) :

$$MC = \frac{\Delta TC}{\Delta y} = \frac{\text{Change in Total Cost}}{\text{Additional output}}$$

c) Marginal Input Cost (MIC or VMI)

or Marginal Factor Cost (MFC).

$$\text{MFC or MIC} = \frac{\Delta TC}{\Delta x} = \frac{\text{Change in Total Cost}}{\text{Additional input}}$$

d) Marginal Revenue (MR)

$$MR = \frac{\text{Change in Total Revenue}}{\text{Additional output}} = \frac{\Delta TR}{\Delta y}$$

Here it is explained by the table and production function curve (PFC) .

Table :
Relationship between Total, Average and Marginal Products :

Unit of fertilizer (x)	x	Total Physical Product TPP (y) or T.P.	y	Average Physical Product APP or A.P (y/x)	Marginal Physical Product MPP or M.P (y/x)		Remarks
0	0	0	0	0	0	I	Increasing at Increasing Marginal return
1	1	2	2	2.0	2		
2	1	5	3	2.5	3		
3	1	9	4	3.0	4		
4	1	14	5	3.5	5	II	Increasing at constant Marginal return
5	1	19	5	3.80	5		
6	1	23	4	3.83	4		
7	1	26	3	3.71	3	III	Increasing at decreasing rate or marginal return
8	1	28	2	3.5	2		
9	1	29	1	3.22	1		
10	1	29	0	2.9	0		
11	1	28	-1	2.54	-1	IV	Decreasing at increasing rate
12	1	26	-2	2.16	-2		

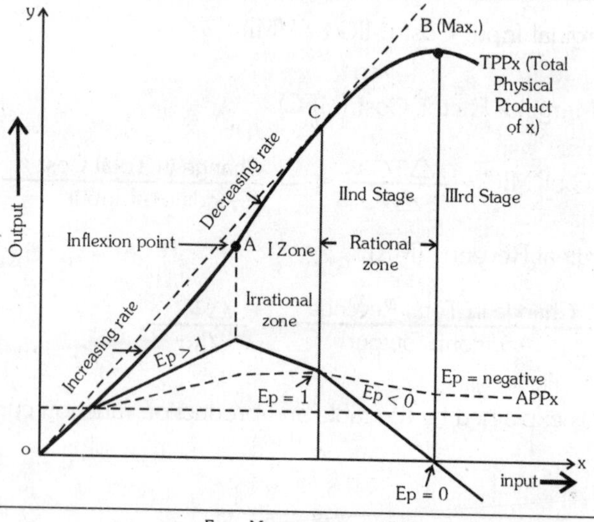

Production Function Curve

These are the inferrence drawn from the Production -function Curve :-

1) Total Physical Product of x (TPP_x) rises at increasing rate of return; MPP_x rising and Production function curve is concave upward.
2) Beyond Inflexion Point 'A' ; TPP_x rises but at diminishing rate and MPPx starts to decline.
3) TPP_x is highest at 'B' or remains constant; and MPP_x =O.
4) TPP_x decreases then MPP_x becomes negative.
5) At Point 'D' ; $MPP_x = APP_x$ and APP_x is maximum.
6) When MPP_x > APPx (more than APPx); then APPx is increasing.
7) When MPPx < APP_x (less than APP_x) ; then APP_x is decreasing but never becomes negative.

Elasticity of Production :

The percentage change in output due to percentage change in input is called Elasticity of Production (E_p). It is the ratio of the percentage change in output to the percentage change in Input.

$$E_p = \frac{\text{Percent Change in Output}}{\text{Percent Change in Input}} \quad \frac{\%\Delta Q}{\%\Delta X}$$

$$= \frac{\frac{\Delta y}{x} X100}{\frac{\Delta x}{x} x100} = \frac{\frac{\Delta y}{x}}{\frac{\Delta x}{x}} = \frac{\Delta y}{\Delta x} = \frac{y}{x}$$

$$= MPP_X \, \frac{y}{x} \quad (\text{Where} = \frac{\Delta y}{\Delta x} MPP_X) \qquad = MPP_X \frac{1}{\frac{y}{x}}$$

$$= MPP_X \frac{1}{APP_X}$$

$$= \frac{MP_X}{AP_X}$$

$$E_p = \frac{MP_X}{AP_X}$$

Inference from the formula :

1) When MP = AP, then E_p =1.

2) When MP > AP, then E_p > 1.

3) When MP < AP, then E_p < 1.

4) When MP = 0, then E_p = 0.

Profitability is increasing upto the increase in APPx = MPPx and there is no harm to the farmer. It is the minimum level of input- use (i.e.fertilizer use in this case). At this stage the farmer is in the First stage . Since in the first stage, profitability is undoubtly increasing, therefore the first stage is called Irrational zone. Most of the Indian farmers are under stage I. But the farmers of Punjab, Haryana and Western U.P. are under stage II i.e. Rational zone where the farmer has to conscious to get highest profitability using the Cost Principle. When MPPx becomes zero and negative afterwards, the production is not profitable and the stage is third stage i.e. Irrational zone.

Under the rational zone i.e. IInd stage, the farmer has to calculate proper and optimum level to get maximum profitability.

Cost Principle

To measure the optimisation condition, there are two methods :-

1) First Method :=

Additional Revenue = Additional Cost

⇒ Added Revenue = Added Cost

⇒ Additional Return = Additional Expenditure

⇒ Value of Marginal Product (VMP) = Value of Marginal Input (VMI)

⇒ Marginal Revenue = Marginal Cost

⇒ **MR = MC** or **VMP = VMI**

⇒ $Py.\Delta y = Px.\Delta x$, where

$$\frac{Py}{Px} = \frac{\Delta x}{\Delta y}$$

Py = Price of Product 'y'

Px = Price of of Input 'x'

Δy = Marginal Product 'y'

Δx = Marginal Input 'x'

$$Py = Px.\frac{\Delta x}{\Delta y} = Px.\frac{1}{\frac{\Delta y}{\Delta x}}$$

$$Py = Px.\frac{1}{MPP_x} \Rightarrow \frac{Px}{MPPx}$$

$$py = \frac{P_x}{MPP_x} \text{ and } MPPx = \frac{Px}{Py}$$

At the point where marginal revenue (MR) is equal to marginal cost (MC) ; that point is the opimum point where the farmer gets maximum net profit

2) Second Method

$$MPPx = \frac{Px}{Py} \Rightarrow \frac{\Delta y}{\Delta x} = \frac{Px}{Py}$$

The proportional marginal change in y to x, is equal to their opposite price ratio. It might be depicted by the cost principle curve.

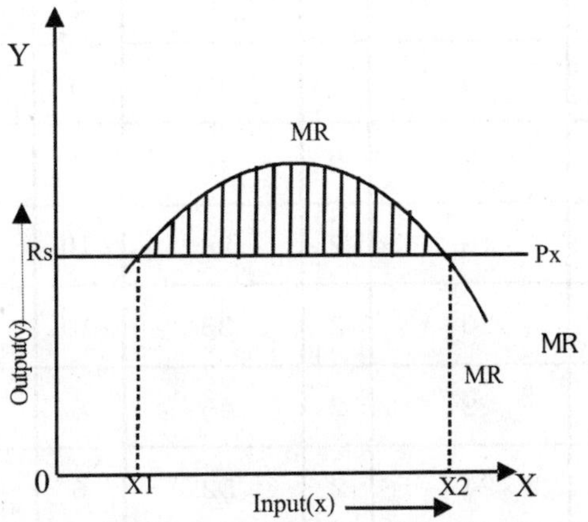

Cost principle curve

In the above graph (cost principle curve) marginal return (MR) curve becomes equal twice to marginal cost (MC) i.e. at x_1 and x_2.

First time MR =MC is in the First stage (i.e.irrational zone) of the Production function curve (PFC) and Second time MR = MC is in the second stage (i.e. rational zone). In the second stage of PFC, MR curve starts to decline and cuts the price line of x at the point x_2 where again MR =MC which denotes the optimum marginal rate of input. The shaded area of the MR Curve denotes net profit .

To get maximum profit in agri-business, three points should be essentially considered in the II stage (rational zone) of the PFC :-

(i) MR = MC

(ii) MC Curve must be rising (see Different Cost Curve)

(iii) Total Revenue > Total Cost means total revenue must be greater than the total cost.

It becomes more clear from the cost principle table.

Table : Cost Principle

Input	OutPut	Total Cost (TC) Rs. 2.0/Per Unit TFC = Rs 10/-	MC	Total Revenue (TR) TR = Py.y @ Rs.2.0/ unit	MR	Profit (TC-TFC)	
0	0	10(TC=T-FC)	0	0	0	-10	
1	2	12	2	4	4	-8	
2	5	14	2	10	6	-4	
3	9	16	2	18	8	2	
4	14	18	2	28	10	10	
5	19	20	2	38	10	16	
6	23	22	2	46	8	24	
7	26	24	2	52	6	26	
8	28	26	2	56	4	30	
9	29	28	2	58	2	30	MC=MR
10	29	30	2	58	0	28	
11	28	32	2	56	-2	24	
12	26	34	2	52	-4	22	

Law of Equi - marginal Return
or
Principle of Opportunity Cost
or
Alternative Cost Principle

According to this law, each unit of the limited or unlumited resources is to be invested in different enterprises in such a way that a farmer gets maximum marginal return.How a farmer can distribute each unit of input among two or more than two different enterprises, is studied and calculated as follows :-

Suppose a farmer has 100 kg fertiliser . How can he distribute it on different crops say wheat, gram and barley so that he gets maximum profit?

Table : Fertilizer and Marginal Return

Fertilizer (kg)		Marginal Returns in Rs. (MR)		
Total	Additional	Wheat	Gram	Barley
10	0	100 III	120 I	80 V
20	10	90 IV	110 II	70 VIII
30	10	80 VII	80 VI	60 XI
40	10	70 X	70 IX	50 XIII
50	10	60 XII	50 XIV	40
60	10	50 XV	40	30
70	10	40	35	20
80	10	30	25	15
90	10	25	20	10
100	10	20	10	5

It is clear from the above table that after the investment of initial 10 kg fertilizer, MR from wheat is Rs. 100/-, from gram is Rs. 120/- and from barley is Rs. 80/- In such case first dose of fertilizer (I) should be applied in the field of gram. Similarly IInd dose in gram, IIIrd dose in wheat, IVth dose in wheat and V^{th} dose in barley and so on according to the table . After 10 th dose, the fertilizer is exhausted . If the cost of fertilizer is Rs. 5/- per kg, means 10kg fertilizer (one dose) costs Rs. 50/- If the resource (here amount of fertilizer is 100 kg costing Rs. 500/-) is limited, a farmer can not invest after the exhaustion of his input (say 10th dose i.e. 100kg fertilizer). But in case of unlimited resouree (i.e. Plenty of fertilizer) the farmer inevests it until MR = MC. Here MC= Rs.5 X10kg = Rs 50/- ; it means the farmer invests the input doses upto XVth dose in wheat where he gets MC = MR = Rs. 50/- . And it is the situation where we see **Equi Marginal Return** in case of **unlimited resources** i.e. Rs. 50/- from wheat, Rs. 50/- from gram and Rs. 50/- from barley equal to marginal cost.

In the case of limited resource i.e. 100 kg fertilizer, the maximum return would be :-

From Wheat : 100 + 90 +80 + 70 = Rs. 340/-

From Gram : 120 + 110 + 80 + 70 = Rs. 380/-

From Barley : 80 + 70 = Rs. 150/-

Total Return = 340 + 380 + 150 = Rs. 870/-

In the case of unlimited resources i.e.plenty of fertilizer, the maximum return would be :-

From Wheat : 100 + 90 +80 +70 + 60 + 50 = Rs. 450/-

From Gram : 120 + 110 + 80 + 70 + 50 = Rs. 430/-

From Barley : 80 + 70 + 60 + 50 = Rs. 260/-

Total Return = 450 + 430 + 260 = Rs. 1140/-

Problem :

A farmer wishes to invest his Rs. 1000/- on the crop, poultry and dairy in the instalment of Rs. 200/- and the total returns from the different enterprises are given in the table. Suggest the farmer how he can distribute his money among different enterprises so that he gets maximum return ?

Total Cost	Total Return		
	Crop	Poultry	Dairy
0	0	0	0
200	500	300	400
400	900	500	700
600	1200	600	900
800	1400	650	1000
1000	1500	680	1050

Solution :

Table: showing distributors of Rs. 1000/- on different Enterprises

Total Cost	Additional Cost (MC)	Crop		Poultry		Dairy	
	MC	Total Return	Maginal Return	Total Return	MR	Total	MR
0	–	0	–	0	–	0	–
200	200	500	500 I	300	300 IV	400	400 II
400	200	900	400 III	500	200	700	300 V
600	200	1200	300	600	100	900	200
800	200	1400	200	650	50	1000	100
1000	200	1500	100	680	30	1050	50

Note :

Other factors viz Diversification , Inclusion of legumes in the cropping system, local conditions etc. are taken into consideration in the invest ment on different enterprises in case of equal marginal returns . e.g. IInd and IIIrd allocations and similarly IV th and Vth allocations in the above table.

Substitution between Inputs

or

Factor -Factor Relationship

or

Rate of Technical Substitution

or

Principle of Least Cost Combination

This law is represented in algebraic form as :

$$Y = f(x_1, x_2)$$

Where y = output and $x_1,\ x_2$ are different inputs.

This principle explains : How the different units of the two or more than two variable factors of production of any commodity can be used so that a farmer gets same quantity of product at the comparatively lowest cost .

isoquant curve /isoproduct /production Indifference curve :

isoquant = iso	+	quant
↓		↓
Equal		quantity

isoquant curve is the curve/path showing various combinations of factors for producing the same quantity of output.

It becomes clear from the following table and graph.

Table : Production Indifference of Wheat

Yield in quintal	X_1	ΔX_1	X_2	ΔX_2	$\frac{\Delta X_1}{\Delta X_2}$ (rate of substitution)
40q	100	-	0	-	-
40	90	10	1	1	10 Where Y = 40 q fixed
40	75	15	2	1	15 x_1 = amount of fertilizer in kg.
40	55	20	3	1	20 x_2 = No. of irrigation
40	30	25	4	1	25
40	0	30	5	1	30

To produce 40 q wheat , either 100kg fertilizer and no irrigation or no fertilizer and 5 no. of irrigation or any other combination in between these two extremes is required. With the application of 1 irrigation,10kg fertilizer application is reduced to produce the same output. Now we have to see whether application of 10kg fertilizer is cheaper or 1 irrigation is cheaper and this is explained by principle of least cost combination. Isoquant curve may be upto infinity in number. Different isoquant curves show the diferent levels of production which are obtained from different combinations of two resources . Four isoquant curves are shown in figure.

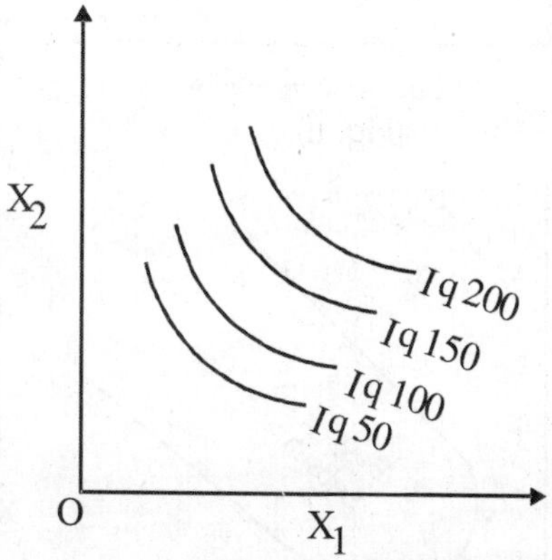

Fig. :Different Isoquant Curves

The shape of isoquant curve depends upon the extent of substitutability (i.e. rate of substitution) of the two inputs :-

i) When two inputs are **perfect substitutes,** the shape of the isoquant is straight line.

ii) If two inputs are good substitutes, the shape of the isoquant is **slightly curved & convex to the origin**.

iii) If two inputs are poor substitutes; isoquant curve is having **steep curvature.**

iv) If two inputs are to be used in fixed proportion i**.e.absolutely non substitutable** ; **isoquants are right angled**.

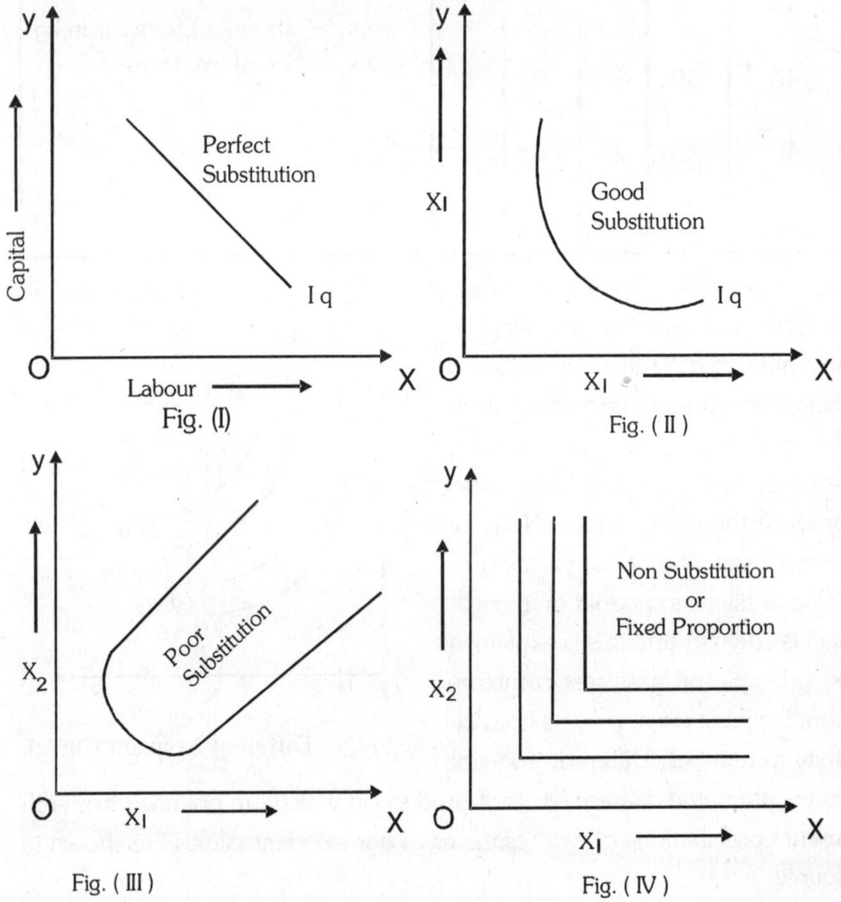

Fig. shapes of iso-quant curves

Types of Factor - Substitution :

a) Increasing rate of factor substitution (IRFS) :- If one factor of production factors' combination is increased ; the Marginal Rate of Substitution (MRS) of other factor is also increased e.g. the increase in number of irrigation (x_1) increases the no . of intercultures (x_2) .

b) Decreasing rate of Factor Substitution (DRFS) : - In agriculture, these is generally decreasing marginal rate of factor substitution because no two factors of production are perfectly substitutable. For example as the no. of irrigation increases, the requirement of fertilizer is reduced for producing the same output.

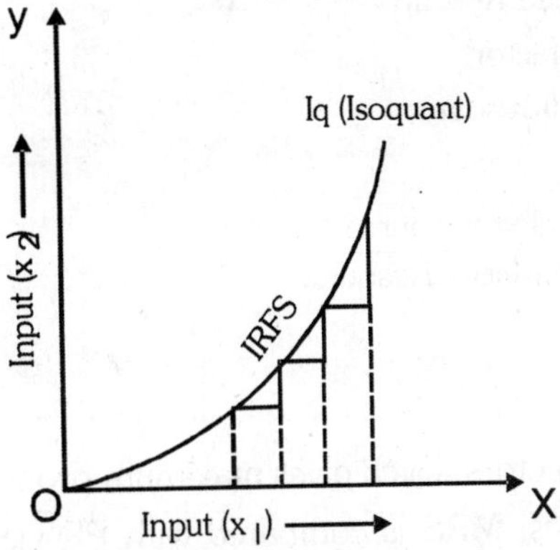

Fig. (A) IRFS

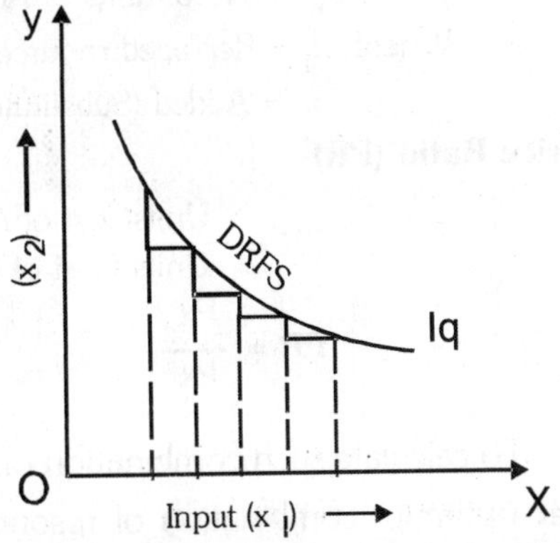

Fig. (b) DRFS

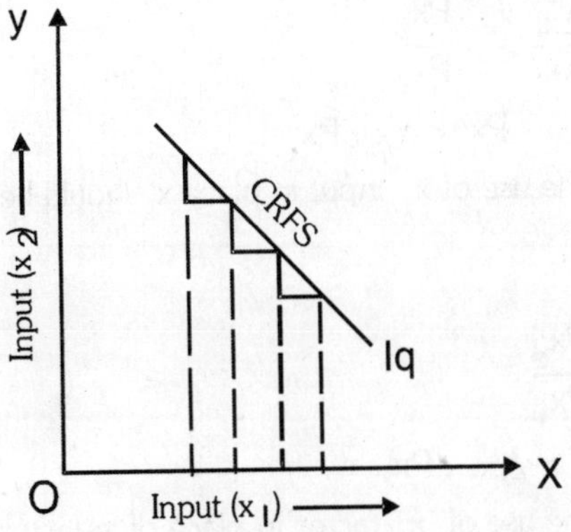

Fig. (c) CRFS

c) **Constant rate of Factor substitution (CRFS) :-** There is constant marginal rate of substitution . e.g. perfect substitution .

Principal of Least Cost Combination :=

Marginal Rate of substitution :

The per unit increase in one factor decreases the unit change in the second factor ; is known as the **Marginal Rate of Technical (or Factor) Substitution (MRFS) .**

$$MRFS\ x_2\ \text{for}\ x_1 = \frac{\Delta x_1}{\Delta x_2}\ ,\quad MRFS\ x_1\ \text{for}\ x_2 = \frac{\Delta x_1}{\Delta x_2}$$

$$MRFS = \frac{\text{No.of units Replaced Resource}}{\text{No.of units of Added Resource}} = \frac{\Delta x_1}{\Delta x_2}$$

Where, x_1 = Replaced reource (factor)

x_2 = Added (Substituted) resource

Price Ratio (PR)

$$PR = \frac{\text{UnitsCost of Added Resource}}{\text{Units Cost of Replaced Resource}}$$

$$\mathbf{PR} = \frac{Px_2}{Px_1}$$

To calculate such combination of factors which gives maximum profits **i.e.** optimum combination of resources; MRS is compared with PR viz. MRS<PR; MRS> PR, or MRS = PR

1) MRFS < PR

$$\frac{\Delta x_1}{\Delta x_2} < \frac{Px_2}{Px_1}$$

$$\Delta x_1 . Px_2 < \Delta x_2 . Px_1$$

In such condition, the use of x_1 input in place x_2 should be increased until we get MRS = PR

2) MRFS > PR

$$\frac{\Delta x_1}{\Delta x_2} > \frac{Px_2}{Px_1}$$

$$\Delta x_1 . Px_2 > \Delta x_2 . Px_1$$

In such condition, the use of x_2 factor in place of x_1 factor should be increased until we gate MRS = PR

3) MRFS = PR

$$\frac{\Delta x_1}{\Delta x_2} = \frac{Px_2}{Px_1}$$

$$\Delta x_1 \cdot Px_2 = \Delta x_2 \cdot Px_1$$

It is the optimum combination to get maximum profit and such equalibrium point is called as **Least Cost Combination.**

To calculate least cost combination in factor - substitution, there are two methods .

Simple Arithmetic Method :

Let x_1 = Fertilizer in kg; x_2 = No.of irrigation

Px_1 = Rs. 2/- per kg fertilizer

Px_2 = Rs. 50/- per irrigation

y = 40 q/ha

S. No.	Combination of Factors		Total Cost in Rs.	Yield
	x_1(kg)	x_2(no)	$(Px_1.x_1 + Px_2.x_2)$	y
1	100	0	(2x100) + 0 = 200	40q/ha (fixed)
2	90	1	(2x100)+(50x1) = 230	,,
3	75	2	(2x75)+(50x2)=250	,,
4	55	3	(2x55) +(50x3)=260	,,
5	30	4	(2x30)+(50x4)=260	,,
6	0	5	(2x0)+(50x5)+250	,,

In the above table, first condition is the cheapest costing Rs. 200/- only. Therefore the first condition is optimum combination of resources to produce same yield.

b) **Graphic Method**

$$\frac{\Delta x_1}{\Delta x_2} = \frac{Px_2}{Px_1} \text{ where } \frac{Px_2}{Px_1} \text{ is fixed for acertain period.}$$

$\frac{Px_2}{Px_1}$ is drawn on the graph where it is called **Price Line** or **Iso - cost line** or **iso-expenditure line**.

Iso - cost line is the tangent to iso - quant curve where $\frac{\Delta x_1}{\Delta x_2} = \frac{Px_2}{Px_1}$

How to draw Price line?

Suppose a farmer wishes to invest Rs. 100/- in the agriculture.

Px_1 = Rs. 2/- per kg
Px_2 = Rs. 50/- per irrigation
If the farmer invests the whole money (i.e. Rs. 100/-) on x_1 (i.e.fertilizer), he will use maximum 50 kg of x_1 therefore to draw price line, we take maximum point 50 on Oy - axis. Similarly if the farmer invests the whole money on x_2 (irrigation), he will use maximum 2 no. of x_2 (irrigation) in such condition, we take maxumum point 2on ox-axis . Now both the points (point 50 on oy -axis and point 2 on ox -axis) are connected with a straight line which will be the price line for input's costs . Such price line is known as Iso -cost line or Iso- expenditure line. Each and every point of Iso- cost line gives the same expenditure.

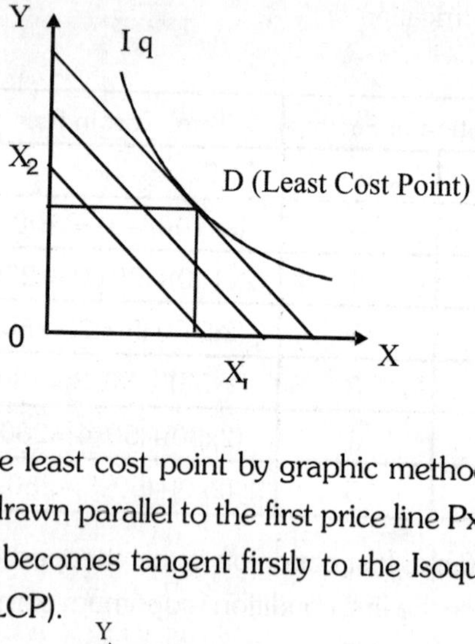

To determine the least cost point by graphic method, different other price lines are to be drawn parallel to the first price line Px_2 Px_1. The point where the price line becomes tangent firstly to the Isoquant curve,will be the least cost point (LCP).

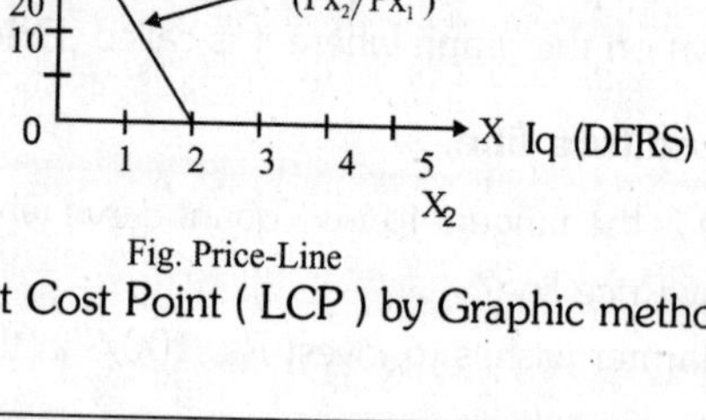

Fig. Price-Line
Fig. : Determination of Least Cost Point (LCP) by Graphic method

Left side of the price line shows lower cost and right side of the price line shows higher cost.The Least Cost point depends upon the slope of the price line. If the farmer has sufficient resources,he can shift to right side and follow other price lines.And thus he gets different least cost points viz. C_1, C_2, & C_3 on different Isoquant Curve (IC). The change in price changes the all least cost points but at a certain time there is only one Expansion Path. The line passing through the least cost points viz. C_1, C_2, C_3 is called **Expansion Path** or combination of Iso -cost lines. Tne size of the expansion path depends on the production function. The poor farmer uses C_1 point (first least cost point), middle farmer uses C_2 point and the big farmer uses C_3 point or any other least cost point (LCP) of right side.

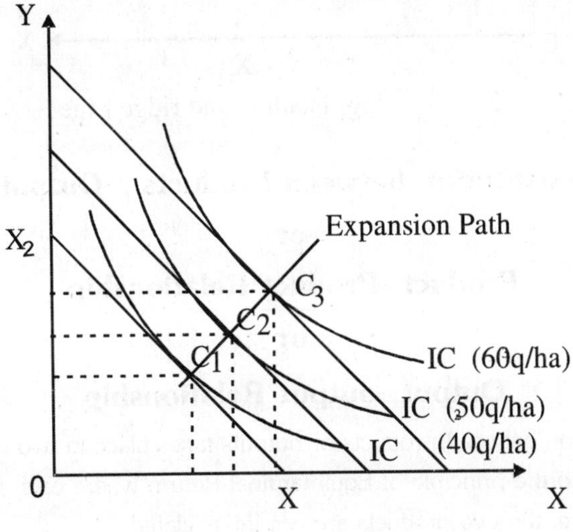

Fig. : Expansion Path having different LCPs.

Isocline and Ridge Line :

isocline = iso + cline

↓ Equal ↓ Inclined

Isocline is the line passing through the equal slopes of the different isoquant curves. Expansion path is a particular isoclne passing through the different least cost points . For expansion path , price line is a necessary .

Ridge line is the line passing through the maximum limits of the different isoquant curves i.e. parallel to the respective axis and thus isocline varies in between these two ridge lines.

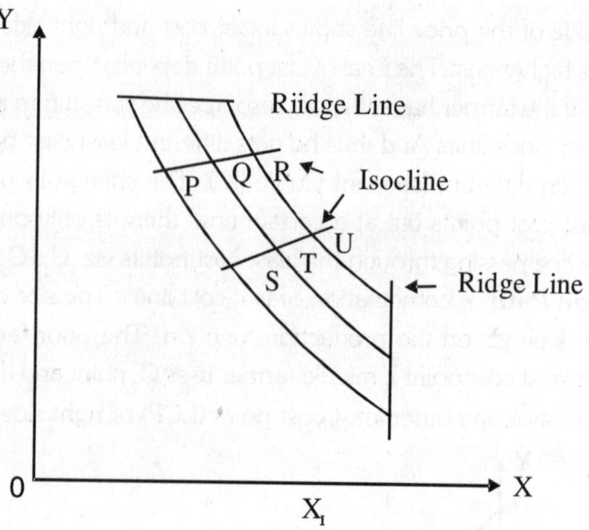

Fig. Isocline and ridge Line

Substitution between Products / Outputs

or

Product -Product Relationship

or

Output -output Relationship

The substitution between products or outputs takes place in two ways :-

i) According to the principle of Equimarginal Return where each enterprise is independent i.e. the two products are not inter-related.

Input ________________ Products

x ________________ y_1, y_2 , y_3 (Different Enterprises)

ii) According to the principle of product -product relationship where the different products are interdependent or inter -related.

Input ________________ Products

x ________________ y_1, y_2 (inter - dependent products)

Farm Products Relationship :

1) joint Products :- Joint products are results from the same production function.Generally all agricultural products are joint products . In agriculture there is one Bye-product with each main product e.g. straw with wheat grain, Cow -dung with milk etc. For a short period, there is only one product but in a long period, there is substitution between products . In the joint products the farmer emphasises on the main product.

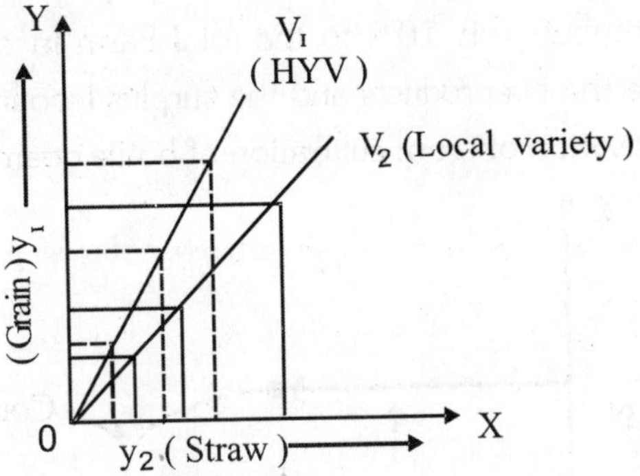

Fig. Joint Products

2) Complementary Relationship :- The change in the level of one product changes the level of other product in the same direction. It means the increase or decrease in the production level of one product increases or decreases the other's respectively. Such products are called complementary products e.g. mixed cropping of Wheat + Gram where the yield of wheat increases due to nitrogen -fixation by gram, the increase in the number of cattles increases the quantity of dung manure. But two enterprises are not always complimentary over-all possible combinations of the two.And such relation always gives the way to competition e.g. Excess forage area reduces the grain production . The complementary relationship after a certain point becomes competitive i.e. after the points B and C in the given graph.

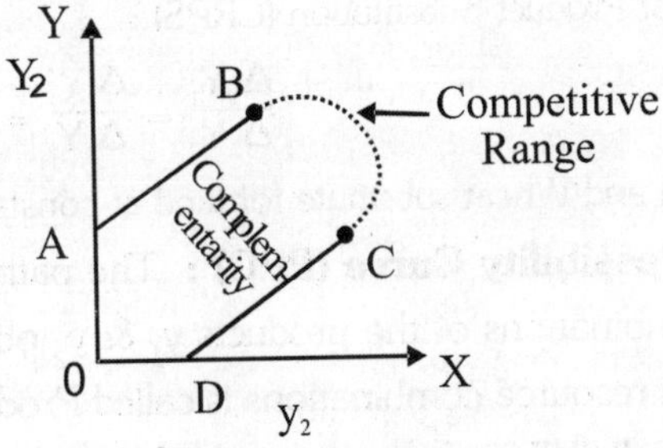

Fig. : Complementary Relationship

3) **Supplementary Relationship :-** The increase or decrease in the production level of one product does not affect the production level of the other product . Such type of relationship between product and product is called supplementary relationship. Here one enterprise is subsidiary enterprise and its contribution is 10% to the total Farm income . Subsidiary enterprises utilise the byeproducts and the surplus labour e.g. Rearing of livestock and cultivation of crop; cultivation of bottle goard with sugarcane etc.

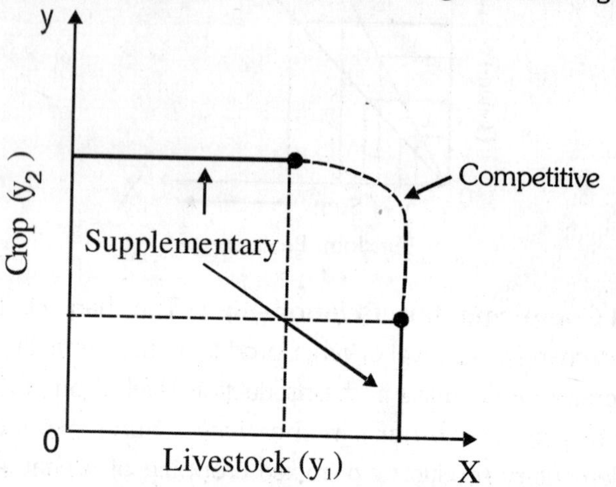

Fig. : Supplementary Relationship

4) **Competitive Relationship :-** When the increase in the production level of one product necessarily decreases the production level of the other product, such relationship is called competitive one. And when two products are competitive, they may substitute at constant rate, increasing rate or decreasing rate.

a) Constant Rate of Product Substitution (CRPS) :

$$\frac{\Delta_1 Y_1}{\Delta_1 Y_2} = \frac{\Delta_2 Y_1}{\Delta_2 Y_2} = \frac{\Delta_3 Y_1}{\Delta_3 Y_2}$$

For example , Gram and Wheat substitute for land at constant rate.

Production Possibility Curve (PPC) : The path or locus passing through various combinations of the products y_1 & y_2 obtained with fixed level of resources or resource combinations is called Production Possibility Curve. PPC is also called Transformation or Iso- revenue curve/Iso -return curve/ Iso -income curve /or opportunity curve. When PPC is a straight line then it is called production possibility line (PPL).

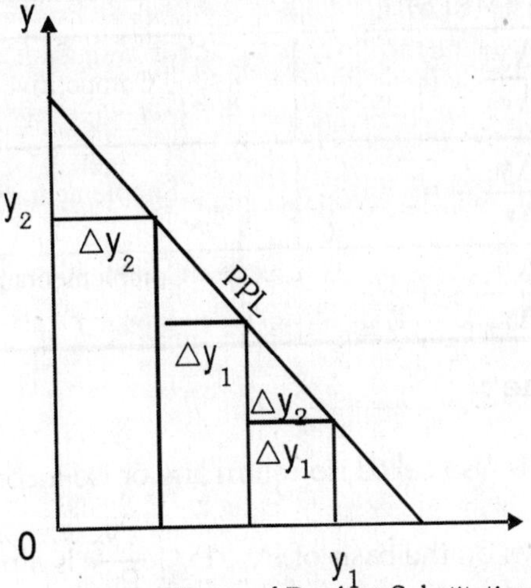

Fig. : Constant Rate of Product Substitution (CRPS)

b) Decreasing Rate of Product Substitution (DRPS) : Two Products within a limited range may substitute at decreasing rate e.g.the substitution between Dairy and crop in short duration.

C) Increasing Rate of Product Substitution (IRPS) :- The increase in the level of one product decreases the level of other product substantially e.g. the substitution between labour and capital ; Rice & Maize. Such type of substitution is common in agriculture sector.

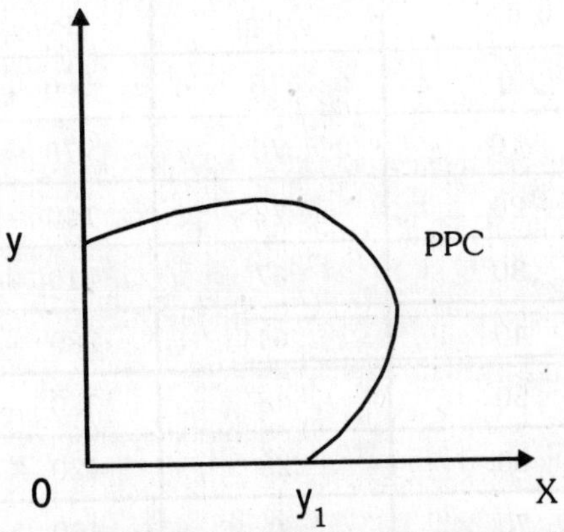

Fig. : Increasing Rate of Product Substitution (IRPS)

The above relationships are summerised below

Marginal Rate of Product Substitution (MRPS)	Enterprise Relationship
$\frac{\Delta y_1}{\Delta y_2}$ or $\frac{\Delta y_2}{\Delta y_1} < 0$	Competitive
$\frac{\Delta y_1}{\Delta y_2}$ or $\frac{\Delta y_2}{\Delta y_1} = 0$	Supplementary
$\frac{\Delta y_1}{\Delta y_2}$ or $\frac{\Delta y_2}{\Delta y_1} > 0$	Complementary

Iso - Revenue Line :

Iso- revenue line is also called iso-return line or iso-income line . This line is to be drawn on the basis of Py_2/Py_1. $\frac{Py_2}{Py_1}$ is a price line but here price indicates the price of the product/output. Therefore it is called iso-revenue line.

There are two methods to get optimum combination of the two products viz. airthmetic calculation method and graphic method.

i) Simple Airthmetic method :

Let,

Py_1 =Rs. 7/- per kg for y_1; Py_2 =Rs. 10/- per kg for y_2

S. No.	Combination of products		Total Income in Rupees
	y_1 (kg)	y_2 (kg)	$Py_1.y_1 + Py_2.y_2 =$
1	0	78	0 + 78 = 780
2	10	76	70 + 760 = 830
3	20	72	140 + 720 = 860
4	30	67	210 + 670 = 880
5	40	61	280 + 610 = 890
6	50	48	350 + 480 = 830
7	60	28	420 + 280 = 700
8	70	0	490 + 0 = 490

from the above table, it is obvious that 5th combination is the optimum to get highest return/income.

ii) Graphic method :- The highest return from optimum product combination is calculated by production possibility curve (PPC) and Iso-revenue line (Py_2/Py_1)

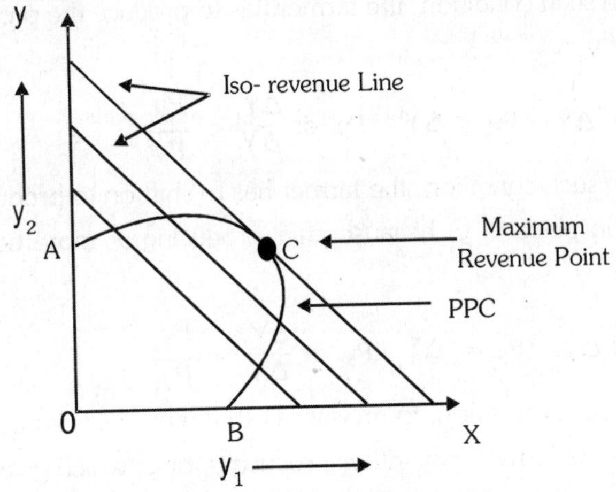

Fig. : Maximum Revenue Point

In the graph, point 'C' is the maximum revenue point where Iso-revenue line is lastly tangent to PPC, Just opposite to find out least cost point.

Here Expansion path is the path or locus passing through the different maximum revenue points of the different production possibility curves.

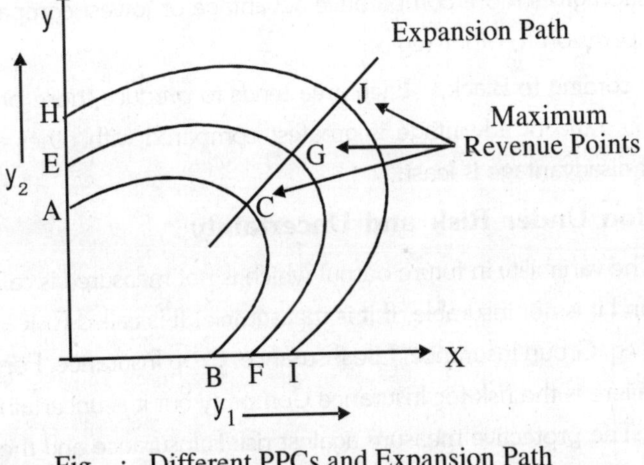

Fig. : Different PPCs and Expansion Path

When the price of the product y_1 is increased, the farmer has to produce the product y_1 more in place of y_2 . There may be three forms or conditions on the basis of marginal rate of product substitution (MRPS) :-

i) $\Delta Y_1 \, . \, Py_1 > \quad . \, Py_2$ or $\frac{\Delta Y_1}{\Delta Y_2} > \frac{Py_2}{Py_1}$

In such condition, the farmer has to produce the product y_1 more in place of y_2.

ii) $\Delta Y_1 \, . \, Py_1 < \Delta Y_2 \, . \, Py_2$ or $\frac{\Delta Y_1}{\Delta Y_2} < \frac{Py_2}{Py_1}$

In such condition, the farmer has to shift on to produce the product $y_2$more in place of y_1 because after producing y_2 more he will get more revenue.

iii) $\Delta Y_1 \, . \, Py_1 = \Delta Y_2 \, . \, Py_2$ or $\frac{\Delta Y_1}{\Delta Y_2} = \frac{Py_2}{Py_1}$

In such condition, to produce both the products is equally profitable. It means the farmer may either produce y_1 or y_2 which gives equal return.

Principle of Comparative Advantage

When the capital is marginal, pereent profit (advantage) or profit per rupee becomes more important, although all the principles of farm management are based on the concept of marginal capital. According to this principle, the farmer has to produce such things in each region whose production gives more comparative advantage or lowest comparative loss in comparasion to other regions.

According to Black, "Each area tends to produce those products for which its ratio of advantage is greatest compared with other area or its ratio of disadvantage is least."

Decision Under Risk and Uncertainty

The variability in future output which is not measured is called uncertainty and it is not insurable. If it is measurable, It is called Risk and Risk is insured eg. Group Insurance, Life Insurance, Crop Insurance. For example, Crop failure is the risk for Insurance Company but it is uncertainty for the farmer. The protective measure against risk is Insurance and the measure against uncertainty is Diversification or growing of crops with low variability in output. To insure the crop, Crop Insurance Scheme has been launched. The precursor of Crop Insurance Scheme was Benjamin Franklin. Two types of Crop Insurance are : (i) Compulsory and (ii) Optional . In 1973, to cover Cotton, Wheat, Groundnut and Potato; a kind of insurance under the auspices of General Insurance Corporation (GIC) was initiated first time in India.

National Agricultural Insurance Scheme (NAIS) :--

National Agricultural Insurance Scheme was announced in June,1999 by Sri Atal Bihari Bajpayee the prime minister of India.Some lacunae and limitations,present in the Comprehensive Crop Insurance Scheme (CCIS) initiated in 1985,have been removed in this scheme.Under CCIS some crops viz. Wheat,Rice and Oilseeds were only covered and thus Commercial Crops like Sugarcane,Potato and Cotton were bereft from CCIS.Only rainfed crops were insured and only those farmers were benefitted who availed the loan.The insurance value was maximum Rs. 10,000/- and was taken in the form of loan from the institutional sources.Whereas NAIS is incomparable and under which all the farmers including commercial and horticultural farmers would be benefited.This scheme provides the protective net to the farmers in the case of Crop-failure due to natural hazards and insectpests and diseases.This scheme is initiated from the Rabi season of the year 1999-2000 and is planned for minimum five years.

Types and Systems of Farming

Types of Farming :

According to Johnson, "When farms in a group are quite similar in the kinds and proportions of the crops and livestock that are produced and in the methods and practices followed in production;That group is described as Type of Farming".

There are many factors which affect the types of farming :--

i) Product relationship

ii) One crop season

iii) Length of Sowing Season

iv) Risk and Uncertainty

v) Long investment & size economy

vi) Business skill

vii) Relative price of farm product

viii) Resource of farmer

ix) Transport facility

x) Farm size

xi) Land Value & Technology Development

xii) Religious belief & Social background.

On the basis of methods and practices followed in the production,kinds and proportion of crops and livestocks;farming is divided into Five types viz.Diversified farming,specialised farming,mixed farming,Dry farming and Ranching farming.

[1] **Diversified Farming :--** A farm on which the income from a single product is less than or not equal to 50%(< 50%)of the total farm income,is called a diversified or general farm and such farming is known as diversified farming. Substence farming or marginal farming is a diversified farming.

One proverb is prevalent related to diversified farming--"A good farmer is one who does not put all the eggs in one basket."

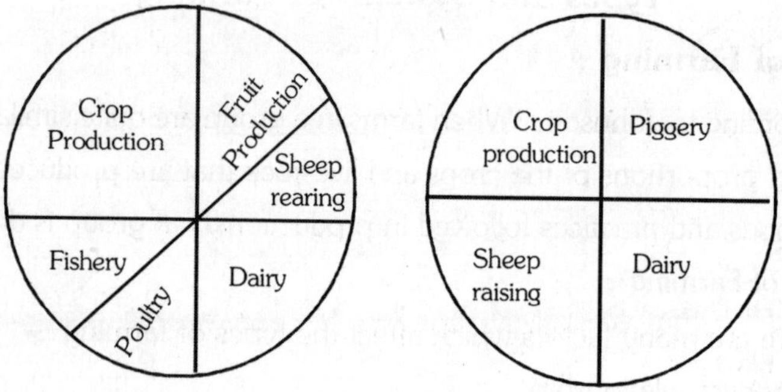

Fig. : Diversified farming

Advantages :--

i) Better use of

a) Land :--by suitable crop rotation

b) Labour :--more employment of family labour,

c) Capital :--profitable use of equipment.

ii) Business risk is reduced due to a crop failure.

iii)Regular and quicker returns are obtained from various enterprises

Disadvantages of diversified farming :--

i) Dificulties in marketing without the help of cooperatives.

ii) Farm supervision is limited due to diversified occupation.

iii) Better equiping of farm is not possible due to expensive implement and machinery for each enterprise.

iv) There are chances of undetected leaks in farm business.

Subsistence/Marginal Farming :--

In such farming farmer has surplus to sell in the market except on occasions when he has to forced sale to get some cash.

Characteristic :--

i) Farm holding are tiny with greater population on land

ii) Resource structure is helplessly poor.

iii) They produce only family consumption and nothing for marketing.

iv) Price elasticity of production is small.

v) Complimentary relationship between enterprises and their byeproduct for cattle maintenance without considering profit or loss.

[2] Specialised Farming :--A farm on which the income from a single product or one enterprise is more than or equal to 50% (≥50%) of the total farm income is called specialised farm and such farming is known as specialised farming following are the conditions for specialised farming :-

i) Where there are special market outlets.

ii) Where economic conditions are fairly uniform for a long period.

iii) Where one enterprise is much affected by abnormal weather conditions.

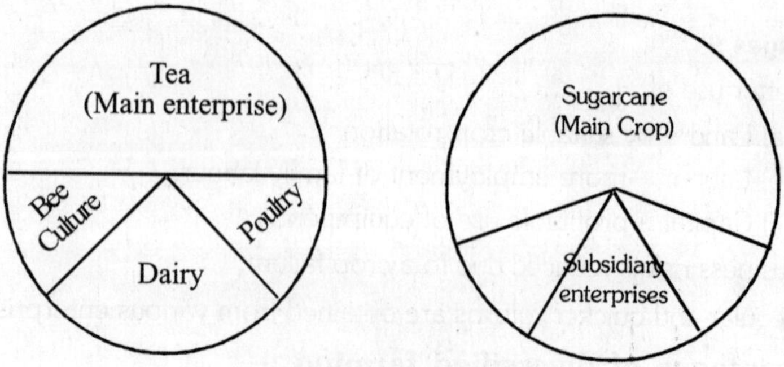

Fig. : Specialised Farming

The nomenclature of specialised farm is done on the name of main crop/ special crop or enterprise whose farm income exceeds 50% of the total farm receipt.For example where farming of tea,sugarcane and potato individually contributes more than or equal to 50% of the total farm income, is called tea garden,sugarcane farm, potato farm respectively.

Advantages :--

i) Better use of land e.g. black soil is particularly suitable for cotton growing.

ii) Better marketing, better grading, processing, transportation and financing of the produce.

iii) Better farm management and less chance of wastage.

iv) Less equipments are required.

v) Labour efficiency is increased.

vi) Farming is according to modern scientific methods.

vii) Efficiency of skill is increased.

Disadvantages :--

i) Greater risk of failure,

ii) Productive resources i.e.land,labour & capital,are not fully utilised.

iii) Soil fertility can't be maintained due to lack of suitable crop rotation..

iv) Bye-product of farm can't be utilised due to lack of livestocks.

v) Farm returns are generally received once in a year.

vi) General knowledge of farm enterprises become limited.

[3] Mixed Farming :-- In such type of farming crop production is combined with livestock raising or dairying.In our country livestocks are inter-related to agriculture due to partial farm mechanisation.The livestock enterprise is complimentary to crop production so as to provide a balanced productive systems of farming.When the livestock begins to compete with the crop for the same resource,the relationship changed from complimentary to competitive nature.The income from the subsidiary enterprise does not exceed 10% and the subsidiary enterprises utilises byeproducts.Mixed farming is a type of diversified farming where livestocks and crop production are inter-dependent on each other.

[4] Dry Farming :-- Dry farming is the profitable production of the useful crops without irrigation on the land which receives less than or equal to 50 cm (i.e. 20 inches) annnual rainfall.The main problem of such farming area is to conserve soil moisture.Natural factors force the farmer to adopt dry farming.Therefore there is no question for comparative profit or loss.The people of such region are struggling their life for livelihood.Following practices are to be adopted for successful dry farming :--

i) Timely field preparation for receiving and conserving soil moisture i.e. summer ploughing.

ii) Timely and proper inter-culture during the growth of crop.

iii) Organic manuring for increasing water holding capacity.

iv) more plant to plant spacing.

v) Adopt mixed cropping.

[5] Ranching Farming :-- Ranch land is such land where agricultural practices are not adopted i.e. land is not utilised for raising crops but the natural vegetations used for grazing the livestock.Who kept such livestock is called Rancher and the farming is called Ranching.Ranchers have their no land and make use of such public land.Ranch land is present in Australia, America, Tibet, China and hilly tracts of India (i.e. area of Bikaner and Jammu & Kashmir). Australian sheep ranching farm is well known in the world.

Summary :

Farming

- **Types**
  - Diversified farming
  - Specialised farming
  - Mixed farming
  - Dry farming
  - Ranching
- **Systems**
  - Individual or peasent farming
  - Capitalistic farming
  - State farming
  - Corporate farming
  - Joint farming
  - Collective farming
  - Cooperative farming
    - Coop.Better farming
    - Coop.Joint farming
    - Coop.Tenant farming
    - Coop.Collective farming

Systems of Farming

Accrding to Johnson,"the combination of products on a given farm and the method or practices that are used in the production of the products is known as the systems of farming."

In Indian condition,systems of farming is concerned in terms of the ways of economic and social functioning.Factors affecting the system of farming are :--

i) Farm tenancy

ii) Farm leases;ownership of land;desire to have group farming.

iii) Economic use of land.

iv) Incentive to cooperative.

On the basis of the ways of economic and social functioning,farming has the following systems :--

[1] Individual/Peasant Farming :--Farmers follow the agriculture practice in their own way and the farmer himself is manager/organiser of his farm and entire family of the farmer helps in making decision. The main objective is to fulfil the family need and not maximisation of the profit.about 70% Indian farmers have been practising this farming system. Land tax is paid to the state govt.by the farmer.

[2] Capitalistic Farming :- Capital is the important factor of production and the main objective is to maximise the profit in such farming. Capitalists use the improved technologies and methods of agriculture on their big farms. Such ststem of farming is prevalent in America and Britain. In India capitalistic farming is restricted to tea, coffee and Rubber gardens.

[3] State Farming :- In this system farms are managed by the govt and the agricultural workers are paid wages generally on monthly basis. The main objective is not always to maximise the profit. Such farming system is practised to carry out farm research work and demonstration as well as to increase the amount of quality seeds. e.g. Research farm, Seed farm, Demonstration farm etc.

[4] Corporate Farming :- This farming is completely managed by operating body of the corporate sector. Here large acreage of land and big amount of capital are required. Such farming system is to be seen in USA, and some parts of India viz. Maharashtra & Tamilnadu.

[5] Joint Farming :- Here two or more than two farmers carry out the agricultural operations Jointly by pooling their agricultural resources and lastly divide the produce according to the privious fixed ratio.

[6] Collective Farming :- In collective system of farming the ownership of the agricultural assets is invested to the society but not to the individual farmer. All the members of the collecctive farm is equally treated and the members are divided into labour-brigade. Leader of the each labour-brigade is selected. The farm is managed by the elected committee. Govt. takes a certain amountof the produce from each collective farm on the rate of priviously announced price. This system is popular in communist countries like Russia, China. There are three main forms of collective farming viz. Toz, Kholkhos and Commune.

[7] Cooperative Farming :- All the farmers or members pool their land, labour and capital on the voluntary basis and perform the farming operations together to get mutual benfit. On the basis of the ownership of land and agricultural operation method, cooperative farming has the following four different systems :-

Different systems of Cooperative farming	Ownership of land	Agricultural Operation method
a. Cooperative Better Farming	Individual	Individual
b. Cooperative Joint Farming	Individual	Collective
c. CooperativeTenant Farming	Collective	Individual
d. Copoerative Collective Farming	Collective	Collective

a) Cooperative Better Farming :- With a certain prupose or objective, farmers join together and perform all operations together after forming a society. Farm is managed by this society . At the last of the year each member gets dividend of the total profit. Here all the farmers retain their individual ownership . All the small holdings too will get the facilities of farm technology.

b) Cooperative Joint Farming :- Members pool at their land while retaining its individual ownership and perform agricultural operations jointly. Farm is managed by the elected council of the society. Each member gets divided in the ratio of his land ownership and labour shared . Any of the members give up the society at anytime. There is homogeneity in crops cultivated .

c) Cooperative Tenant Farming :- A society is registered consisting of a number of farmers. The whole land is divided into equal or unequal small plots. These smaller holdings are then leased to individual members of the society . The society provides facilities of credit, seed, manure and implements and undertakes marketing of the produce. Every member pays a fixed rent for his holding but the produce of his holding is his own and entirely at his disposal.

d) Cooperative Collective Farming :- Members pool all their land, animals, other natural resources and equipments together into a common ownership . Farm is managed by elected council and modern farm technology is used. Such farm is highly mechanised and usually a large scale farm. Each member will be paid wages and a share in the surplus produce.

Farm Planning & Budgeting

Farm Planning : -

Planning means taking decision in advance. To prepare a set of operations before performing the work is called Planning . Farm Plannig is to prepare a set of operations in advance to get maximum satisfaction from the available farm resources. Farm planning is for a certain period. Tandon and Dhondyal defined the farm planning in this way --- " Farm planning is a process for deciding in the present what to do in future about the combination of crops & livestock to be raised through rational use of resources ."

The planning of the operations and their execution is the secret of economic success. Planning is a forward-looking approach. Farm Plan is an integrated, coordinated and advance programme of actions which seek to present an opportunity to cultivators to improve his level of income. The main objective is to maximise the annual net income sustained over a long period of time and the ultimate objective of any farm planning is the improvement in the standard of living of the farmer, But the objective of farm planning is not always to maximise net income. The farmer also prepares a suitable farm plan to fulfil their family need. The farmer selects a suitable altemative among the different alternatives available on the farm.

On the basis of the time-frame farm planning is of two types :-

1) Annual Planning :- When the farm plan is made for one year only and its account is maintained annualy , such planning is called annual planning.

2) Long-range Planning :- Here plan is prepared for the long period to say 5-10 years. Long -range planning is good for farm business.

Characteristics of good farm plan : - Under Indian condition,followings are the characteristics of good farm plan :-

i) Good farm plan provides for efficient use of farm resources.
ii) balanced combination of enterprises.
iii) Avoid excessive risks.
iv) Provides flexibility.
v) Utilises farmer's knowledge, training and experiance and takes account of their likes and dislikes.
vi) efficient marketing facilities .
vii) Programme of obtaining, using & repaying the credit.
viii) uses of up -to -date modern agricultural methods and practices.

Thus Farm Planning is a process of making decisions regarding the organisation and operation of a farm business so that it results in a continuous maximization of net returns of a farm business. And farm plan is a programme of the total farm activity of a farmer drawn up in advance.

Farm Budgeting :

Farm budgeting is a process of estimating costs, returns and net profits of a farm or a particular enterprise during the plan- period.Thus farm budgeting is a method of analysing plans for the use of agricultural resources at the command of decision maker. So planning and budgeting go side by side.

Farm budgeting is of two types :-

[1] **Enterprise ≡ Partial Budgeting** :- Enterprise budgets are the input output relationship for individual enterprises. Enterprise budgets are used to estimate inputs required , costs incurred and expected returns from a particular enterprise. Enterprise budgets provide useful information regarding the resources requirements and the relative profitability of different enterprises

Partial budgeting refers to estimating the outcome or returns for a part of business . partial budgeting is a method of making a comparative study of the cost -and -return analysis resulting from a change in a part of the business organisation. This change may be made through a careful selection from among alternative methods of production or practices, the choice of which is based on the opportunity cost of relative profitability and does not affect the total farm organistation vitally. This budgeting does not consider the complimentary and supplementary relationships amongst themselves which are quite common among farm enterprises at low level of production.

Partial or enterprise budgeting is used where the change in the activity under study would not affect the farm organisation vitally. Under suh budgeting only variable costs are evaluated and only marginal costs and marginal returns are estimated here.Enterprise or partial budgeting is needed :--

1) In the adoption of new enterprise in place of old one.
2) In the increasing the size of ongoing enterprise on the farm
3) In buying or bringing any new machinery.
4) In the adoption of new agriculture technique.
5) In the use of any new input.
6) In the small changes in farming system and cropping practices.
7) In the change of fertiliser,irrigation or other agricultural patterns.

[2] Total or Complete or full Budgeting :--

The cost-and -return analysis of the whole farm as a single unit is estimated in the case of drastic changes of the farm business is called complete budgeting.Complete budgeting involves complete reorganisation of the farm business.It considers all the crops,livestock producing methods and estimate costs and returns for the farm as a whole.Here both the costs i.e.variable and fixed costs are included in the computation of full farm budgeting.

Complete farm budgeting is needed in

1) before starting farming on a new farm.
2) Comparative cost-and -return analysis of alternative farm plans.
3) In the drastic changes in the farm organisation and farm operations i.e. in the case of complete re-organisation of the farm business e.g.adoption of new crop rotation, new methods of crop production and livestock rearing.

4) Sudden change in the prices of farm inputs & or farm outputs.

5) In the major change in the availability of one or more production resources.

Algebraic formula of full farm budget :--

$R = (Q_1P_1 + Q_2P_2 + ... + Q_nP_n) - (X_1P_1 + X_2P_2 + X_mP_m) - FC$

Where R = Net Profit, Q = Quantity of produce, X = Input factor, P = Price, FC = Fixed Cost.

⇒ Net Profit of Farm business = Total receipts from all produces – total costs of input factors – fixed cost.

Table : Difference between Partial & Complete budgeting :

Partial Budgeting	Complete Budgeting
1) It considers a few alternatives do not affect the organisation vitally.	1) It takes care of all the alternatives.
2) Partial budgeting does not indicate the break -even point as to when to start one practice and abandon another.	2) Full budgeting does indicate the break-even-point.
3) It does not consider all the alternatives open to a farmer within the restraints ofhis present resources.	3) It considers all the alternatives open to a farmer
4) It is the estimation of cost-and - return analysis for a part of farm business.	4) It is the estimation of cost-and return analysis of the whole farm business as a single unit.
5) It considers variable costs only.	5 It considers variable and fixed both costs.
6) It is used in the minor changes in the the farm business.	6) It is used in the drastic change farm business.
7) It is estimated in the context of a available farm structure.	7) In it several alternative farm plans to be economically evaluated.

8) Partial budgeting is needed in the partial changes in the farm business e.g. in the adoption of new enterprise in place of old one.	8) Full budgeting is needed in the case of complete reorganisation of the farm business (e.g. adoption of new crop rotation) or before starting farming on a new farm.
9) It fails to consider all the relevant factors in maximising net returns to the farm as a whole.	9) It considers all the relevant factors.
10) It does not allow substitution between resources.	10) It allows substitution between resources.
11) It does not consider the complementary and supplementary relationship between different enterprises but it simply assumes to be competitive.	11) Competitive, complementary and supplementary relation-ships are taken care of in full budgeting.
12) It is simple,quick and easy.	12) It requires more time and efforts and more basic data in accurate form.

❑❑❑

27 Objective Model Questions

1. Agro-Climatic Regional Planning was initiated in India with the inception of :
   (a) First five year plan
   (b) Grow more food campaign.
   (c) In late eighties.
   (d) The Seventh Five year plan.
2. Rice production is highest in areas having :
   (1) High rainfall
   (2) Canal irrigated lands.
   (3) Bright sunny days during and after anthesis.
   Choose the correct Combination using the codes given below :
   (a) 1,2 and 3 (b) 1 and 2
   (c) 2 and 3 (d) 1 and 3.
3. The ecosystem determining agro-ecological zones has two components viz. Biotic and abiotic.Abiotic components comprise of :
   (a) Light,water and animals.
   (b) Minerals,plants and soil organisms.
   (c) Materials like water, minerals,salts, atmospheric gases, energy like light, heat and stored energy.
   (d) Water, wind and plants.
4. Ability of a system (say Agriculture)to maintain inter-regional equity and enhance,if required,the flow of its products and services by way of linkages with other systems without damaging its own long-term potential is called.
   (a) an efficient system. (b) a sustainable system.
   (c) a productive system. (d) none of the above.

5. Crop mixtures have proved biologically most dynamic under.

(a) water stress conditions (b) Irrigated conditions.

(c) high soil fertility conditions (d) large holding conditions.

6. In Jute growing areas the usual alternate crop is

(a) Sugarcane (b) Cotton

(c) Rice (d) Wheat

7. Which one of the following terms is used to denote the relative land area under sole crop required to produce the same yield as obtained under a mixed or an intercropping system at the same management.

(a) Land utilisation index (b) Cropping intensity

(c) Land equivalent ratio (d) Relay cropping

8. Pulses fit well in cropping systems as they are :

(a) Short duration crops.

(b) Long duration crops.

(c) Disease-resistant crops.

(d) Moisture stress- resistant crops.

9. As per principle of crop rotation pulses should be included in rotation because:

(a) They increase total pulse production.

(b) They release land quickly for the second crop.

(c) They help in maintenance of soil fertility.

(d) They increase the yield of succeeding cereal crop.

10. Which one of the following pairs is NOT correctly matched ?

(a) Relay cropping : Sowing pulse in rice crop prior to harvest.

(b) Inter cropping : Wheat after rice.

(c) Multi-tier cropping : Coconut - Pepper- Cacao -Pineapple.

(d) Sole cropping : Potato

11. Which One of the following is the principal 'N' fixing algale community of blue-green algae?

(a) Rhizobia (b) Anabaena

(c) Azotobacter (d) Azospirillium

12. Groundnut pegs when developed in the soil form.

(a) roots (b) stems

(c) tubers (d) fruits

13. Fire curing is followed in :

(a) Chewing type of tobacco (b) Bidi tobacco

(c) Hookah tobacco (d) Cheroot tobacco

14. Consider the following physiological stages of wheat :

(1) Crown root initiation (2) seedling

(3) Tillering (4) Ear emergence.

The correct sequence of the above physiological stages is :

(a) 2,3,1 and 4 (b) 2, 3, 4, and 1

(c) 2,1,3 and 4 (d) 1, 2, 3and 4

15. Low quality of Jute fibre is attributed to

1. Discolouration of fibre.
2. Very short fibre strands.
3. Reduction of cellulose in fibre.
4. Reduction of cell wall strength.

Over-retting decreases the fibre quality due to

(a) 3 & 4, (b) 1 & 3, (c) 1 & 4, (d) 1 & 2

16. Which of the following two crops are resposible for almost 75% of pulse production in India?

(a) Gram and moongbean

(b) Pigeonpea and moongbean

(c) Moongbean and lentil

(d) Gram and pigeonpea

17. Triticale is a cross between :

(a) Wheat and barley (b) Barley and rye

(c) Wheat and rye (d) Wheat and oat

18. An international crop research institute called CIMMYT in Mexico deals with two crops grown widely in India,these crops are :

(a) Rice and Jute (b) Rice and Wheat

(c) Maize and wheat (d) Maize and millet

19. Which one of the following occupies the highest volume percent in a fine sandy loam surface soil in good condition for plant growth?

(a) Organic matter

(b) Pore space which includes air and / or water

(c) Mineral matter

(d) Others including macro and micro flora and fauna

20. Pedology considers soil as a :

(a) Natural body

(b) Purely a synthesised one

(c) Medium for plant growth

(d) Powdered rock mass of the earth's crust.

21. Chlorites are basically silicates of :

(a) Magnesium with some iron and aluminium.

(b) Iron with some magnesium & aluminium

(c) Aluminium with some iron and magnesium.

(d) Iron and aluminium.

22. The cation exchange capacity of a mineral soil containing predomi nately kaolinitic clay minerals generally ranges between :

(a) 3-15 m.e. per 100 gm soil

(b) 15-40 m.e. per 100 gm. soil.

(c) 80-100 m.e. per 100 gm. soil

(d) More than 100 m.e. per 100 gm. soil

23. In soil science micelle stands for :

(a) Colloidal particle carrying positive charge.

(b) Colloidal particle carrying negative charge.

(c) Micropores in between textural particles.

(d) Negatively charged silt particles with high CEC.

24. Soils high in swelling clays which crack widely upon drying resulting in shrinking,shearing and soil mass movement are characteristics of :

(a) Ultisols (b) Entisols

(c) Podzols (d) Vertisols

25. Total porosity of a soil with 1.325 g/ cc B.D. and 2.65 g/ cc P.D. is (B.D. :Bulk density,P.D. :Particle density).

(a) 30% (b) 40%

(c) 50% (d) 60%

26. Which one of the following micro-organisms is responsible for maximum nutrient cycling in the soil?

(a) Bacteria (b) Actinomycetes

(c) Fungi (d) Algae

27. Which one of the following statements is correct ?

(a) Rhizobia fix nitrogen in soil non-symbiotically.

(b) Azotobactors fix nitrogen symbiotically in association with legumes.

(c) Blue green algae fix nitrogen in paddy fields under submerged conditions.

(d) Clostridia fix nitrogen symbiotically in association with legumes.

28. When organic matter with wide C : N ratio (70:1) is added to the soil, the net effect is the disappeearance of mineral nitrogen due to

(a)Mineralisation (b) Immobilisation

(c) Nitrogen transformations (d) Nitrogen reduction.

29. The addition of excess soluble phosphatic fertilizers to soil may affect plant growth by :

(a) Inhibiting the growth of microorganisms.

(b) Precipitating iron and aluminium.

(c) Reducing the availability of zinc.

(d) Reducing the availability of calcium.

30. As plants grow to maturity,the concentration of plant nutrients changes.What kind of change occurs in respect of N, P,K,Ca and Mg ?

(a) Concentration of all the five elements decreases.

(b) N and P decrease, K, Ca and Mg increase.

(c) N, P and K decrease, Ca & Mg increase.

(d) N, P and K increase,Ca & Mg decrease

31. In order to assess the immediate fertility level of a given soil for potas sium, which one of the collowing forms of potassium present in soil would one test for?

(a) Total K (b) Exchangeable K

(c) Water soluble K (d) K held in the silt

32. What culture should be given priority on groundnut cultivation:

(a) Azospirilla (b) Mycorrhiza

(c) Phosphobacteria (d) Rhizobia:

33. Which one of the following concentrated organic manures has maxi mum phosphorus content?

(a) Fish meal (b) Stera meal

(c) Poultry manure (d) Guano

34. Match List I with List II and select correct answer using the codes given below the list :

List I	List II
(Classes of fertilizers)	(Example of different classes)
A. Organic fertilizer	1. Urea
B. Inorganic fertilizer	2. Isobutylidene diurea.
C. Biofertilizer	3. Ammonium sulphate
D. Slow release fertilizer.	4. azolla.
	5. 2- chloro-6-pyridene.

Codes

(a)	A	B	C	D
	1	2	3	5
(b)	A	B	C	D
	1	3	4	2
(c)	A	B	C	D
	2	1	5	4
(d)	A	B	C	D
	5	1	4	3

35. What is the effect of application of phosphatic fertilizer to a green manure leguminous crop on the availability of phosphorus to the suc ceeding crop ?

(a) Greater amount of phosphorus is taken up and converted into organic form, which is released on decomposition of green matter.

(b) Green manure increases the availability in inorganic form to the next crop.

(c) It does not affect the availability of phosphorus to succeeding crop.

(d) It fixes the phosphorus through humus and reduces the avail-ability.

36. The nitrogenous fertilizer manufactured in India which occupies the first place in production today is :

(a) Ammonium sulphate (b) Ammonium Phosphate

(c) Urea (d) Calcium ammonium nitrate.

37. Which one of the following fertilizer is least hygroscopic :

(a) Urea (b) Ammonium sulphate

(c) Diammonium phosphate (d) Calcium ammonium nitrate.

38. Diammonium phosphate requirement for supplying 45kg N and 115 kg P_2O_5 to one hectare peanut crop is-

(a) 100kg (b) 150kg

(c) 200kg (d) 250kg

39. Zinc, Iron , Copper, Boron, Manganese and molybdenum are called micro nutrients because they are :

(a) Nutrients of microscopic significance.

(b) Nutrients which produce microscopic effects.

(c) Essential nutrients but are required in minute quantities .

(d) Nutrients required by micro-organisms.

40. The strongest chemical agent for breaking dormancy is-

(a) Potassium nitrate (b) Gibberellin

(c) Ethylene (d) Cumerin

41. The optimum cardinal temperature points for germination of rice seeds are :

(a) 18-20°C (b) 20-22°C

(c) 30-32°C (d) 37-39°C

42. Interveinal chlorosis of the younger leaves of the plants is the charac teristic symptom of

(a) Boron deficiency (b) Zinc deficiency

(c) Iron deficiency (d) Phosphorus deficiency

43. The rice variety containing "Dee - geo - woo - gen" is-

(a) Indrasan (b)Basmati

(c) Tilak (d) IR- 8

44. Which one of the following soils is best for vegetable cultivation ?

(a) Sandy (b) Sandy loam

(c) clay loam (d) clay

45. The ill efects of submergence of roots of deciduous fruit plants in water,for very long time,is due primarily to :

(a) Lack of nutrients (b) Lack of aeration

(c) Excess of moisture (d) Poor absorption of moisture

46. Which one of the following groups of crops is most suitable and recom mended for intercropping in the young orchards ?

(a) Short duration fodders

(b) Short duration fruit crops

(c) Poplars and eucalyptus

(d) Short duration legume vegetables.

47. The main objective of growing intercrops with the main fruit crop is to

(a) Help the fruit crop to grow better

(b) Improve the soil fertility

(c) Check soil erosion

(d) Get additional income

48. In relation to crop rotation which one of the following statements is incorrect ?

(a) Legume may be followed by non-legume

(b) Vegetable susceptible to a particular pest should be followed by a resistant one.

(c) Heavy feeders should be followed by low feeders.

(d) Deep rooted crop should be followed by same type of crop.

49. The two important principles of extension education are :

(a) Participation and leadership

(b) Participation and class room teaching.

(c) Leadership and classroom teaching.

(d) Leadership and subject-matter knowledge.

50. Which one of the following categories the discipline of extension edu cation belong ?

(a) Pure science
(b) Applied biological science
(c) Applied behavioural science
(d) Art.

51. Which one of the following extension programmes linked farm plan with credit for farmers ?

(a) Integrated Rural Development programme.

(b) Community Development preogramme

(c) Intensive Agricultural Development programme.

(d) Etawah pilot project.

52. Puppet show in villages as a method for transfer of technology is an example of :

(a) Mass media
(b) Group media
(c) Individual media
(d) Tribal media

53. The volume (in litres) of one cumec flow of water in an hour is :

(a) 2.6×10^6
(b) 3.6×10^6
(c) 4.6×10^6
(d) 5.6×10^6

54. A farmer is to select suitable crops for a drought prone unirrigated area. Which one of the following groups of crops should be chosen for mixed cropping ?

(a) Sesame, maize, oats
(b) Sorghum Wheat
(c) Sesame,Sorghum/Pearl millet,niger
(d) Maize,upland rice,berseem.

55. Which one of the following can be suitably grown as a mixed crop with wheat ?

(a) Cabbage
(b) Cotton
(c) Jowar
(d) Mustard

56. Consider the following statements,Relay cropping implies :

1. Sowing of second crop when the first crop has attained maturity and is nearing harvest.
2. A situation where two or more crops occupy a habitat with spatial annidation.
3. Overlapping of a part of the life span,of the two crops without any competition of the above statements.

(a) 1,2 & 3 are correct
(b) 1 & 2 are correct
(c) 2 & 3 are correct
(d) 1 & 3 are correct.

57. *Imperata cylindrica* is associated with :

(a) Sugar beet
(b) tea (weed)
(c) lentil
(d) cereals

58. Match list 1 (crops) with list II (weedicides) and select the correct answer using the codes given below the lists :

List I	List II
A. Wheat	1. Simazine
B. Sugareane	2. Isoproturon
C. Chickpea	3. Thiobencarb
D. Rice -Paddy	4. Fluchloralin.

Codes :

(a) A - 1, B - 2, C - 4, D - 3.

(b) A - 2, B - 1, C - 4, D - 3.

(c) A - 1, B - 2, C - 3, D - 4.

(d) A - 2, B - 1, C - 3, D - 4.

59. According to modem concepts, the major function of inter-row tillage is :

(a) Moisture conservation through soil mulching.

(b) Improved soil aeration .

(c) Improved soil granulation.

(d) Moisture conservation through weed destruction.

60. 'Ratooning ' is practised as a matter of routine in :

(a) Sugarbeet and sugarcane

(b) Sugarcane and maize.

(c) Sugarcane & Napier grass

(d) Sugarcane only.

61. DAPOG method of crop raising is associated with :

(a) Inter culture operation

(b) Nursery raising

(c) Seed-bed preparation

(d) Intercultivation

62. Where adequate irrigation water is available four to six irrigations are applied to the wheat crop. At which of the following stages of growth is the first irrigation recommended ?

(a) Crown-root initiation stage

(b) Tillering stage

(c) Critical stage of growth

(d) Flowering stage.

63. In high P-fixing soils, water soluble phosphatic fertilizers are applied to the crops essentially as :

(a) Broad cast

(b) Placement below the seed

(c) Placement above the seed

(d) Top dressing.

64. Spreading varieties of groundnut are preferred in areas receiving late rains in the Kharif season, because :

(a) They are of longer duration

(b) Their water requirements are high

(c) Their seed is dormant

(d) Their seed is bold.

65. Which one of the following parts of the tobacco plant synthesises nicotine ?

(a) Root (b) Stem

(c) Branch (d) Leaf

66. In wet season, the initiation of panicle primordia in rice variety of 120 days duration takes place in :

(a) 60 to 65 days after sowing

(b) 65 to 70 days after sowing

(c) 70 to 75 days after sowing

(d) 75 to 80 days after sowing.

67. Match list I (fertilizer) with list II (equivalent acidity) and select the correct answer using the codes given below the listes :

List I	List II	Codes
A. Urea	1. 148	(a) A-5, B-2, C-3, D-4
B. Ammonium Sulphate	2. 128	(b) A-4, B-3, C-2, D-1
C. Ammouium Chloride	3. 110	(c) A-2, B-5, C-4, D-3
D. Anhydrous Ammonia	4. 84	(d) A-4, B-1, C-2, D-3
	5. 64	

68. Consider the following states/ U.T. :

1. Pondicherry 2. Punjab

3. Andhra Pradesh 4. Haryana

As per the 1991 -92 data, the correct sequence of the DECREASING order in respect of the consumption of N + P_2O_5 + K_2O fertilizer nutri ents (in Kg / ha/year) is :

(a) 1, 2, 4, 3 (b) 2, 1, 3, 4

(c) 1, 2, 3, 4 (d) 2, 1, 4, 3.

69. Which one of the following oil-cakes, in addition to its manurial value, acts also as a nitrification inhibitor ?

(a) Mustard (b) Groundnut

(c) Neem (d) Sesamum (til)

70. Which one of the following organic manures has the narrowest C/ N ratio ?

(a) Biogas slurry (b) Farm-Yard manure

(c) Compost (d) Groundnut cak

71. Biofertilizer is a

(a) Mixture of organic manure and microorganisms.

(b) Mixture of inorganic fertilizer & microorganisms.

(c) Culture having the desired strain of micro organisms.

(d) Decomposed compost enriched with mineral solubilisig bacteria.

72. Super-digested compost is called so because :

(a) It is prepared by allowing it to decompose beyond the normal pro cess of decomposition.

(b) It contains plant nutrients more than that present in ordinary com post.

(c) Superphosphate is added to the compost during its preparation.

(d) Its P_2O_5 content is more than that present in the ordinary com post.

73. Match list I with List II and select the correct answer using the codes given below the lists :

List I	**List II**
(Name of fertilizer)	(Type of fertilizer)
A. Ammonium Sulphate	1. Biofertilizer
B. Diammonium Phosphate	2. Straight fertilizer

C. Algae 3. Organic manure

D. F. Y. M. 4. Complex fertilizer

Codes :

(a) A-2,	B - 4,	C-3,	D–1
(b) A-3,	B - 2,	C- 4,	D-1
(c) A-2,	B - 4,	C- 1,	D- 3
(d) A-4,	B - 2,	C- 3,	D- 1.

74. Bray suggested certain modification in the Mitscherlich equation. The modified equation is written as :

(a) (dy/ dx) = (A-Y) C

(b) A = B (I -Y) / Y.

(c) Log (A - y) = log A - Cx.

(d) log (A-y) = log A - Clb -Cx.

75. The mineral chlorite has an extra layer of :

(a) Montmorillonite (b) Illite

(c) gibbsite (d) Brucite

76. Which one of the following minerals is the main source of boron in soils?

(a) Pyrolusite (b) Tourmaline

(c) Haematite (d) Hemimorphite .

77. Criteria of essentiality of nutrients for plants were given by

(a) Arnon (b) Stout

(c) Russel (d) Liebig

78. Apatite is a group of minerals which mainly contains

(a) Sulphur (b) Phosphorus

(c) Nitrogen (d) Titanium

79. Match List i (micro nutrients) with List II (typical deficiency symptoms) and select the correct answer using the codes given below the lists.

List I	**List II**
A. Manganese	1. Speckled yellow of sugar beet
B. Molybdenum	2. Write bud of maize
C. Zinc	3. Whiptail in cauliflower
D. Copper	4. Reclamation desease of cereals

Codes :

(a) A - 1, B - 2, C -3, D - 4

(b) A - 1, B -3, C - 2,D - 4

(c) A - 2, B -4, C - 3, D - 1

(d) A -2, B -3, C- 4, D- 1

80. Chlorite belongs to the :

(a) 2 : 2 type (b) 1 : 1 type

(c) 2 : 1 type (d) amorphous type

81. The soil layer of maximum leaching is indicated by the symbol :

(a) A_1 (b) A_2

(c) B_1 (d) B_2

82. Which of the following pairs are correctly matched?

1. Humus	Ligno-protein complex
2. Humin	Insoluble in dilute alkali
3. Fulvic acid	High molecular weight.
4. Humic acid	Insoluble in dilute acid.

Select the correct answer using the codes given below :

Codes :

(a) 2, 3 & 4

(b) 1, 2 & 3

(c) 1, 2 & 4

(d) 1, 3 & 4

83. Consider the following statements :

Soil is a medium for plant growth because :

1. It provides mechanical support and nutrients to the plants .

2. It contains beneficial micro organisms .

3. Plants cannot be grown without soil of these statements :

(a) 1 alone is correct (b) 2 & 3 are correct

(c) 1 & 3 are correct (d) 1, 2 & 3 are correct

84. The horizons which make up the profile of a forest soil would include.

(a) A, B & C (b) A, C & O

(c) B & O (d) A, B, C & O

85. Essential elements are those :

(a) Which could be replaced by other elements with similar properties.

(b) Which are required in large quantities .

(c) Whose deficiency would prevent the completition of plant's life cycle.

(d) Which would promote plant growth.

86. Molybdenum is required by plants because it is a cofactor for

(a) Cytochrome Oxidase (b) Nitrate reductase

(c) Nitrite reductase (d) Phosphatase

87. Match List I (crops) with List II (Varieties) and select the correct answer using the codes given below the lists :

List I	**List II**	**Codes**
A. Cotton	1. Ratna	(a) A- 4, B-2, C-1, D-
B. Potato	2. Sujata	(b) A-2, B-4, C-1, D-3
C. Rice	3. Girija	(c) A-4, B-2, C-3, D-1
D. Wheat.	4. Kufri sinduri	(d) A-2, B-4, C-3, D-1

88. Khaira disease of rice can be controlled by spraying

(a) Copper sulphate (b) Manganese sulphate

(c) Borax (d) Zinc sulphate.

89. Consider the following categories of adopters of an improved practice

1. Early adopters 2. Early majority

3. Late majority 4. Laggards

5. Innovators

90. The basic unit of development under the Integrated Rural Develop ment Programme is a:

(a) District

(b) Community development block

(c) Village

(d) Family

91. The following of the four important reports on extension development

1. The Royal commissions Report.
2. Agricultural Administration Team Report
3. Grow More Food Enquiry Committee Report
4. Nalagarh Committee Report

The correct chronological order of these reports is :

(a) 1, 3, 2, 4 (b) 3,1,2,4

(c) 3,1,4,2 (d) 1,3,4,2

92. The sources of information are classified into the following three groups on the basis of their origin within and outside the social system

1. Personal localite 2. Personal cosmopolite

3. Cosmopolite.

The correct sequence of these groups in terms of their use is :

(a) 1, 2, 3 (b) 1, 3, 2

(c) 2, 3, 1 (d) 2, 1, 3

93. A high percentage of cropped area in India is covered by rice be cause :

(a) There is adequate rainfall

(b) Rice can be grown in different seasons.

(c) Rice is adaptable to most types of land and agroclimate.

(d) A large part of the population of India is rice-eating.

94. Which one of the following approaches has been adopted by the plan ning commission in delineating India into agro-climatic zones?

(a) Project approach

(b) Sectoral approach

(c) Crop production approach.

(d) Holistic approach.

95. Consider the following statements :

Agro-climatic zoning of cultivable land is based on :

1. Water deficit condition

2. Water surplus condition

3. Geographical location of the region of the above statements.

(a) 1, 2, & 3 are correct (b) 1 & 3 are correct

(c) 2 & 3 are correct (d) 1 & 2 are correct.

96. Organic matter content of 'histosols'and 'hydromorphic'soil is

(a) 30-40% (b) 50-57%

(c) 60-70% (d) 2-18%

97. Consider the following statements regarding the causes of low crop productivity of sandy soils :

1. Sandy soils have very low water retention capacity.
2. Sandy soils easily fix applied phosphatic fertilizer.
3. Applied plant nutrients easily leach out through drainage.
4. Sandy soils have very low organic matter content.

Of these statements :

(a) 1, 2 & 3 are correct (b) 1, 2 & 4 are correct

(c) 1, 3 & 4 are correct (d) 2, 3 & 4 are correct.

98. The cation exchange capacity (in units of milli equivalents per 100 gram of soil)of Kaolinite clay varies between :

(a) 3 & 15 (b) 16 & 40

(c) 41 & 80 (d) 81 & 100

99. The phosphorus content as P in plough layer of one hectare of soil containing 0.1% P_2O_5 is.

(a) 430 kg (b) 860 kg

(c) 1290 kg (d) 1500 kg.

100. Match LIst I (Principal soils of India) with List II(Toxonomic soil order) and select the correct answer using the codes given below the lists:

List I	List II
A. Black soils of Maharashtra	1. Alfisol.
B. Laterite soils of Kerala	2. Entisol
C. Red soils of chhotanagpur	3. Ultisol
D. Gangetic alluvial soils	4. Vertisol.

Codes :

(a) A - 3, B - 4, C - 1, D - 2

(b) A - 3, B -4, C -2 , D - 1

(c) A -4, B - 3, C - 2, D - 1

(d) A- 4, B - 3, C -1, D -2.

101. Match List I (soil bacteria) with List II (microbiological reactions) and select the correct answer using the codes given below the list.

List I	List II
A. Nitrobacter	1. Symbiotic of atmospheric nitrogen.
B. Azotobacter	2. Non-symbiotic fixation of atmospheric nitrogen.

C. Nitrosomonas — 3. Conversion of ammonia into nitrate in soil

D. Rhizobium — 4. Oxidation of nitrite to nitrate.

Code :

(a) A- 4, B - 2, C - 3, D - 1

(b) A - 2, B - 4, C - 1, D- 3

(c) A - 4, B - 2, C - 1, D - 3.

(d) A - 2, B - 4 C - 3, D - 1

102. 'Whiptail' symptom in cauliflower is due to the deficiency of :

(a) Copper (b) Moly bdenum

(c) Iron (d) Zinc.

103. The number of essential elements required for the growth of most higher plants is :

(a) 3 (b) 6

(c) 8 (d) 16

104. Match List I with List II and select the correct answer using the codes given below the lists :

List I (Nutrient element)	**List II** (Function in the plant)
A. Nitrogen	1. Reduction of nitrate.
B. Phosphorus	2. Root growth.
C. Potassium	3. Acceleration of vegetative growth.
D. Molybdenum	4. Translocation of Photo synthates .

Codes :

(a) A - 3, B - 2, C - 4, D -1

(b) A - 4, B - 2, C - 1, D - 3

(c) A - 4, B - 3, C - 2, D -1

(d) A - 3, B -1, C - 4, D -2.

105. 'Crop logging' is a method of

(a) Soil fertility evaluation.

(b) Plant analysis for assessing requirement of nutrients for crop production.

(c) Assessing crop damage.

(d) Testing suitability of fertilizers.

106. In a Compost pit, the heat of combustion of organic matter raises the temperature when decay is occuring rapidly . During decay, the temperature rises by :

(a) 75 - 90^{0}C (b) 60 - 80^{0}C

(c) 5- - 72^0 C (d) 40 - 60^{0}C

107. Which of the following pairs of rhizobium spp. and the host genera on which the rhizobium spp. subsists, are correctly matched?

1. *R. leguminosorum*..Pisum
2. *R. Japonicum*..Glycine
3. *R. meliloti*..Medicago.

Select the correct answer using the codes given below :

Codes

(a) 1, 2 & 3 (b) 1 & 2

(c) 1 & 3 (d) 2 & 3

108. Poor nitrification rate of non-edible mahua (*Madhuca latifolia*) cake is due to

(a) the presence of oil

(b) Wide C : N ratio

(c) Low nitrogen content

(d) The presence of certain alkaloids

109. Which of the following pairs is correctly matched ?

Fertilizers	Nutrient content
(a) Urea	42.0 (% N)
(b) Diammouium phosphate	20.0% (% P)
(c) Ammonium sulphate	21.5(%N)
(d) Muriate of Potash	47.5(%K)

110. The neutralising values of the following liming materials are such that

(a) $CaO > Ca(OH)_2 > CaCO_3 > CaMg(CO_3)_2$

(b) $CaO > Ca(OH)_2 > CaMg(CO_3)_2 > CaCO_3$

(c) $Ca(OH)_2 > CaO > CaMg(CO_3)_2 > CaCO_3$

(d) $CaCO_3 > CaO > Ca(OH)_2 > CaMg(CO_3)_2$

111. The nitrogenous fertilizer of choice for tea is :

(a) Calcium ammonium nitrate

(b) Urea

(c) Ammonium sulphate

(d) Ammonium nitrate.

112. Transport of phosphate from the cytosol (p^H about 7.0) into vacuoles (p^H nearly 5.5) involves.

(a) $(H_2PO_4)^-$ alone (b) $(HPO_4)^{2-}$ alone

(c) $(PO_4)^{3-}$ alone (d) $(H_2PO_4)^-$ and $(HPO_4)^{2-}$

113. Diminished carbohydrate status of the plant placed under high nitrogen supply is due to :

(a) Lower photosynthetic activity

(b) Vigorous protein synthesis

(c) Increased transpiration ratio

(d) Increased vegetative growth

114. Consider the following symptoms :

1. Younger leaves show signs of deficiency first.
2. High amide and nitrate nitrogen.
3. Effect more pronounced on shoots when compared to roots and cruciferous plants develop narrow leaf lamina.
4. Blockage of the synthesis of cystine, methionine etc. Development of the above mentioned symptoms is attributable to the deficiency of :

(a) Calcium (b) Magnesium

(c) Nitrogen (d) Sulphure

115. Match List I (Symptoms) with List II (causes) and select the correct answer using the codes given below the lists :

List I	List II
A. Khaira disease of rice	1. Phosphorus deficiency
B. Browning of cauliflower	2. Molydenum deficiency
C. Purple colouration of Maize leaves.	3. Potassium deficiency
D. Firing of tobacco leaves.	4. Zinc deficiency

Codes :

(a) A - 2, B - 4, C - 3, D - 1

(b) A - 2, B - 4, C - 1, D - 3

(c) A -4, B - 2, C - 3, D - 1

(d) A - 4, B - 2, C - 1, D - 3

116. Consider the following international crop improvement institutes:

1. International centre for Maize and wheat Improvement, Mexico.
2. International crops Research Institute for the semi-Arid Tropics, India.
3. International Rice Research Institute, Philippines.
4. International Centre for Potato, Peru.

The correct chronological order in which these institutes were established is :

(a) 1, 3, 2, 4
(b) 3, 1, 2, 4
(c) 3, 1, 4, 2
(d) 1, 3, 4, 2

117. Match List I (Parasite) with List II (Nature of the parasite) and select the correct answer using the codes given below the lists :

List I	List II
A. Striga	1. Semi- parasite on stem.
B. Cuscuta	2. Semi- parasite on root
C. Loranthus	3. Total parasite on root.
D. Orobanche	4. Total parasite on stem

Codes :

(a) A- 1, B - 2, C -3, D - 4

(b) A - 2, B -4, C - 3, D - 1

(c) A - 2, B - 4, C - 1, D - 3

(d) A - 1, B - 2, C - 4, D - 3

118. While working as an extension specialist in a village, one finds that the package of practices are Not being followed despite the best communication methods one had employed for the purpose one finds the following reasons for non- adaptivity :

1. 60% are laggards .
2. 20% innovative but orthodox.
3. 20% are old and orthodox.

The best remedial measure would be to :

(a) adopt 1 & 3

(b) adopt 2 alone

(c) avoid 3

(d) adopt 2 and make them team leaders of

119. Consider the following institutions :

1. Zila parishad
2. Gram panchayat.
3. Village School
4. Village Co-operative

The basic rural institutions would include :

(a) 1, 2 & 3
(b) 2, & 4
(c) 2, 3 & 4
(d) 1, 3 & 4

120. Which of the following statements are correct ?

1. Rainfed agriculture contributes about 44% to the total food production in India.
2. In india, the area under wheat lies some where between 23 and 24 million hectares .
3. India's Productivity in rice is about 1.75 tonnes/ha .
4. According to the National Wastelands Development Board, an area of about 75 million hectares is under Wastelands .

Select the correct answer using the codes given below :

Codes :

(a) 2, 3 & 4
(b) 1, 3 & 4
(c) 1, 2 & 4
(d) 1, 2 & 3

121. Which one of the following types of crop cultivations will be most appropriate and effective in forests ?

(a) Extensive cropping
(b) Intensive cropping
(c) Mixed cropping
(d) Inter-cropping

122. In wheat, seminal roots :

(a) Induce extra strength against lodging.

(b) are responsible for nourishing the plants during early stages of the crop.

(c) Take care of the plants during the later stages of the crop.

(d) appear during the vegetative phase of the crop.

123. The latest botanical name of green gram is :

(a) *Vigna radiata* (L) wilezek

(b) *Vigna mungo* (L) Kepper

(c) *Phaseolus mungo* (L)

(d) *Phaseolus vulgaris* (L)

124. Consider the following statements :

Puddling of soil for transplanting paddy is done to :

1. Control weeds

2. Reduce rate of infiltration.

3. Improve soil structure.

4. Increase aeration. of these statements :

(a) 1, 2, 3 & 4 are correct.

(b) 2 & 4 are correct .

(c) 1 & 2 are correct .

(d) 1, 3 & 4 are correct.

125. Match list I (crop) with List II (optimum time of sowing in North India) and select the correct answer using the codes given below the lists :

List I	**List II**
A. Cotton	1. First fortnight of November
B. Barley	2. Second fortnight of October
C. Gram	3. Whole of the month of May
D. Groundnut	4. First fortnight of July.

Codes :

(a) A - 3, B - 1, C - 4, D - 2

(b) A - 3, B - 1, C - 2, D - 4

(c) A - 1, B - 3, C - 2, D - 4

(d) A - 1, B - 3, C - 4, D - 2.

126. Match List I (crop) with List II (weed specific to the crop) and select the correct answer using the codes given below the lists :

List I	**List II**
A. Rice (low land)	1. Phalaris minor
B. Maize	2. Echinochloa crusgalli
C. Wheat	3. Lantana camara
D. Uncultivated land	4. Euphorbia hirta.

Codes :

(a) A - 2, B - 3, C - 1 D - 4

(b) A - 1, B - 2, C - 4, D - 3

(c) A - 1, B - 3, C - 2, D - 4

(d) A - 2, B - 4, C - 1, D -3

127. Consider the following crops :

1. Wheat 2. Potato

3. Maize 4. Moong

The correct sequence of sowing these crops in a cropping sequence planned under irrigated conditions for north-western plains of India would be :

(a) 4, 3, 1, 2 (b) 4, 3, 2, 1

(c) 3, 4, 1, 2 (d) 3, 4, 2, 1

128. Which one of the following is a pair of pre-emergence herbicides ?

(a) Paraquat and butachlor

(b) 2, 4-d and thiobencarb

(c) Dalapon and triazine

(d) Fluchloralin and pendimethalin .

129. Irrigation efficiency is very low in irrigated farming in India mainly because :

(a) Large irrigated area is under canal irrigation system.

(b) A major part of the soil types is not suitable for irrigation.

(c) Of the poor quality of irrigation water.

(d) Of the adverse climatic conditions.

130. Which one of the following irrigation methods is best suited for an undulating topography?

(a) Flow irrigation (b) Check basin method

(c) Sprinkler irrigation (d) Furrow irrigation

131. Which one of the following is Not a component of the moisture potential in soils.

(a) Matric potential (b) Turgor potential

(c) Gravity potential (d) Osmotic potential

132. The crops suitable for cultivation in saline sodic soil would include.

(a) Soyabeen and Rice (b) Soybean & Sugarbeet

(c) Sugarbeet & Rice (d) Rice & Maize

133. The Rice crop tolerates submergence because :

(a) The plant does not need oxygen.

(b) The plant has mechanism to transport oxygen through leaves.

(c) Bacteria keep the rhizosphere Oxidised .

(d) Oxygen penetrating to root zone through air movement is sufficient.

134. The unconsolidated material on the underlying rock is called.

(a) Regolith (b) Soil

(c) Solum (d) Earth

135. If a soil has a bulk density of 1.50 mg/m^3 and a particle density of 2.65 mg/m^3, then the percentage pore space of that soil will be :

(a) 35.0 (b) 38.5

(c) 43.4 (d) 48.5

136. Active soil forming factors would include :

(a) Vegetation and topography

(b) Time, topography and climate

(c) Time and vegetation.

(d) Vegetation and climate.

137. The black cotton soils of central India have been derived from :

(a) Granites (b) Granodiorites

(c) Basalt (d) rhyolite

138. Which one of the following minerals is least weatherable in soil?

(a) Olivine (b) Quartz

(c) Calcite (d) Biotite

139. Match List I (Microbial reactions) with List II (Soil organisms) and select the correct answer using the codes given below the lists :

List I	List II
A. Sulphate reduction	1. Thiobacillus.
B. Sulphate oxidation	2. Desulfotomaculum
C. Nitrite Oxidation	3. Rhodospirillum
D. Nitrogen Fixation	4. Nitro bacter.

Codes :

(a) A - 1, B - 2, C - 3, D - 4

(b) A - 2, B -1, C - 4, D - 3

(c) A - 2, B - 1,C - 3, D - 4

(d) A - 1, B - 2, C - 4, D - 3.

140. The following figure shows the relationship between nutrient concentration in the plants and yield of crops :

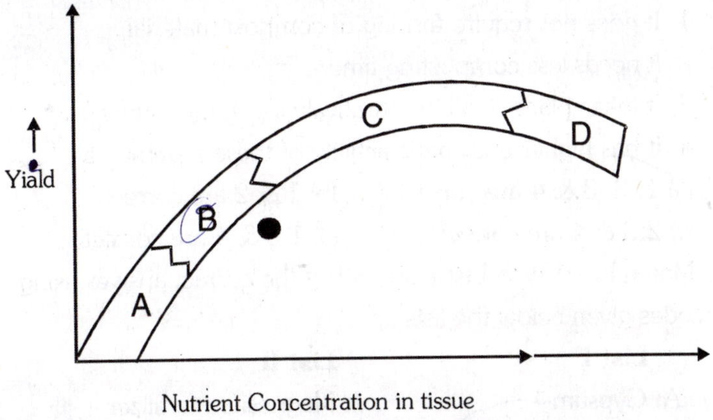

Which portion (s) of the curve fall (s) in hidden hunger zone?

(a) A & B (b) B & C

(c) B alone (d) A alone

141. The best estimate of the available p in phosphatic fertilizers is obtained from the :

(a) Water soluble fraction

(b) Citrate in soluble fraction

(c) Citrate soluble fraction

(d) Water soluble and citrate soluble fractions

142. Given that :

= Increase of yield per unit of growth factor.

A = Maximum yield.

Y = Yield without growth factor.

C = A constant.

Mitscherlich equation for evaluation of soil fertility is given by :

(a) $\frac{Dy}{Dx} = (A-y)C$ (b) $\frac{Dx}{Dy} = (A-y)C$

(c) $\frac{Dx}{Dy} = \frac{(A-y)}{Dx}$ (d) $\frac{Dy}{Dx} = (A+y)C$

143. Consider the following statements :Vermicompost is considered to be better because :

1. It does not require turning of compost material.
2. It needs less composting time.
3. It takes place at a comparatively low temperature.
4. It has higher enzymatic activity of these statements :

(a) 1, 2, 3 & 4 are correct (b) 1 & 2 are correct

(c) 2,3 & 4 are correct (d) 1,3 & 4 are correct

144. Match List 1 with List II and select the correct answer using the codes given below the lists :

List I	List II
A. Gypsum	1. Magnesium fertilizer
B.17 : 17 : 17	2. Micronutrient fertilizer
C. Serpentine	3. Calcium fertilizer
D. Chelates	4. Compound/Complex fertilizer.

Code :

(a) A - 3, B - 2, C - 1, D - 4.

(b) A - 3, B - 4, C - 1, D - 2.

(c) A - 1, B - 2, C - 3, D - 4.

(d) A - 1,B - 4, C - 3, D - 2.

145. Sulphate of potash is preferable to muriate of potash (MOP) for tobacco because.

(a) The chlorine in the MOP afect the burning quality of tobacco.

(b) The chlorine in the MOP increases the nicotine content of tobacco.

(c) The application of MOP results in the luxury con sumption of potassium.

(d) The sulphate ion helps to increase the availability of other nutrients.

146. Match List I (Reactants used for the preparation of fertilizers) with Lst II (fertilizer Produced) and select the correct answer using the codes given below the lists :

List I	List II
A. Rock phosphate, Phosphoric acid	1. Urea
B. Rock phosphate, phosphoric acid, nitric acid, ammonia.	2. Ammouium Sulphate.
C. Ammonia, gypsum, Co_2	3. Nitrophosphate
D. Ammonia, CO_2	4. Concentrated super phosphate.

Codes :

(a) A - 4, B - 3, C - 1, D - 2

(b) A - 4, B - 3, C - 2, D - 1

(c) A - 3, B - 4, C - 2, D - 1

(d) A - 3, B - 4, C - 1, D - 2.

147. Entry of potassium ions into root hair in soils having a low potas sium ion content in the soil , is mediated through.

(a) Ion exchange mechanism.

(b) Mass flow phenomenon.

(c) Donnan equalibrium process.

(d) Utilisation of metabolic energy.

148. Match List I (source of character) with List II (character exploited) and select the correct answer using the codes given below the lists :

List I	List II
A. Norin - 10	1. Resitance to *Puccinia recondita* causing leaf rust in wheat.
B. Opaque - 2	2. Semi- dwarfing habit in rice.

C. Dee-gee-woo- gen — 3. Protein content in maize.

D. Lr genes — 4. Short stature in wheat.

Codes :

(a) A - 4, B - 3, C - 1, D - 2

(b) A - 4, B - 3, C - 2, D - 1

(c) A - 3, B - 4, C - 1, D - 2

(d) A - 3, B - 4, C - 2, D - 1.

149. A high yielding semi- dwarf variety of rice was First developed at:

(a) CRRI, Cuttack (b) IARI, New Delhi

(c) ICRISAT, Hyderabad (d) IRRI, Manila

150. The most popular extra-early maturing variety in pigeon pea is:

(a) UPAS - 120 (b) Gulbarga

(c) Type - 7 (d) Type - 17

151. While adopting a new innovation in agriculture, farmers go through the following four stage in the mental process :

1. Awareness 2. Intereset

3. Action 4. Satisfaction

The correct sequence of these four stages is :

(a) 1, 2, 4, 3 (b) 2, 1, 3, 4

(c) 1, 2, 3, 4 (d) 2, 1, 4, 3

152. The fruits of research and development findings could reach the poorest of the poor through :

(a) Research personel themselves

(b) NGO's

(c) A strong and devoted extension angency.

(d) agricultural universities & Colleges .

153. In a communication process, Information flows through the following four elements :

1. Message 2. Communicator

3. Channel 4. Audience

The correct sequence of flow of information is :

(a) 2, 3, 1, 4 (b) 3, 2, 1, 4

(c) 1, 3, 2, 4 (d) 1, 2, 3, 4

154. For providing inputs like quality seeds, fertilizers and pesticides, the agency that is present at the Village panchayat samiti level is

(a) NABARD (b) Nationalised banks

(c) Cooperative society (d) Insurance companies.

155. Among the states growing cotton as rainfed crop, the one which obtains the highest yield in the country is :

(a) Maharashtra (b) Haryana

(c) Gujrat (d) Punjab

156. North-east monsoon causes rainfall in the south-eastern part of the peninsular India during :

(a) November - December (b) February - March

(c) September - October (d) May - June.

157. Which one of the following pairs of states dominates at present in pigeon pea cultivation in India, on account of favourable agroclimatic conditions ?

(a) Bihar and Haryana.

(b) U.P. and Maharashtra

(c) M.P. & Rajasthan.

(d) Karnataka & Andhra Pradesh.

158. Cropping intensity and gestation period of the crops grown are :

(a) Inversely related (b) Directly related

(c) Exponentially related (d) Not inter related

159. Keeping in view the available technology for Kharif and Rabicrops, Which one of the following patterns would be more in demand for crop planning ?

(a) Multiple cropping

(b) Mono culture

(c) Location specific cropping

(d) Mixed cropping

160. At present, the farming system of India has become.

(a) export oriented (b) Technology oriented

(c) Market oriented (d) Domestic need oriented

161. The anti- quality constituent in young seedlings of forage sorghum is:

(a) Coumarins (b) Terpenes

(c) Acetic acid (d) Hydrocyanic acid.

162. Which one of the following species belongs to pop maize?

(a) Everta (b) Amylacea

(c) Indurata (d) Ceratina.

163. High yield of maize is obtained in regions where the atmospheric. temperature during the growing season is

(a) High during the day and low during the night.

(b) High during the day and high during the night

(c) Low during the day and high during the night

(d) Low during the day and low during the night

164. Seed Potato from the hilly regions of India is preferred because :

(a) It is cheaper

(b) Good varieties are grown in those regions .

(c) It has better germination percentage.

(d) It is disease free.

165. Which of the following pairs of common name and scientific name are correctly matched ?

1. Proso millet (cheena) ------------------ *Panicum miliaceum.*
2. Kodo millet ---------------------------- *Paspalum scorbiculatum.*
3. Italian millet -------------------------- *Panicum antidotale.*
4. Finger millet --------------------------- *Eleucine coracana.*

Select the correct answer using the codes given below :

Codes :

(a) 1, 2 & 3 (b) 1, 3 & 4

(c) 1, 2 & 4 (d) 2, 3 & 4

166. The dwarf wheat varieties should be :

(a) Sown at the same depth at which the tall wheat varieties are sown.

(b) Sown at a depth deeper than that for tall wheat varieties.

(c) Sown at a depth shallower than that for tall wheat varieties.

(d) Broadcast & incorporated in the top soil.

167. Which one of the following chemicals is used to treat seed potato to break its dormancy?

(a) Ethrel (b) Thiourea

(c) Indole acetic acid (d) Naphthalene acetic acid.

168. The interculture operation in standing broadcast rice crop is commonly known as :

(a) Thinning (b) Spudding

(c) Rotation (d) Beushening

169. Consider the following statements . In paddy, fertilizer application should be directed to the :

1. Oxidised zone. 2. Reduced zone

3. Surface.

(a) 1 alone is correct (b) 2 alone is correct

(c) 2 & 3 are correct (d) 1 & 3 are correct

170. The following operations are practised in the cultivation of tobacco

1. Priming 2. Topping 3. Desuckering.

The correct sequence of these operations in the field is :

(a) 3, 2,1 (b) 3,1, 2

(c) 2, 3, 1 (d) 2, 1, 3

171. Consider the following statements :

A mulching material with high albedo value is preferred because :

1. It reflects back more of the incident radiation.

2. It reduces considerably the down ward transmission of heat.

3. It has an evaporative cooling effect of these statements ..

(a) 1 & 3 are correct (b) 1 & 2 are correct

(c) 2 & 3 are correct (d) 1, 2 & 3 are correct.

172. If the land utilisation Index (LUI) of a crop is 0.50, then the duration of that crop will be :

(a) 183 days (b) 172 days

(c) 200 days (d) 135 days

173. Match List I (common names of weeds) with List II (their scientific names) and select the correct answer using the codes given below the lists :

List I	**List II**
A. Nut grass (Motha)	1. *Echinochloa crusgalli*
B. Bermuda grass (Doob)	2. Cyperus rotundus .
C. Yellow nut sedge (pila motha)	3. *Cynodon dactylon*
D. Barnyard grass (sawan)	4. *Cyperus esculentus*

Codes :

(a) A - 1, B - 3, C - 4, D - 2
(b) A - 2, B - 3, C - 4, D - 1
(c) A - 2, B - 4, C - 3, D - 1
(d) A - 1, B - 4, C - 3, D - 2

174. If a farmer practices 'Sugarcane- Cotton - Wheat,' the cropping intensity at his farm will be :
(a) 100% (b) 150%
(c) 200% (d) 300%

175. Which one of the following triazines is preferred in dryland maize:
(a) Simazine (b) Atrazine
(c) Metribuzin (d) All

176. For getting 10% advantage, the LER should be :
(a) 1.00 (b) 0.90
(c) 1.10 (d) 110

177. If the nitrogen content of a soil is 0.56% Organic matter (%) content of that soil will be :
(a) 1.72 (b) 5.6
(c) 9.5 (d) 0.95

178. Which one of the following actinomycetes is responsible for N- Fixation?
(a) Anabaena (b) Azolla
(c) Frankia (d) Clostridium

179. The 'India Farming' is published from :
(a) ICAR (b) IARI
(c) CAZRI (d) ICRISAT

* The following items consist of two statements, one labelled the Assertion A' and the other labelled the 'Reason R' you are to examine.

These two statements carefully and decide the assertion A and the Reason R are individually true and if so, whether the Reason is a correct explanation of the Assertion. Select your answer to these items using the codes given below.

Codes :

(a) Both A & R are true and R is the correct explanation of A.

(b) Both A & R are true but R is Not a correct explanation of A.

(c) A is true but R is false.

(d) A is false but R is true.

180. Assertion (A) : Since independence the share of agriculture in na tional income has been decreasing steadily.
Reason (R) : The percentage of population dependent on agriculture has by and large remained constant since independence.

181. Assertion (A) : During the green revolution both the yield per hect are and area under wheat increased sharply.
Reason (R) : India is following the former soviet pattern of agricultural development.

182. Assertion (A) : Extension is taken as essential and major part of tech nology transfer.
Reason (R) It is incorrect to equate Agricultural extension with the term 'Technology Transfer'that includes the additional functions of input supply and agri-services.

183. Assertion (A) :Pulses when included in intercropping with cereals act as complimentary crops.
Reason (R) :Pulses with tap root system do not complete for nutrients and moisture with cereals having fibrous root system.

184. Assertion (A) :Symbiotic nitrogen fixation occurs in the nodules of leguminous plants.
Reason (R) Nitrate reductase activity occurs in nodules.

185. Assertion (A) :Soil moisture moves from wet to dry zone.
Reason (R) :Soil moisture moves in response to a potential gradient.

186. Assertion (A) :In sandy soil,the nutrient availability is low.
Reason (R) :The water holding capacity of a sandy soil is poor.

187. (A) :Assertion :Berseem is the best intercrop for citrus orchards.
Reason (R) : Berseem enriches the orchard soil. (a)

188. Assertion (A) : The basic thrust of the lab to land programme is increased productivity,additional income availability and greater employment opportunity. (b)
Reason (R) :The lab to land programme has now been integrated with Krishi Vigyan Kendra in the country :Growth duration of latesown wheat is less than that of the crop sown at the optimal date.

189. Assertion (A) :Late sown wheat crop requires less amount of fertilizer than does the crop sown at the optimal date. (a)
Reason (R) :Growth duration of late-sown wheat is less than that of the crop a sown at the optimal date.

190. Assertion (A) :Potato seeds are treated with one percent aqueous solution of thiourea for an hour before sowing. (a)
Reason (R) : Potato seeds have a dormancy period.

191. Asertion (A) :A deep soil profile regulates the free movement of nutrients for a good crop growth. (a)
Reason (R) : Eluviation is the result of illuviation process favouring accu-mulation of nutrient elements and colloided content in B hrizon.

192. Assetion (A) :Shale is an igneous rock. (d)
Reason (R) : Shales were formed by sedimentation of fine clay in sea-bed in the remote past.

193. Assertion (A) :Granular structure is a characteristic of grassland soils.
Reason (R) :Granular soils are rich in organic matter content. (b)

194. Assertion (A) :Low availability of phosphorus is a problem in both acidic and alkaline soils. (b)
Reason (R) :Soil p^H regulates the availability of different soil nutrients.

195. Assertion (A) : Farmyard manure is beneficial in sandy as well as clayey soils. (b)
Reason (R) : Farmyard manure contains all the three major nutrients namely on as average 0.5, 0.25 and 0.5% of N, P_2O_5 and K_2O respectively.

196. Assertion (A) : An extension worker may reach a large number of people through mass media at low cost per unit of coverage.
Reason (R): The mass- media enable extension workers to greatly increase their, teaching intensity. (a)

197. Assertion (A) : In India, area under wheat is more than what is necessary in a normal year, while area under oil seeds and pulses is still below the required levels. (a)
Reason (R) : Area under pulses and oilseeds have been shifted over to the production of wheat in recent years on account of higher prices and improved productivity potential of high yieldding variety wheat seeds.

198. Assertion (A) : Alkali soils need more puddling for the cultivation of rice when compared to normal soils. (a)
Reason (R) : Alkali soils have more exchangeable sodium on the exchange complex than the normal soil.

199. Assertion (A) : Organic manures with wider C/N ratio when added to the soil decompose rapidly in soil and increase the plant available nitrogen. (d)
Reason (R) : Organic acids are produced during decomposition of organic manures .

200. Assertion (A) : The complex and mixed fertilizers with the same nutrient contents are equally effective in crop production.
Reason (R) : The complex fertilizers are granular, free flowing and (b) easy to apply whereas fertilizer mixtures are physical blends of two or more straight fertilizers mechanically mixed.

201. Assertion (A) : Soil organic matter plays a pivotal role in sustaining soil productivity and fertility. (a)
Reason (R) : Plant nutrients are slowly released from soil organic matter.

202. Assertion (A) : Cultivation of rice in the non-traditional areas in India has increased. (c)
Reason (R) : Food habits of the people of non -rice growing areas have changed.

203. Assertion (A) : Seed drmancy in rice is an important element in maintaining the viability of seed.

Reason (R) : Seed dormancy in rice prevents seed germination in the panicle while the crop is still in the field.

204. Assertion (A) : Green manuring in soil results in better utilisation of phosphorus by subsequent crops.

Reason (R) : Decomposition of green manure releases carbon dioxide which dissolves in water to form carbonic acid which in turn helps dissolution of phosphorus.

205. Assertion (A) : The botanical name of noble cane is *Saccharum officinarum.*

Reason (R) : There is no difference between *Saccharum officinarum* & *S. Spontaneum.*

206. Soils with high P^H are generally deficient in .

(a) Zn & Mn (b) B & Fe

(c) Cu & Mo (d) Ca & Mg

207. Which one of the following plant nutrients is useful in increasing resistance to diseases and insect - pests ?

(a) Calcium (b) Phosphorus

(c) Nitrogen (d) Potassium

208. Use of materials like sulphur, polyform UF, isobutylidene diurea etc. improves fertilizer use efficiency.When such additives are used ,the N- fertilizers are termed as.

(a) Prilled urea.

(b) Controlled release nitrogenous fertilizers.

(c) Slow nitrifying N - fertilizers

(d) Complex fertilizers.

209. For preparing a mixed fertilizer, 100kg of ammonium sulphate and 125 kg of single super phosphate were mixed 25kg of filler was added to this mixture. In terms of N - P_2O_5 - K_2O, the fertilizer grade of this mixture will be .

(a) 5 -5- 0 (b) 5 - 10-0

(c) 8-8-0 (d) 10 - 10-0

210. In respect of a sugarcane crop to be grown in an acidic soil having high P-fixing capacity,the fertilizer of choice woud be
(a) triple superphosphate
(b) Finely ground rock phosphate
(c) Coarse rock phosphate
(d) superphosphate

211. Match list I (Biofertilizer) with list II (properties) and select the correct answer using the codes given below the lists :

List I	**List II**
A. Azospirillium	1. Helps in phosphorus uptake.
B. Azotobacter	2. Floating aquatic fern.
C.VAM	3. Grows on and inside the roots of graminaceous crops.
D. Azolla	4. Free living nitrogen fixiing bacterium.

Codes :
(a) A-4, B-3, C-2, D-1 (b)A-3, B-4, C-2,D-1
(c) A-3, B-4,C-1,D-2 (d) A-4,B-3,C-1,D-2

212. A soil with 20 me/100 g CEC and 5 me/100 g exchangeable H will have a base saturation of
(a) 25.0% (b) 75.0%
(c) 80.0% (d) 133.3%

213. what would be the concentration of nutrients in plants when they are suffering from hidden hunger?
(1) Above deficiency symptom zone.
(2) Below profitable yield zone
(3) Below deficiency symptom zone
(4) Above profitable yield zone
Select the correct answer using the codes given below :
(a; 1 & 4 (b) 2& 3
(c) 1 & 2 (d) 3 & 4

214. There is a net increase in the mineralisation of organic nitrogen in soil when
(a) C : N ratio is 20 to 30 : 1
(b) C : N ratio is below 20 : 1
(c) C : N ratio is above 30 : 1
(d) Nitrogen is taken up by the crops.

215. Hydraulic conductivity of soils varies directly with

(a) Capillary porosity

(b) Non-capillary porosity

(c) Total porosity

(d) Water holding capacity

216. Tolerance to soil salinity of the given set of crops is such that

(a) Barley > wheat > upland > paddy > beans

(b) Barley > wheat > beans > upland paddy

(c) Wheat > barley > beans > upland paddy

(d) Wheat > barley > upland paddy > beans

217. Which one of the following pairs is not corrcetly matched ?

(a) Expanding type of clay minerals	-------------	Higher waterholding capacity
(b) High calcium content in soil	-------------	Iron chlorosis
(c) Water inundated soils	-------------	High nitrogen mineralisation
(d) Sandy soils	___________	Split application of fertilizers.

218. Phosphorus fixation refers to the phenomenon of :

(a) Absorption of phosphorus by the plant and synthesis of biological compounds .

(b) Reaching of Phosphorus from soil by percolating waters.

(c) Conversion of soluble phosphorus into plant unavailable forms.

(d) Anion exchange with colloidal constituents of soil.

219. Consider the following materials :

(a) Montmorillonite (b) Humus

(c) Kaolinite (d) Illite

The correct sequence of these materials in decreasing order of their CEC is

(a) 1, 2, 3, 4 (b) 2, 1, 3, 4

(c) 2, 1, 4, 3 (d) 1, 2, 4, 3

220. Which one of the following sets of soil conditions is more suitable for plant growth ?

(a) Sandy soils with high aeration and low water holding capacity.

(b) Clay soils with high water holding capacity and low aeration.

(c) Organic soils with high water holding capacity and low aeration.

(d) Loamy soils with medium aeration and medium water holding capacity.

221. On the basis of total mass, the relative abundance of the given elements in the earth's crust is such that .

(a) Si > O_2 > Fe > Al (b) O_2 > Al > Si > Fe

(c) O_2 > Si > Fe > Al (d) O_2 > Si > Al >Fe

222. Which one of the following systems needs irrigation throughout the year?

(a) Diara land cropping system.

(b) Intercropping system

(c) Mixed cropping system.

(d) Multiple cropping system.

223. A farmer is to select a herbicide for controlling Cyperus spp. and dicot weeds on his sugarcane crop. The crop is followed, after its toon, by dhaincha in the kharif and lentils in the rabi . In this case , the herbicide of choice would be.

(a) Atrazine (b) dalapon

(c) diuron (d) 2, 4-D

224. Match list I (Herbicides) with list II (chemical group) and select the correct answer using the codes given below the lists :

List I	**List II**
A. Dalapon	1. Quaternary ammonium compound
B. 2, 4, 5 -T	2. Symmetrical triazenes
C. Simazine	3. Phenoxy compounds
D. Paraquat	4. Aliphatic acids.

Codes :

(a) A - 3, B - 4, C - 1, D - 2 (b) A - 4, B - 3, C - 1, D - 2

(c) A - 4, B - 3, C - 2, D - 1 (d) A - 3, B - 4, C - 2, D -1

225. When irrigation water is available for two irrigations only, the wheat crop should receive these at :

(a) CRI and late tillering stages

(b) CRI and boot stages

(c) CRI and flowering stages

(d) CRI & milk stages .

226. Dehaulming in potato is done to obtain

(a) higher yield (b) quality seed tuber

(c) higher starch content (d) higher protein content

227. Cropping pattern is defined as

(a) the proportion of area under different crops

(b) growing different crops in a recurrent succession

(c) The porportion of area under different crops at a particular point of time.

(d) a method of growing different crops.

228. Crop rotation is defined as a

(a) system of growing different kinds of crops on the same land.

(b) system of growing different crops in succession on the same land.

(c) method of growing different crops.

(d) system of growing the same crop on different fields.

229. Match list I (Host legume) with list II (Rhizobium sps) and select the correct answer using the codes given below the lists.

List I	**List II**
A. Fenugreek	1. *R. meliloti*
B. Clovers	2. *R. Phaseoli*
C. Peas	3. *R. trifolii*
D. Beans	4. *R. Leguminosarum*

Codes :

(a) A - 1, B - 2, C - 4, D - 3 (b) A - 1, B - 3, C - 2, D - 4

(c) A - 3, B - 1, C - 2, D - 4 (d) A - 1, B - 3, C - 4, D - 2

230. Match list I with list II and select the correct answer using the codes given below the lists :

List I (Crops)	**List II** (Term used in relation to the crop)
A. Corn.	1. Retting

B. Cotton — 2. Tasselling

C. Peanut — 3. Ginning

D. Jute — 4. Pegging

Codes :

(a) A - 1, B - 2, C - 3, D - 4 (b) A - 2, B - 3, C - 4, D - 1

(c) A - 2, B - 3, C - 1, D - 4 (d) A - 1, B - 3, C - 4, D - 2

231. Wheat is of

(a) Mediterranean origin (b) Indian origin

(c) Mexican origin (d) Chinese origin

232. The popular variety, Ganga Safed -2 of maize is a

(a) Single cross hybrid (b) double cross hybrid

(c) 3-way cross hybrid (d) double top cross hybrid

233. Consider the following factors :

(a) Soil characteristics

(b) Mean annual rainfall.

(c) Potential evapotranspiration

(d) Number of drought days in a year

The planning commission in India has divided the country into different agro- climatic regions based on factors.

(a) 1, 2, & 3 (b) 1, 2, & 4

(c) 1, 3, & 4 (d) 2, 3, & 4

234. The main purpose of T & V programme is to

(a) Identify research problems

(b) Bring about coordination in the administration of agricultural extension.

(c) Update the technical knowledge and skills of agricultural exension workers

(d) Help farmers in the adoption of latest innovations in agriculture.

235. An extension worker practically explains Japanese method of Paddy cultivation to agroup of farmers and also show its superiority over the traditional method. In the process, which of the following communication methods have been used by him ?

(a) Method demonstration (b) Field visit

(c) Result demonstration (d) Study tour

Select the correct answer using the codes given below :

Codes :

(a) 1, 2, 3 & 4 (b) 1 & 3

(c) 2 & 3 (d) 1, 2 & 4

236. Consider the following statements :

Extension in the long means change in

(a) Knowledge (b) Skill

(c) Work (d) Attitude

Of these statements

(a) 1, 2, & 3 are correct (b) 1, 3 & 4 are correct

(c) 1, 2, & 4 are correct (d) 2, 3 & 4 are correct

237. Electrical conductivity of saturation extract of saline-sodic soils in m mhos per cm is :

(a) 2.0 (b) 2.0 - 3.0

(c) 3.5 (d) more than 4.0

238. Deficiency symptoms of calcium on plants are first noticed at :

(a) Lower leaves (b) Middle leaves

(c) Terminal leaves (d) All leaves

239. When organic matter is added to the soil :

(a) Bulk density is increased

(b) Bulk density is decreased

(c) Bulk density does not change

(d) None of these

240. The availability of molybdenum is more at :

(a) p^H - 5.0 (b) p^H - 7.0

(c) p^H - 7.5 (d) p^H - 9.0

241. ' Seed plot technique' is followed in :

(a) Wheat (b) Paddy

(c) Potato (d) Bajra

242. Azotobacter fixes atmospheric nitrogen

(a) Symbiotically

(b) Non - symbiotically

(c) Both symbiotically and non - symbiotically

(d) None of the above is correct

243. The soil is well flocculated if it has more of

(a) Sodium (b) Zinc

(c) Potassium (d) Calcium

244. Rock phosphate is used in

(a) Neutral soil (b) Acidic soil

(c) Alkaline soil (d) Saline soil

245. Montmorillonite clay has silica and alumina layers respectively in the following proportion:

(a) 2 : 1 (b) 1 : 2

(c) 3 : 2 (d) 2 : 3

246. Which one of the following microorganisms is involved in nitrifica tion in the soil:

(a) Pseudomonas (b) Micro - coccus

(c) Thio - bacillus (d) Nitrosomonas

247. Which one of the following is a growth retardant

(a) 2, 4, 5 - T (b) Cycocel

(c) CEPA (d) TIBA

248. Orobanche is a

(a) Total root parasite (b) Total stem parasite

(c) Partial root parasite (d) Partial stem parasite

249. Television, radio , fairs and exhibitions are the media of

(a) Group communication (b) Individual communication

(c) Community communication (d) Mass communication

250. Central Soil Salinity Research Institute is located at :

(a) Jodhpur (b) Karnal

(c) Dehradoon (d) Hyderabad

251. Highest number of sugar mills in India is located in :

(a) Maharastra (b) Tamil Nadu

(c) Uttar Pradesh (d) Bihar

252. The term extension was first used by .

(a) America (b) England

(b) India (d) Japan

253. Serialise the following stages of adoption process :

(a) Interest (b) Evaluation

(c) Trial (d) Awareness

Pick out the answer from the following .

(a) A B C D (b) C A B D

(c) D A C B (c) C B A D

254. Sustainable agriculture can be achieved by :

(a) More resources (b) More subsides

(c) More teachnological support (d) Better prices of the produce

255. Which one is not a system of farming

(a) State farming (b) Collective farming

(c) Cooperative farming (d) Dry - farming

256. Wheat is a :

(a) C_3 plant (b) C_4 plant

(c) both (d) None of the above

257. Paddy has :

(a) 3 stamens (b) 4 stamens

(c) 5 stamens (d) 6 stamens

258. Cultivation of hybrid rice is most popular in :

(a) Japan (b) Korea

(c) China (d) India

259. International Crop Research Institute for semi-arid tropics is situated at :

(a) Jodhpur (b) Bharatpur

(c) Hyderabad (d) Colombo

260. The translocation of sugars from source to sink takes place through :

(a) Sieve tubes of the phloem (b) Companion cells

(c) Xylem vessels (d) Medullary rays

261. The rate of water absorption by plants is decreased due to deficiency of :

(a) Boron (b) Potassium

(c) Iron (d) Manganese

262. The secondary plant element is :

(a) Nitrogen (b) Calcium

(c) Zinc (d) Boron

263. Pungency in mustard oil is caused by :

(a) Amino acid (b) Erusic acid

(c) Glucosilates (d) None of these

264. Seed viability is tested by use of :

(a) Bromide (b) Iodide

(c) Chloride (d) None of these

265. The weight of a half mole of water is :

(a) 0.5 gram (b) 1.0 gram

(c) 9.0 gram (d) 18.0 gram.

266. 'Azolla' (a nitrogen fixing organism) is in symbiotic association with:

(a) Fern and blue green algae

(b) Broyophyta & blue green algae

(c) Fungi and blue green algae

(d) Nostoc and blue green algae

267. The core metal in chlorophyll is

(a) Fe (b) Mg

(c) Mn (d) Zn

268. Phosphorous is taken by plants in the form of :

(a) Phosphorus chloride (b) Phosphoric acid

(c) Phosphorus sulphite (d) Phosphorus carbonates

269. Which one is not correctly matched :

Crop	Seed rate (kg / ha)
(a) Gram	80
(b) Til	4
(c) Mustard	10
(d) Sunflower	12

270. Origin of soybean is :

(a) India (b) USA

(c) Japan (d) China

271. Green revolution in wheat is basically attributed to

(a) Greater use of seeds of high yielding varieties

(b) Higher use of N, P, K fertilisers

(c) Greater availability of irrigation

(d) Greater use of plant protection chemicals .

272. Chromosome number in bread wheat is :

(a) 22 (b) 32

(c) 42 (d) 62

273. Whiptail in cauliflower is caused due to :

(a) deficiency of iron

(b) deficiency of boron

(c) deficiency of molybdenum

(d) deficiency of sulphur

274. SAR of alkali soil is

(a) less than 15 (b) more than 15

(c) less than 13 (d) more than 13

275. Urea hydrolysis in the soil produces :

(a) Uric acid (b) Ammonium carbonate

(c) Ammonium carbonate (d) Amides

276. Paddy crop can be successfully grown under water-logged conditions because :

(a) Paddy roots do not require O_2

(b) paddy roots have capacity to absorb atmospheric O_2 directly

(c) Needed O_2 is supplied from the foliar parts of paddy to the roots.

(d) Needed O_2 is supplied by the microorganisms present in the soil.

277. Sunflower oil is rated as a good quality oil because it contains :

(a) Low quantity of saturated fatty acids

(b) Low quantity of unsaturated fatty acids

(c) High quantity of saturated fatty acids

(d) High quantity of unsaturated fatty acids

278. What is the form in which zinc is absorbed by the plants :

(a) Zn + (b) Zn++

(c) Zn +++ (d) None

279. Which group of microbes increases Phosphate solubility.

(a) Azotobacter (b) Pseudomonas

(c) Clastridium (d) Rhizobium

280. Maturity in sugarcane can easily be determined by hand refracto meter when its reading is :

(a) 10 - 15 (b) 17 - 18

(c) 20 - 22 (d) > 22

281. Main objective of puddling in paddy is to :

(a) Destroy insect - pests

(b) Create impermeable layer in the soil

(c) Mix the fertilizer in the soil

(d) None is true

282. The size of the clay particle in soil falls in one of the ranges :

(a) 0.25 - 0.10 mm (b) 0 . 10 - 0.05 mm.

(c) 0. 05 - 0.002mm (d) < 0.002 mm

283. Potato is propagated through :

(a) Tuber (b) Micro - tuber

(b) TPS (True potato seed) (d) All of these

284. Ammonia is converted into nitrate by :

(a) Fungus (b) Algae

(c) Earthwarms (d) Bacteria

285. Out of the total cropped area of India , forest occupy only :

(a) 26% (b) 33%

(c) 35% (d) 19.5%

286. The first Director General of Indian Council of Agricultural Research was :

(a) Dr. M.S. Swaminathan (b) Dr. N.S. Randhava

(c) Dr. G.S. Randhava (c) Dr. B. P. Pal

287. The fertilizer containing nitrogen in amide form is :

(a) Urea (b) CAN

(c) Ammonium Sulphate (d) Ammonium Phosphate

288. Near the canal banks soils become :

(a) Alkaline (b) Acidic

(c) Neutral (d) Well drained

289. Azotobacter is useful in fixing nitrogen from air in :

(a) Non - leguminous crop like rice & wheat

(b) Leguminous crops

(c) Soybean

(d) None of these

290. Most of the plants obtain nitrogen from the soil in the form of :

(a) Nitrite (b) Nitrate

(c) Free nitrogen gas (d) Nitric acid

291. Which of the following is a micronutrient :

(a) Calcium (b) Zinc

(c) Nitrogen (d) Phosphorus

292. Which of the plant growth regulator is responsible for breaking dormancy :

(a) Auxin (b) Cytokinini

(c) Gibberellic acid (d) 2, 4- D

293. Deficiency of which of the following element causes general yellowing and poor growth of the plant ?

(a) Phosphorus (b) Nitrogen

(c) Calcium (d) Boron

294. How many elements have so far been found essential for plant growth ?

(a) 20 (b) 16

(c) 30 (d) 18

295. The plants growing in shady places adopt themselves by having following modification for more efficient carbon assimiation :

(a) Leaves become broader

(b) Leaves become narrower

(c) Stem is modified into thorn

(d) Leaves become leathery

296. Which of the following Agricultural Universities was started in 1960?

(a) C.S. Azad University of Agriculture & Tech, Kanpur

(b) Punjab Agricultural University Ludhiana

(c) U.P. Agricultural University, Pant Nagar

(d) Rajendra Agri . University , Pusa, Bihar

297. Extension workers should be well versed with economic principles :

(a) Never

(b) Partially

(c) Without it they can 't give effective advice from business point of view.

(d) This is not necessary

298. Subsistence farming is characterised by

(a) Enough surplus for sale.

(b) Distres sale

(c) Can hire large no. of labourers

(d) The resource structure is rich

299. The agro - climatic zones in which India is divided are :

(a) Ten (b) Fifteen

(c) Twelve (d) Six

300. The intensity of following one year crop rotation ' maize-potato -onion' is :

maize - potato - onion

(a) 100% (b) 150%

(c) 200% (d) 300%

301. Which variety of the following relates to oilseeds

(a) CSH -4 (b) Janak

(c) Ganga - 5 (d) Varuna

302. Which of the following is commonly used for green manuring in India ?

(a) Dhaincha (b) Lentil

(c) Pea (d) Potato

303 . Most effective phosphatic fertilizer for crops in acidic soil is :

(a) DAP (b) Rock phosphate

(c) SSP (d) None

304. Under T & V system who is mainly responsible to make regular contact with farmers ?

(a) Dy. Director (Ext) (b) Kisan Sahayak/ Sevek

(c) SDO (d) BDO

305. Which weedicide is used to kill weeds of broad leaves in legumes?

(a) 2, 4 - D (b) 2, 4, 5- T

(c) 2, 4 - DB (d) None

306. Windbreaks are planted in orchards towards :

(a) South - west (b) North - west

(c) North - east (d) East - south

307. Foundation seed is produced from :

(a) Nucleus seed (b) Breeder seed

(c) Quality seed (d) Certified seed

308. Pulse crop which does not fix nitrogen from atmosphere easily is

(a) Arhar (b) Pea

(c) Soybean (d) Rajma

309. If only three irrigations are available indicate critical stages of wheat when these will be applied __

(a) Crown root initiation , flowering , doughing stage

(b) CRI, tillering, doughing stage

(c) Tillering , jointing, milking

(d) Critical root intiation, late jointing & dough stage

310. What is the content of organic matter in alluvial soil -

(a) 0.86 - 6.53 (b) 0.34 - 0.77

(c) 0.28 - 1.10 (d) 5. 9 - 8.7

311. The group of organisms which convert ammonia into nitrites and nitrites to nitrates are respectively -

(a) Nitrosomonas, nitrobactor

(h) Pseudomonas, Azotobactor

(c) Streptomyces, Bacillus, Anthracis

(d) Micromonospora, Nocardia

312. Which of the following elements is not essential element for plants but proves to be beneficial for some ?

(a) Copper (b) Boron

(c) Sodium (d) Iodine

313. The most efficient use of phosphorus is achieved by

(a) Broadcasting at the sowing time

(b) Top dressing 21 days after sowing

(c) Basal placement at the time of sowing

(d) Foliar spray

314. Which of the following nutrient is cycled in the soil through atmosphere?

(a) Phosphorus (b) Potash

(c) Zinc (d) Nitrogen

315. C.F.T.R.I. is situated at

(a) Mysore (b) New Delhi]

(c) Lucknow (d) None of these

316. Origin of groundnut is :

(a) India (b) Brazil

(c) Africa (d) Japan

317. The zinc deficiency in paddy plants causes :

(a) Blight disease (b) Ergot disease

(c) Smut disease (d) Khaira disease

318. Maize - toria -wheat-moong rotation is termed as a system of

(a) mixed cropping (b) relay cropping

(c) multiple cropping (d) inter cropping

319. Coarse grained crops are grown on :

(a) high rainfall areas (b) water-logged areas

(c) lowland areas (d) upland areas

320. Which one of the following crops is grown in all seasons :

(a) Sugarcane (b) Groundnut

(c) Lentil (d) Maize

321. Growing plants of different heights in the same field at the same time is known as :

(a) Multiple cropping (b) Multi - storeyed cropping

(c) Relay cropping (d) Mixed cropping

322. Water use efficiency is the highest in case of :
(a) Flood irrigation (b) Border irrgation
(c) Sprinklar irrigation (d) Drip irrigation

323. Potato tuber is modified form of :
(a) Root (b) Stem
(c) Leaf (d) Flower

324. The edible portion of sweet potato is :
(a) modified root (b) modified stem
(c) leaf (d) fruit

325. Berseem is a fodder crop and is recommended for sowing in the months of -
(a) March - April (b) June -July
(c) October - November (d) January - February

326. Among the following types of soil water which one is the most useful for the plants :
(a) Capillary water (b) Hygroscopic water
(c) Gravitational water (d) Combined water

327. Richest source of Nitrogen is :
(a) Dried blood (b) Raw bone meal
(c) Fish manure (d) Dry activated sludge

328. Geological erosion is also known as :
(a) Accelerated erosion (b) Natural erosion
(c) Soil erosion (d) Water erosion

329. Which one of the following is not an element of communication
(a) message (b) feed - back
(c) channel (d) fidelity

330. Browning of cauliflower can be corrected by use of :
(a) Ammonium molybdate (b) Zinc sulphate
(c) Boric acid (d) Urea

331. Vijay, Amber, Sona & Kisan are composite varieties of :
(a) Wheat (b) Rice
(c) Maize (d) Jowar

332. Planting of succeeding crop before harvest of standing crop is known as :
(a) Mixed cropping (b) Relay cropping
(c) Intercropping (d) Multiple cropping

333. Soil fertility is reduced due to :
(a) Continuous cropping (b) Imbalanced use of fertilizers
(c) Over irrigation (d) Poor drainage

334. In sugarcane - Wheat intercropping, Sugarcane is planted in the month of :
(a) October (b) February
(c) March (d) April

335. Berseem seeds are inoculated with :
(a) Azolla culture (b) Azotobacter culture
(c) Rhizobium culture (d) None of these

336. Scorching or burning on margins of bottom leaves manifests the deficiency of :
(a) Nitrogen (b) Zinc
(c) Copper (d) Potassium

337. 'Diara' lands are :
(a) Areas flooded by sea
(b) Areas created by sand storms
(c) Areas on both sides of canal
(d) Areas located on either side of river which get flooded every year.

338. Which one is an insect trap crop :
(a) Arhar (b) Paddy
(c) Bhindi (d) Potato

339. Carbon is an essential element for plant growth.Plants take it from:
(a) Soil (b) Water
(c) Air (d) Light

340. Winter maize is recommended to be sown slopes of ridges :
(a) Southern (b) Northern
(c) Eastern (d) Western

341. Motha belongs to the family :

(a) Graminae (b) Malvaceae

(c) Cyperaceae (d) Liliaceae

342. Extension education focuses upon development of :

(a) An individual (b) a family

(c) Asocial group (d) A community

343. The category of seed being the costliest one is :

(a) Foundation (b) Trutaful

(c) Certified (d) Breeder

344. Potassic fertilizer containing maximum quantity of potash is :

(a) Muriat of Potash

(b) Sulphate of Potash

(c) Potsassium nitrate

(d) Potassium ammonium nitrate

345. The extension work must be based on needs and interests of :

(a) People (b) Leader

(c) Official (d) Govt.

346. The plants meet their carbon requirements by absorbing :

(a) Soil minerals

(b) Soil organic matter

(c) Carbon monoxide from atmosphere

(d) Carbon dioxide from atmosphere

347. Salt tolerant crop is :

(a) Cowpea (b) Pea

(c) Sugarbeet (d) Muskmelon

348. Which one of the following pairs is correctly matched :

(a) Red soil -- Calcareous

(b) Alkali sol -- P^H - 9

(c) Black soil -- Rich in kaolinite

(d) Laterite soil -- P^H - 8

349. Nitrogen is a major nutrient because :

(a) It is available in soil more

(b) It is constituent of protein

(c) It is available in atmosphere more

(d) It is required by plants more

350. Peat soils are generally :

(a) Black,heavy & acidic
(b) Red, light & acidic
(c) Black, light & basic
(d) Black,heavy & neutral.

351. A branch of science that deals with the study of origin of the soil, its classification and description is known as :

(a) Edaphology
(b) Ecology
(c) Pedology
(d) Geology

352. Soil- amendment used to reclaim usar (barren) land is :

(a) Calcium sulphate
(b) Sodium sulphate
(c) Potassium sulphate
(d) Calcium carbonate

353. The salt affected soils are divided in how many classes depending upon electrical conductivity and sodium contents of their solution extract.

(a) Two
(b) Three
(c) Four
(d) Five

354. Which micro- organism is present in large quantities in soil :

(a) Algae
(b) Bacteria
(c) Fungi
(d) Protozoa

355. Sugar content shall be highest if the refractometer reading is :

(a) 20
(b) 25
(c) 11 - 12
(d) 17 - 18

356. Origin of maize crop is :

(a) India
(b) America
(c) Africa
(d) England

357. Concept of plant ideotype was first propounded by

(a) C.M. Donald
(b) J.D. Eastin
(c) Meyer & Anderson
(d) M.S. Swaminathan

358. Ratio of grain to stalk in Pigeon pea is :

(a) 1 : 2
(b) 1 : 3
(c) 1 : 10
(d) 1 : 6-8

359. Brix percent in sugarcane juice is the indicator of :

(a) Salt content in juice (b) Sucrose percent in juice
(c) Total solids in juice (d) None of these

360. Which one of the following is the example of parallel intercrop ping _
(a) Maize + Soyabean (b) Sugarcane + Mustard
(c) Potato + Mustard (d) Sugarcane + Potato

361. Need based and skill oriented vocational training to practicing farmers and employed extension personnel is provided through:
(a) Training & visit (b) Krishi Vigyan Kendra
(b) National Demonstration (d) Lab to Land Prog

362. Albert Mayer is the name associated with :
(a) Etawah pilot Project
(b) Nilokheri Development Project
(c) Farraka Development Project
(d) Shantiniketan.

363. Foundation seed is produced from :
(a) Nucleus seed (b) Breeder seed
(c) Quality seed (d) Certified seed

364. Name a cereal variety developed by mutation breeding in India.
(a) Sonalika (b) Jagannath
(c) Kiran (d) Savita

365. Before transplanting of paddy , puddling is done :
(a) to increase the availability of nutrients
(b) to increase the activity of micro-organisms
(c) to reduce the percolation of water in soil
(d) to increase the percolation of water in soil.

366. Name the first phase of water erosion :
(a) Sheet erosion (b) Rill erosion
(c) Splash erosion (d) None of these

367. 'Operation Flood Programme' relates to :
(a) Conserving soil (b) Controlling flood
(c) Boosting milk supply (d) Funds for flood victims

368. Soils brought by Ganga-Yamuna and Indus rivers and their tributar

ies are deep and are known as :

(a) Bhabar soils (b) Vindhyan soils

(c) Desert soils (d) Alluvial soils

369. The principle male sterile line introduced for hybrid pearl millet production in India was :

(a) Tift 23 A (b) BJ - 104

(c) MHB - 118 (d) CO - 2

370. Which one of the following genes was utilised in the production of high protein composites in maize ?

(a) Rht -1 (b) Dee zee woo zen

(c) Opaque - 2 (d) Norin

371. Match list I (Crop) with list II (Variety) and select the correct answer using the codes given below :

List I	List II
A. Rice	1. UP 262
B. Wheat	2. Hybrid Ganga
C. Maize	3. Red kafir
D. Sorghum	4. IET 4094

Codes :

(a) A-4, B-1, C-3, D-2 (b) A-1, B-4, C-2, D-3

(c) A-4, B-1, C-2, D-3 (d) A-1, B-4, C-3, D-2

372. The objective of TRYSEM is to provide technical skills to the rural youth from families below the poverty line. The target group comprises rural youth between the ages of :

(a) 20 and 25 (b) 18 and 35

(c) 14 and 35 (d) 14 and 30

373. Consider the following statements :

In Interpersonal channels of communication

(a) message flow tends to be one-way

(b) quantum of feedback available is very high.

(c) Communication is face to face .

(d) Speed of reaching a large audience is relatively slow.

Of these statements .

(a) 1,2,3 & 4 are correct (b) 1 & 4 are correct
(c) 2, 3 & 4 are correct (d) 2 & 3 are correct

374. Which one of the following is the special scheme launched by the ICAR for transfer of technology to the farmers ?
(a) All-India Co- ordinated Research Projects
(b) Operational Research Project
(c) Lab to Land Programme
(d) National Demonstration Programme

375. Groundnut is not an economical crop for deep black soils because:
(a) Phosphorus requirement is high
(b) Swelling and shrinkage affect root system
(c) Peg penetration is very poor
(d) Drainage conditions are poor

376. Which of the following are used as fibre crops :
1. *Corchorus capsularis*
2. *Helianthus annus*
3. *Linum usitatissimum*
4. *Sesamum indicum*
Select the correct answer using the codes given below :
(a) 1 & 2 (b) 1 & 3
(c) 2 & 4 (d) 3 & 4

377. Which of the following agro-climatic factors lead to successful sugarbeet production ?
1. Loam or sandy loam soils .
2. Soil salinity between 1.0 & 1.5%
3. Average temp of 30^0 C and above throughout the growing period .
4. Average temperature between 19.5^0 C and 22^0 C throughout the growing period.
Select the correct answer using the codes given below :
Codes :
(a) 2 & 4 (b) 1, 2 & 4
(c) 1, 2 & 3 (d) 1 & 3

378. Which of the following would one find in a submerged rice field ?

1. Ammonical nitrogen 2. Ferrous sulphide 3. Reduced state

Select the correct answer using the codes given below

codes :

(a) 3 alone (b) 1 alone

(c) 1 & 2 (d) 1, 2 & 3

379. Water use efficiency is defined as

(a) dry matter produced per unit of water used by the crop .

(b) econimic yield per unit of irrigation water applied .

(c) ratio of water used by the crop to the irrigation water applied

(d) ratio of net irrigation to gross irrigtion including other losses.

380. Gross water requirement of a crop is the

(a) sum of net irrigation requirement, effective rainfall and other unavoidable losses during irrigation.

(b) total available water which is present in the soil profile during the crop period.

(c) same as total rainfall in the crop period under heavy rainfall areas

(d) evapotranspiration and water used by the crop for its meta bolic activities.

381. Which of the following statements explain the true relationship between soil water potential and moisture content of loam , sandy and clay soil ?

1. Clay soils hold much more water at a given potential than loamy soils .
2. Loamy soils hold much more water at a given potential than clayey soils .
3. Loamy soils hold much more water at a given potential than sandy soils.
4. Clay soils hold much less water at a given potential than loamy soils.

Select the correct answer using the codes given below :

Codes :

(a) 1 & 2 (b) 2 & 3

(b) 3 & 4 (d) 1 & 3

382. Match list I (Herbicide) with list II (chemical group) and select the correct answer using the codes given below the lists :

List I	List II
A. Paraquat	1. Phenoxy alkanoic acid
B. Dalapon	2. Aliphatic acids
C. 2, 4-D	3. Bipyridillium compounds
D. Alachlor	4. Acetanilides

Codes :

(a) A-3, B-2, C-4, D-1 (b) A-3, B-2, C-1, D-4

(c) A-2, B-3, C-1, D-4 (d) A-2, B-3, C-4, D-1

383. Match list I (Crop) with list II (Associated weed) and select the correct answer using the codes given below the lists :

List I	List II
A. Tobacco	1. *Cinchorium intybus*
B. Lucern	2. *Cuscuta sp.*
C. Berseem	3. *Sorghum halepense*
D. Sugarcane	4. *Orobanche sp.*

Codes :

(a) A-4, B-2, C-3, D-1 (b) A - 2, B -4, C - 1, D -3

(c) A-2, B-4, C-3, D-1 (c) A- 4, B-2, C-1, D-3

384. Match list I (Common name of herbicide) with list II (Trade name of the herbicide) and select the corrcet answer using the codes given below the lists :

List I	List II
A. Alachlor	1. Basalin
B. Oxadiazon	2. Stomp
C. Pendimethalin	3. Ronstar
D. Fluchloralin	4. Lasso

Codes :

(a) A -3, B - 4, C - 1, D - 2

(b) A - 4, B - 3, C - 1, D - 2

(c) A - 4, B - 3, C - 2, D - 1

(d) A - 3, B - 4, C - 2, D - 1

385. The ratio of 'organic carbon' to 'organic matter' in soils is

(a) 1.0 : 1.7 (b) 1.0 : 2.0

(c) 1.7 : 1.0 (d) 20.0 : 1.0

386. A given soil sample contains organic carbon = 0.6 %, phosphorus= 0.1 %, potassium = 1 %

The percent organic matter, P_2O_5 and K_2O contents of the soils is

Organic matter	$\mathbf{P_2O_5}$	$\mathbf{K_2O}$
(a) 1.02	0.29	1.2
(b) 1.02	0.229	1.2
(c) 1.12	2.29	0.83
(d) 1.7	0.86	0.43

387. The bulk density of a soil with particle density of 2.65 g/cc and pore space 40% would be :

(a) 1.29g/cc (b) 1.45g/cc

(c) 1.59g/cc (d) 1.72g/cc

388. Consider the following types of erosion :

1. Gully erosion
2. Splash erosion
3. Normal erosion
4. Rill erosion
5. Sheet erosion

The correct sequence in terms of the INCREASING order of soil loss from the field due to these types of erosions is

(a) 2,3,4,5,1 (b) 2,3,5,4,1

(c) 3,2,4,1,5 (d) 3,2,5,4,1

389. Consider the following statements regarding the diagram given below -

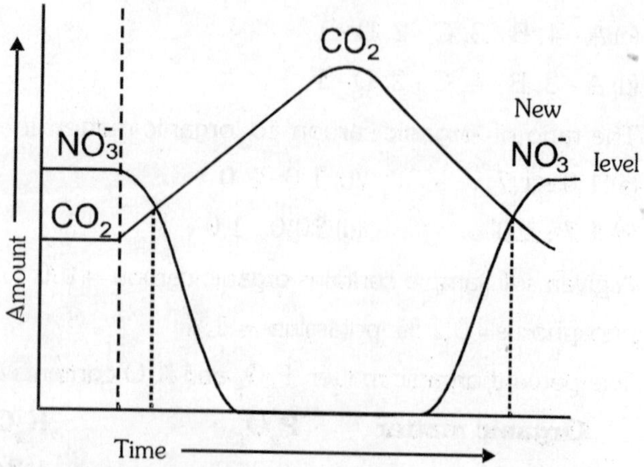

Organic residue with wide C : N ratio is applied

← No_3^- depression period → I

Changes in No_3^- - N level due to microbial activity when fresh organic residue is added to the soil.

1. Nitrate is retained by the soil and hence the nitrate-nitrogen is temporarily immobilised during decomposition.
2. Higher nitrate level observed at the end of decomposition is due to the release of this nutrient by the soil.
3. The peak level of CO_2 and $[NO^-_3]$ observed in the diagram indi cates the maximum activity of microbes developed during de composition. of these statements

(a) 1, 2 and 3 are correct

(b) 1 and 2 are correct

(c) 2 and 3 are correct

(d) 3 alone is correct

390. The primary aim of multiple cropping is to

(a) have higher gross retum

(b) inrease food production

(c) maintain soil fertility

(d) provide technology for achieving the goal of higher level of land use intensity.

391. The main objective of growing the 'catch' crop is to

(a) add more residue to the soil

(b) Suppress weeds

(c) prevent cracking of soil

(d) get an additional income without further investment

392. Silica-sesquioxide ratio is usful in the evaluation of soil in respect of

(a) nutrient availability (b) Silicon content

(c) Iron & aluminium content (c) its age.

393. Consider the following statements :

Cation exchange of soil colloidal clay is due to

(a) Silicon tetrahedral layer

(b) Isomorphious substitution

(c) Hydration of clays

(d) Broken edges of clay lattices of these statements

(a) 1 & 2 are correct (b) 2 & 4 are correct

(c) 3 & 4 are correct (d) 1 & 3 are correct

394. Non-essential but useful nutrients for plants would include

(a) Carbon & Vanadium (b) Vanadium & Chlorine

(c) Vanadium & Sodium (d) Sodium & Chlorine

395. Which of the following are the criteria for determining the esentiality of nutrients elements ?

1. A deficiency of the particular element makes it impossible for the plant to grow normally or complete its life cycle.
2. The deficiency symptoms can be corrected only by supplying the nutrients in question.
3. The nutrient is abundantly present in the soil for rapid plant absorption.
4. The element is directly involved in the plant metabolism.

Select the correct answer using the codes given

Codes :

(a) 1 & 3 (b) 1, 2 & 4

(c) 2, 3 & 4 (d) 1, 2, 3 & 4

396. Which one of the following nitrogen factors would bring down wide C : N ratio material of 100 : 1 to the desirable C : N ratio of 40 : 1 for compost making ?

(a) 1.0 (b) 1.5

(c) 2.0 (d) 2.5

397. Consider the following characteristics of legume root nodule :

(a) Large & elongated

(b) Often clustered on primary root

(c) Pink to red centre.

Among these, the effective root nodules would include

(a) 2 & 3 (b) 1 & 2

(c) 1 & 3 (d) 1, 2 & 3

398. Nitrogen fertilizer use efficiency in rice can be increased by using

(a) sulphur coated urea

(b) blue green algae

(c) Urea super granules

(d) Sulphur - coated urea & urea super granules

399. Following fertilizers are available for preparing a fertilizer mixture :

1. Ammonium sulphate

2. Urea

3. Single super phosphate

4. Basic slag

5. Muriate of potash

A combination of the above-mentioned fertilizers which can be mixed & kept for three months or more is

(a) 1, 4 & 5 (b) 2, 3 & 4

(c) 2, 4 & 5 (d) 1, 3 & 5

400. Match list I (Diagnostic symptoms) with list II (Deficient nutrient) and select the correct answer using the codes given below the lists-

List I	List II
A. Death of apical buds & leathery leaves	1. Zinc
B. Uniform paling of plants without much of stunting	2. Sulphure

C. Uniform paling of top leaves and lower leaves remain green — 3. Molybdenum

D. Interveinal chlorosis of newly formed leaves & reduction in leaf size — 4. Boron

Codes :

(a) A-4, B-3, C-1, D-2 (b) A-3, B-4, C-1, D-2

(c) A-3, B-4, C-2, D-1 (d) A-4, B-3, C-2, D-1

401. Which one of the following nutrients plays the most effective role in controlling the rate of transpiration?

(a) Phosphorus (b) Potassium

(c) Zinc (d) Magnesium

402. Plants roots are

(a) Positively phototropic (b) Negatively hydrotropic

(c) Negatively geotropic (d) Positively geotropic

403. Development of farm sector gives a boost to secondary sector and development of primary and secondary sectors give a boost to the tertiary sector. This is called the 'Product Contribution of Agricul ture''. This concept was coined by

(a) R. H. Tuck (b) Ragner Frisch

(c) Simon Kuznet (d) F.H. Knight

404. Farm management as a resource allocation proposition is related to

(a) Inter-farm management

(b) Intra- farm management

(c) Both inter and intra farm management

(d) Inter- regional management

405. In the figure given below product y_1 is denoted on y-axis and product y_2 on the x-axis. The price of a unit of y_1 & y_2 are Py_1 and Py_2 respectively.

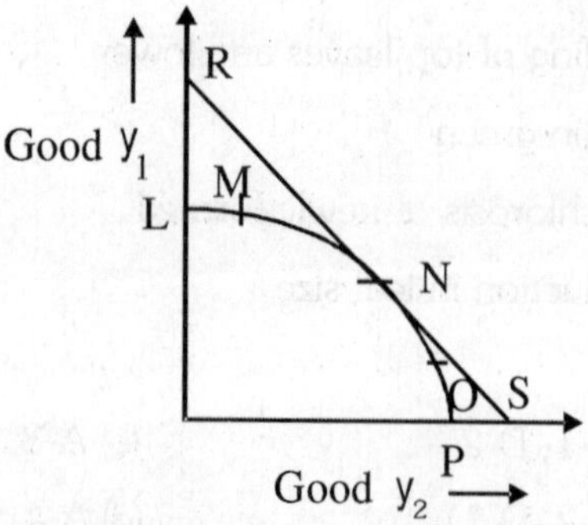

Point N in the figure corresponding to the point where

(a) $\frac{\Delta y_1}{\Delta y_2} = \frac{Py_1}{Py_2}$ (b) $\frac{\Delta y_1}{\Delta y_2} = \frac{Py_2}{Py_1}$

(c) $\frac{y_1}{y_2} = \frac{Py_2}{Py_1}$ (d) $\frac{\Delta Py_1}{\Delta Py_2} = \frac{y_2}{y_1}$

406. The paid-out costs of a farmer is Rs.500 and fixed costs is Rs.250.If the gross returns from farm business is Rs.3,700 the capital turn over would be :

(a) 17.42 (b) 9.80

(c) 4.93 (d) 3.33

407. absence of interaction between resources and activities in the process of production is called.

(a) Linearity (b) Non-linearity

(c) Quasi-non linearity (d) Quasi-linearity

408. In the short run,Average Cost,Average Variable Cost & Marginal Cost curves are 'U' shaped due to the operation of

(a) Law of equimarginal returns

(b) Substitution between two factor inputs

(c) Law of variable proportion

(d) Time comparison principle

409. The size of holding which provides full employment to the farm family and provides a reasonable standard of living could be termed as:

(a) economic holding (b) optimum holding

(c) basic holding (d) family holding

410. Farm system of the Western Himalayan Region of India is mostly dominated by

(a) arable farming (b) plantation crops

(c) pasture (d) horticulture

411. Consider the following four steps involved in the study of agricul tural production economics :

1. Determine and define the conditions which provides for optimum use of resources.
2. Delineate means and methods for changing the existing use of resources to the optimum level.
3. Determine the extent to which the existing use of resources deviates from the optimum.
4. Analyse the factors or forces which are responsible for the existing production patterns and resource use.

The proper sequence of the steps in such study should be

(a) 1,2,3,4 (b) 1,3,4,2

(c) 4,1,3,2 (d) 2,3,1,4

412. Though all categories of farmers are interested in higher returns from their enterprises,what they really try to maximise is

(a) net profit

(b) profit as percentage of investment

(c) farm business income

(d) gross profit

413. Given,

$Y=25\ X_1^{0.35}X_2^{0.49}$, where

Y=output,

X_1=input factor 1,

X_2=input factor 2,

the nature of returns to scale will be

(a) increasing (b) decreasing

(c) constant (d) fluctuating

414. The given diagram shows break-even point of bullock power and tractor power.In this context,which of the following decisions

would be termed rational?(Y-axis in the diagram refers to expendi ture on bullock or tractor power in thousands of rupees and the X- axis refers to the size of holding in hectares.)

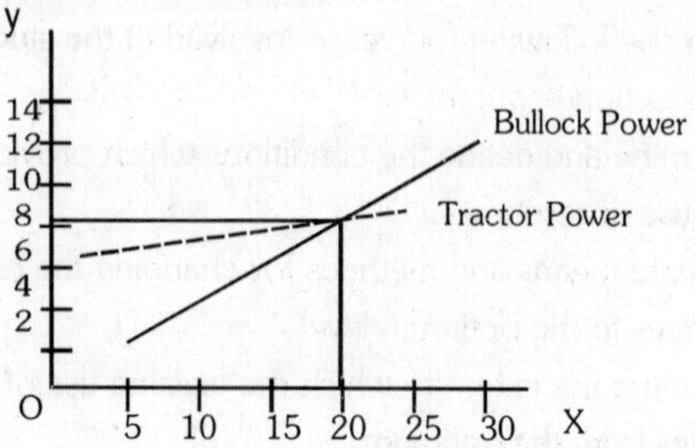

1. Bullock power should be used in place of tractor power.
2. If the size of the holding is 20 hectares,either bullock or tractor power could be used.
3. Tractor power is more economical than bullock power for farm sizes over 20 hectares.

Select the correct answer using the codes given below :

Codes :

(a) 1,2 and 3 (b) 1 and 2

(c) 2 and 3 (d) 1 and 3

415. Consider the following data relating to a farm :

total assets	: Rs. 3000/-
working assets	: Rs. 800/-
current assets	: Rs. 1500/-
current liabilities	: Rs. 200/-
total liabilities	: Rs. 900/-

The net capital ratio in this case would be

(a) 3.33 (b) 4.0

(c) 7.5 (d) 11.5

416. In the given diagram,which one of the following segments illus trates feasible zone in a linear programming problem?

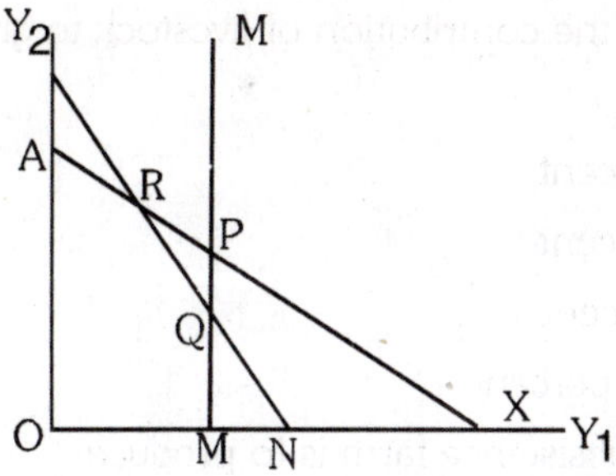

(a) OXRYM (b) RPQ

(c) QMN (d) XNR

417. The term 'operational holding'refers to the

(a) land area owned by a farmer

(b) cropped area cultivated by a farmer

(c) net land area cultivated by a farmer

(d) land area leased in by a farmer

418. Which one of the following formulae gives the correct measure of 'Rate of Capital Turnover'?

(a) (Gross Income / Total Farm Assets) X 100

(b) (Net Income / Total Farm Asets)X100

(c) (Farm business Income / Total Farm Assets)X100

(d) (Capital Investments / Total Farm assets)X100

419. Sufficient condition of farm efficiency is generally refered to as choice indicator because it

(a) is a valid criterion of the assumption of certainty

(b) is consistent with the choices made under perfect competition

(c) is consistent with the welfare increase to the farming population

(d) helps to make allocation compatible with the given objectives

420. Farm budgets are used to

(a) compare economic profitability of different production activities within or among farms

(b) indicate whether a proposed change will be profitable under a given set of factors

(c) explore conditions under which certain farm practices become profitable or unprofitable

(d) do all the above

421. In mixed farming,the contribution of livestock to gross farm income is

(a) at least 10 percent

(b) at least 30 percent

(c) at least 49 percent

(d) more than 50 percent

422. The object of a subsistence farm is to produce

(a) essential food crops for family and market

(b) essential food crops for family only

(c) food and fibre crops for family and market

(d) vegetable crops for family only.

423. In the given figure, product Y_1 is denoted on the y- axis and product Y_2 is denoted on the x-axis. The price of a unit of Y_1 and Y_2 are P_{Y1} and P_{Y2} respectively . Products Y_1 and Y_2 are substitutes in the range of

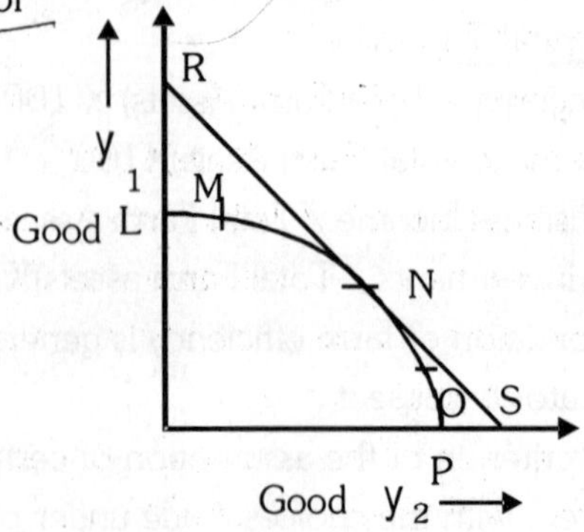

(a) LM
(b) MO
(c) OP
(d) LP

424. The principle of equi-marginal return states that the profits will be the greatest if unit resources are used where.

(a) marginal returns are the highest

(b) average returns are the highest

(c) total returns are the highest

(d) net returns are the highest

425. Consider the following statements :

1. Iso-cost lines are always straight lines .
2. Iso- cost lines further to the right portray higher cost and those nearer to the left portray lower cost .
3. The slope of an Iso- cost line cannot be expressed as the ratio of the price of one input to the price of another input.

The properties of an Iso- cost line would include

(a) 1 and 3 (b) 2 and 3

(c) 1 and 2 (d) 1, 2 and 3

426. Match List I with List II and select the correct answer using the codes given below the lists :

List I	List II
A. Factors	1. Technical complements
B. Resources	2. Technical substitutes
C. Irrational resource combinations	3. Technical complementarity
D. Farm planning tools	4. Budgets and programming.

Codes :

(a) A-2, B-1, C-3, D- 4 (b) A-1, B- 2, C-3, D-4

(c) A- 1, B-2, C-4, D-3 (d) A-2, B- 1, C-4, D-3

427. If the production function is homogeneous of degree one, then the scale elasticity of output will be

(a) equal to unity (b) less than unity

(c) greater than unity (d) equal to zero

428. If technical progress leads to a rise in labour productivity of 5% then an increase of 5% in the wage rate would imply that.

(a) workers will derive all of the benefits

(b) there will be cost-push inflation.

(c) unit labour costs will remain unchanged

(d) profit per unit will invariably fall

429. The inflection point on total product curve corresponds to the

(a) maximum of marginal product curve

(b) maximum of average product curve

(c) maximum of total product curve

(d) minimum of marginal product curve

430. The given diagrams correspond to

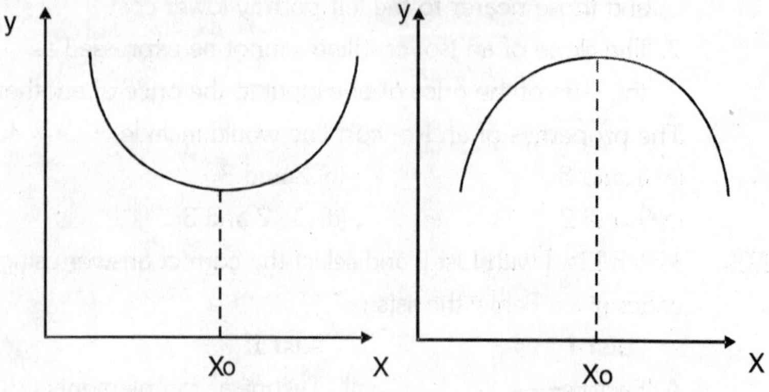

(a) quadratic equation (b) Cobb- Douglas equation

(c) monotoric equation (d) non-monotoric equation

431. At a particular point on a Neo-classical Production Function graph, the magnitudes of TPP, APP and MPP are 18, 6 and 7 respectively. The E_p (elasticity of production) at this point would be

(a) 3 (b) 2.57

(c) 1.16 (d) 0.85

432. Consider the following situations :

1. Output remaining constant from a smaller labour force.
2. An increase in output resulting from an increase in labour, other things remaining constant.
3. Output remaining constant following an increase in the working hours per day.

Example (s) of increased productivity of labour would include

(a) 1 alone (b) 1 & 3

(c) 2 & 3 (c) 1, 2 & 3

433. Which one of the following economic principles helps a farmer to allocate his limited resources to two different enterprises ?

(a) The law of variable proportions

(b) The law of comparative advantage

(c) The law of equimarginal returns

(d)The principle of fixed and variable costs.

434. Which one of the following production functions best explains the returns to scale effect ?

(a) $y = ax_1^{a1}x_2^{a2}$ ---------x_n^{an}

(b) $y = a_o + a_1x_1 + a_2x_2 +$ ----------- $+a_nx_n$

(c) $y / x_1 = a_o + a_1x_1 + a_2(x_2/x_1) +$ -------- $+ a_n(x_n/x_1)$

(d) $y = a_o + a_1x + a_2x^2 +$ ------------ $+ a_nx^n$

435. When instability of prices of a commodity is explained by the cob webb model, which of the following conditions pertaining to the demand and supply functions should hold true?

1. The intercept of the demand function should be greater than the intercept of the supply function.
2. The intercept of the supply function should be greater than the intercept of the demand function.
3. The slope of the demand function should be greater than the slope of the supply function.
4. The slope of the supply function should be greater than the slope of the demand function.

Select the correct answer using the codes given below :

Codes :

(a) 2 & 4 (b) 2 & 3

(c) 1 & 4 (d) 1 & 2

436. Match List I with List II and select the correct answer using the codes given below the lists :

List I	List II
A. Minimum Support Price	1. Farmer
B. Re - finance	2. Govt. of India
C. Price taker	3. NABARD
D. Transportation	4. Time utility
E. Storage	5. Space utility

Codes :

(a) A-2, B-3, C-1, D-5, E-4 (b) A-2, B-3, C-5, D-1, E-4

(c) A-3, B-2, C-1, D-4, E-5 (d) A- 3, B-2, C-5, D-1, E-4

437. Which one of the following principles will be applicable when a farmer is faced with the problem of purchase of heifers versus milch cows on his dairy farm ?

(a) Cost Principle

(b) Opportunity Cost Principle

(c) Comparative advantage Principle

(d) Time Comparison Principle

438. Match List I (Decisions) with List II (Techniques) and select the correct answer using the codes given below the lists :-

List I	List II
A. Capital investment	1. Least Cost Combination
B. Economically Optimum level of resource use	2. Pay - back period
C. Economies of Farm size	3. Linear Programming
D. Optimum enterprise combination under constraining resources	4. Average Cost Curve

Codes :

(a) A-2, B-1, C-3, D-4 (b) A-2, B-1, C-4, D-3

(c) A-1, B-2, C-4, D-3 (d) A-1, B-2, C-3, D-4

439. All categories of Indian farmers are becoming increasingly conscious of the geographic development of crop husbandry. Which one of the following is the main reason behind this ?

(a) Higher labour efficiency

(b) Increased Irrigation efficiency

(c) Suitable ecological efficiency

(d) Accleration in Production efficiency.

440. When marginal rate of factor substitution $\frac{\Delta x_1}{\Delta x_2}$ is more than the inverse price ratio $\frac{Px_1}{Px_2}$, then for better returns one should use

(a) more of x_1 (b) more of x_2

(c) more of both x_1 & x_2 (d) less of both x_1 & x_2

441. Under Product - Product relationship, if a farmer is producing v_1 & v_2 products under non-optional condition, that is, for the most profitable combination of y_1 & y_2, he should

(a) increase the output of y_1

(b) increase the output of y_2

(c) increase the output of y_1 & y_2

(d) decrease the output of y_1 & y_2

442.

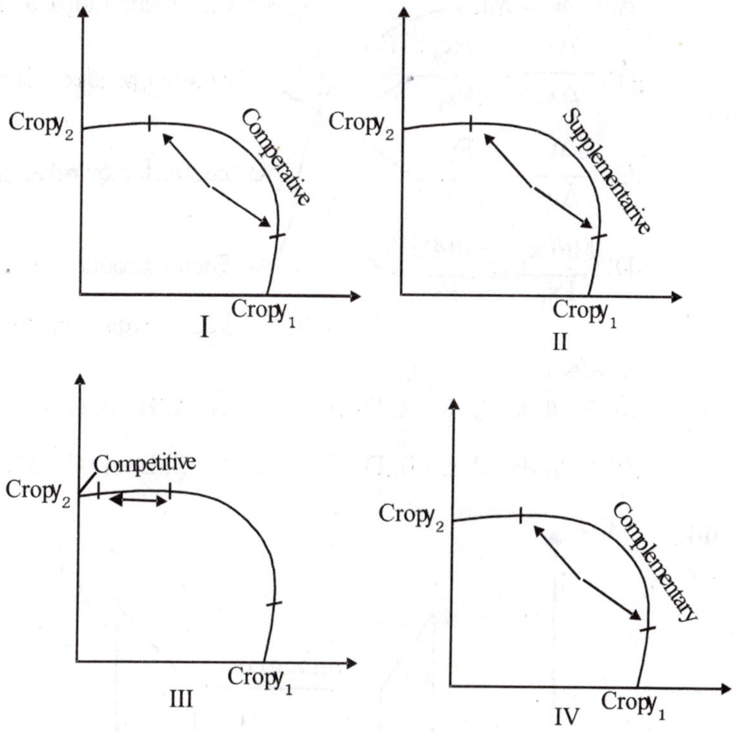

If the data obtained from a particular dose of input in the production of two partially competitive crops are plotted as shown in the above diagrams, then the diagram relating to the most profitable, crop combination would be the one shown in

(a) I　　(b) II

(c) III　　(d) IV

443. Production should be discontinued when the price is

(a) less than the average cost of production

(b) equal to the average cost of production

(c) equal to the marginal cost of production

(d) less than the average variable cost of production.

444. Match List I with List II and select the correct answer using the codes given below the lists

List I	List II
A. VMP = MFC	1. Complementary goods
B. $\frac{\Delta x_1}{\Delta x_2} = \frac{Px_2}{Px_1}$	2. Product-product relationship
C. $\frac{\Delta y_1}{\Delta y_2} = \frac{Py_1}{Py_2}$	3. Factor-factor relationship
D. $\frac{MPPx_1}{Px_1} = \frac{MPPx_2}{Px_2}$	4. Factor-product relationship
	5. Equi-marginal returns

Codes:

(a) A - 4, B - 3, C - 2, D - 1 (b) A - 4, B - 3, C - 2, D - 1

(c) A - 4, B - 3, C - 5, D - 2 (d) A - 2, B - 4, C - 3, D - 5

445.

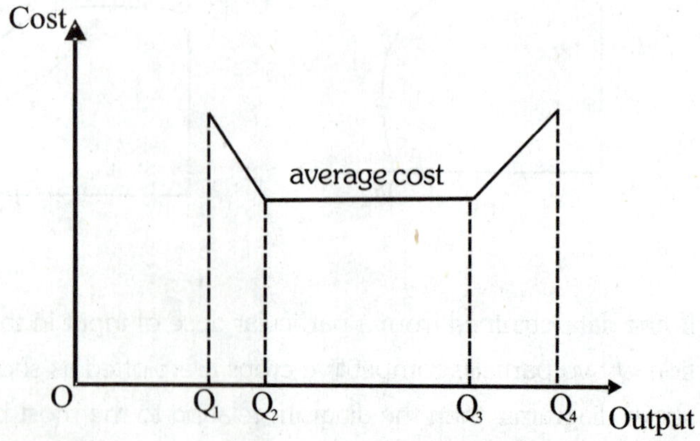

According to the cost-output graph shown in the above figure, increasing returns to scale will be available in

(a) Q_1 to Q_2 (b) Q_1 to Q_3

(c) Q_2 to Q_4 (d) Q_3 to Q_4

446. Match List I (Theory) with List II (Author) and select the correct answer using the codes given below the lists

List I	List II
A. Substitution effect of a price change	1. Slutsky
B. Theory of rent	2. Keynes
C. Liquidity Preference theory	3. Samnelson
D. Revealed Preference theory	4.Ricardo

Codes :

(a) A-4,B-1,C-3,D-2 (b) A-1,B-4,C-2,D-3

(c) A-1,B-4,C-3,D-2 (d) A-4,B-1,C-2,D-3

447. The following data pertains to a linear programming problem involving wheat & mustard as the activities :

Net price per unit of wheat activity = Rs.1.80

Net price per unit of mustard activity = Rs.2.80

Marginal rate of substittution of wheat for mustard = 0.5

Given this situation,each unit increase in mustard activity will

(a) increase profit by Rs. 1.80 (b) increase profit by Rs.1.90

(c) decrease profit by Rs. 3.80 (d) decrease profit by Rs.1.90

448. When marginal product is increasing at increasing rate then :

(a) Total product increase at a decreasing rate

(b) Total product increase at a increasing rate

(c) Average product decreases

(d) None of the above

449. Least cost combination of inputs is achieved when :

(a) $\frac{\Delta x_2}{\Delta x_1} > \frac{Px_1}{Px_2}$ (b) $\frac{\Delta x_2}{\Delta x_1} < \frac{Px_1}{Px_2}$

(c) $\frac{\Delta x_2}{\Delta x_1} = \frac{Px_1}{Px_2}$

(d) None of the above

- Δx_1 & Δx_2 indicate change in x_1 & x_2
- P indicates price.

450. Profit is maximised when :

(a) Marginal cost < Marginal return

(b) Marginal cost = Marginal return

(c) Marginal cost > Marginal return

(d) None of the above

451. The Elasticity of Production (EP) at the starting point of stage II,in Law of Diminishing Return Would be :

(a) > 1.0 (b) 1.0

(c) < 1.0 (d) 0.0

452. Level of optimum production is available in :

(a) I stage of Production (b) II stage of Production

(c) III stage of Production (d) None of the above

453. Which one of the following kind of budgeting would you advise to a new farm planner :

(a) Enterprise budgeting (b) Complete budgeting

(c) Partial budgeting (d) None of these

454. The cost of Production of the crop can be minimised by using economic principle :

(a) Law of diminishing return (b) Law of opportunity cost

(c) Cost Principle

(d) Principle of least cost combination

455. Principle of equi-marginal returns is applied when :

(a) The resources are unlimited

(b) The resource prices are high

(c) The prices of resources are varying

(d) The resources are limited

456. Do the T. C. and the V.C. increase with increase in the output ?

(a) Never (b) Yes

(c) Partially (d) With decrease in output

457. When a new variety of crop is introduced :

(a) Partial budgeting is required

(b) Full budgeting is a must

(c) No budgeting is needed

(d) Only planning is needed

458. The objective of preparation of calender of operations is :

(a) To prepare a cropping scheme

(b) To calculate the cost of production

(c) To assess the need of human and bullock power

(d) To adjust farm enterprises

459. The farm plan would be based on the availability of effficiency of various factors of Production :

(a) True (b) False

(c) Partially true (d) Incomplete statement

460. Farm budget is a tool of farm management analysis :

(a) True (b) False

(c) Never includes cost (d) It is a futile exercise

461. When a farm is managed by Govt. officials it is called a :

(a) Peasant farm (b) Capitalistic farm

(c) Collective farm (d) State farm

462. When the depreciation of an asset is measured by dividing the first cost by its estimated life, it is called :

(a) Diminishing value method

(b) Sum of the years digit method

(c) straight line method

(d) Revaluation method

463. When $\frac{\Delta x_1}{\Delta x_2} > \frac{Px_2}{Px_1}$ then we can :

(a) Employ any combination of x_1 & x_2

(b) Do nothing

(c) Go on substituting x_1 for x_2

(d) Employ x_2 only

464. Which one of the following pairs is correctly matched ?

(a) Common cost ------------- Cost associated with variable resources

(b) Average cost -------------- A study of returns & costs

(c) Marginal cost ------------- Additional cost to produce an additional unit of output

(d) Prime cost --------------- Cost related to fixed resources.

465. Which one of the following pairs is not correctly matched ?

(a) Elasticity of Production greater than unity	-----------	Stage II of classical production function
(b) Negative marginal product	-----------	Stage III of classical production function
(c) Negative elasticity of production	-----------	Stage III of classical production function
(d) Average product decreases but is greater than marginal product	-----------	Stage II of classical production function

466. Which one of the following relationships will provide least cost combination of input use ?

(a) $\frac{\Delta x_1}{\Delta x_2} = \frac{Px_2}{Px_1}$ (b) $\frac{\Delta x_1}{\Delta x_2} > \frac{Px_2}{Px_1}$

(c) $\frac{\Delta x_1}{\Delta x_2} < \frac{Px_2}{Px_1}$ (d) $\frac{\Delta x_1}{\Delta x_2} = \frac{Px_1}{Px_2}$

467. Which one of the following principles is employed in production when resources are scarce ?

(a) Added Cost is equal to added returns

(b) Equi- marginal principle

(c) Comparative advantage

(d) None of the above

468. Break - even point is determined by

(a) MC = MR (b) MC = AVC

(c) MC = AC = AR (d) TR = TC

469. Suppose that Cost A_2 on a farm is Rs. 26,450 and the rent of the leased - in land is Rs. 2,250 . Cos A_1 on this farm will be Rs .

(a) 24, 000 (b) 24, 200

(c) 28, 000 (d) 28, 700

470.. Consider the following statements :

Diversification of farm production helps farmers through stabilising

1. agricultural prices
2. farm incomes
3. agricultural production

of these statements

(a) 1 & 2 are correct (b) 1 & 3 are correct

(c) 1,2 & 3 are correct (d) 2 & 3 are correct

471. A normative farm plan is one which the farmer

(a) has already put into operation on his farm

(b) should adopt during the current five year plan period

(c) had actually adopted in the earlier five year plan periods

(d) should adopt

472. In mixed farming,the contribution of livestock to gross farm income is

(a) at least 10% (b) at least 30%

(c) at least 40% (d) more than 50%

473. During the period 1990 to 1995,the contribution of agriculture sector to the gross domestic product of India has been in the range of

(a) 20 to 30% (b) 30-40%

(c) 40-50% (d) 50-60%

474. Given that

$y = 30X_1^{0.25}X_2^{0.60}$

where y = output

x_1= input factor 1 and

x_2 = input factor 2

The nature of return to scale will be

(a) constant (b) increasing

(c) decreasing (d) fluctuating

475. The given diagram shows isoquants for different combinations of two input factors x_1 & x_2.Consider the following statements in this regard :

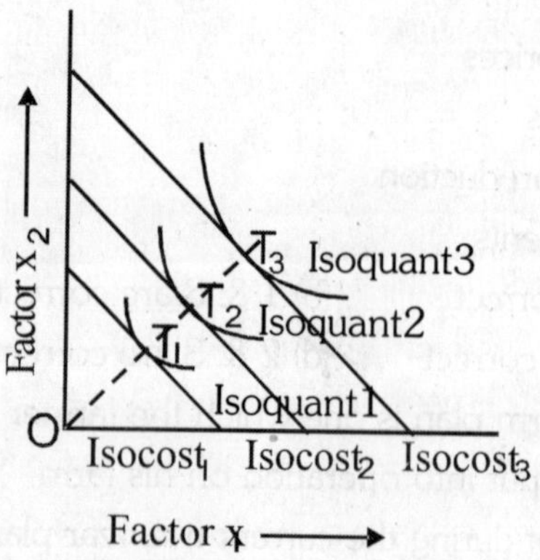

Higher production can be achieved by

1. Moving from isoquant$_1$ to isoquant$_3$
2. moving from isoquant$_3$ to isoquant$_1$
3. increasing the budget outlay.

of these statements

(a) 2 alone is correct (b) 1 alone is correct

(c) 2 & 3 are correct (d) 1& 3 are correct

476. In the given diagram,which one of the following segments represents feasible zone in a lin ear programming problem?

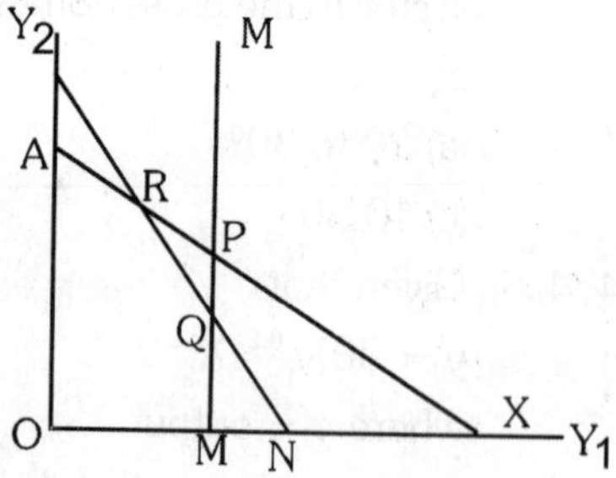

(a) RPQ

(b) QMN

(c) XNR

(d) OARQM

477. Marginal rate of factor substitution $\left(\frac{\Delta x_1}{\Delta x_2}\right)$ is equivalent to

(a) $\frac{Px_1}{Py_1}$ (b) $\frac{Px_2}{Py_1}$

(c) $\frac{Px_2}{Px_1}$ (d) $\frac{Py_1}{Py_2}$

478. Given that

$y = 1245\, X_1^{0.30} X_2^{0.80}$

(a) output increase (b) output decrease

(c) output remains constant (d) data insufficient

479. When one buyer and large seller,the market condition is

(a) Oligopoly (b) Perfect market

(c) monopoly (d) Duopoly

480. The price which necessarily covers the variable cost and also the fixed price is :

(a) The long run price (b) The short run price

(c) The market price (d) The equilibrium price

481. The central Agmark lab is located at :

(a) Mumbai (b) Nagpur

(c) Bangalore (d) Kolkata

482. IBRD is synonymus to :

(a) World Bank

(b) Asian Development Bank

(c) International Monetary Fund

(d) World Trade Organisation

483. MPP (Marginal Physical Product) of a varial input is negative :

(a) Only in I stage (b) Only in II stage

(c) Only in III stage (d) in II & III stage

484. Match List I with List II and select the correct answer using the code given below :

List I (Farm Incomes)	List II (Measure)
A. Family Labour Income	1. G.I. - Cost A_1
B. Farm Business Income (owner farmer)	2. G.I. - Cost A_2
C. Net Income	3. G.I. - Cost B
D. Farm Business Income (tenant farmer)	4. G.I. - Cost C

Code:

(a) A-2, B-1, C-3, D-4 (b) A-1, B-2, C-4, D-3

(c) A-1, B-2, C-3, D-4 (d) A-3, B-1, C-4, D-2

485. Read the following statements and mark your answer according to code given below :

I. For variable input; APP = MPP in the Ist stage

II . For variable input; APP = MPP at the end of Ist stage

III. For variable input; APP = MPP at the start of IInd stage

IV. For variable input; APP = MPP in the IInd stage

Answer Code :

(a) I & III are correct (b) II & III are correct

(c) I & IV are correct (d) II & IV are correct

486. " Mechanisation" in agriculture causes displacement of agricultural labour". Which one of the following statements DOES NOT fully support this viewpoint ?

(a) There is a greater outflow of agricultural labourers to the cities in search of jobs

(b) The total mumber of man-days utilised, infact in creases after mechanisation.

(c) There is a tendancy among agricultural labour to improve their skills.

(d) The marginal productivity of labour is declining.

487. Consider the following statements :

An 'Operational holding' is one which includes all lands used wholly or partially for agricultural production. The improvement in the size of operational holdings has largely been due to

1. extension of irrigation facilities
2. land reclamation schemes
3. consolidation of holdings

of these statements

(a) 1, 2 & 3 are correct (b) 1 & 3 are correct

(c) 2 & 3 are correct (d) 1 & 2 are correct

The following questions consist of two statements, one labelled the 'Assertion A' and the other labelled the 'Reason R'. You are to examine these two statements carefully & select your answers to these questions using the codes given below :

Codes :

(a) Both A & R are true & R is the correct explanation of A.

(b) Both A & R are true but R is NOT a correct explanation of A

(c) A is true but R is false

(d) A is false but R is true.

488. Assertion (A) : The wage rate of agricultural labourers is not equal at all times in the year .

Reason (R) : The seasonality of farm work causes seasonal variations in the opportunity cost of labour.

489. Assertion (A) :Divisibility of input & output factors is an essential requirement of linear programming model.

Reason (R) : Nature of input & output factors has to be continuous in a linear programming model.

490. Assertion (A) : Region II is a rational region of production.

Reason (R) : Region II of a production function has a point of maximum profit.

491. Cooperative society is basic institution for :

(a) . Political growth of the villagers

(b) Social growth of the villagers

(c) Cultural growth of the villagers

(d) Socio- economic growth of the villagers .

492. Read the following statements and mark your answer according to the code given below :

I. C.D is a programme

II. C.D is a cultural event

III. C.D is a social event

IV. C.D is a Project

Answer code :

(a) I & III are correct
(b) I & IV are correct
(c) II & III are correct
(d) II & IV are correct

493. Which one of the following is not correctly matched :

	Approach	**Technique**
(a)	Mass	Exhibition
(b)	Individual	Farm Visit
(c)	Group	Mela
(d)	Mass	Newspaper

494. Match list I with list II and select the correct answer using the code given below :

List I (Aids)	List II (Method of Contact)
A. Model	1. Group contact
B. Slide	2. Mass contact
C. Telephone Call	3. Personal contact
D. Poster	4. Mass contact

Code :

(a) A-2,B-3,C-4,D-1 (b) A-1,B-2,C-3,D-4
(c) A-3,B-4,C-2,D-1 (d) A-3,B-2,C-4,D-1

495. Marketing surplus is
(a) Total Production - Total Family Requirement
(b) Total Consumption - Marginal Consumption
(c) Total Cost - Marginal Cost
(d) Total Production - Total Cost

496. The Food Corporation of India was established in
(a) 1960 (b) 1962
(c) 1965 (d) 1967

497. Tick the correct group of plant nutrients which act as energy exchangers :
(a) K,Na,Ca & Mg (b) C,N,P & S
(c) H & O (d) Fe,Mn,Mo,Cu,B & Zn

498. Which gas is released from paddy field :
(a) CH_4 (b) H_2S
(c) CO_2 (d) NH_3

499. Solubility of rock phosphate can be improved by :
(a) Nitrosomonas (b) Nitrobacter
(c) Azotobacter (d) Bacillus Polymexa

500. Chemical which regulates opening & closing of stomata is :
(a) Kaoline (b) 2, 4 - D
(c) PMA (d) Paraquat

501. Which of the following intercropping system has synergistic effect on component crop :

(a) Autumn planted Sugarcane + Mustard

(b) Autumn planted Sugarcane + Chick pea

(c) Autumn planted Sugarcane + Wheat

(d) Autumn planted Sugarcane + Potato

502. Which one of the following is not correctly matched :

Crop	Test wt. in gm.
(a) Pearl - millet	5-7
(b) Wheat	20-22
(c) Mungbean	34-36
(d) Sunflower	60 -62

503. In case of mungbean seed :

(a) The amount of carbohydrate is higher than that of protein.

(b) The amount of protein is higher than that of carbohydrate.

(c) Protein & carbohydrate are present in equal proportion

(d) There is no carbohydrate

504. Immobilization of sulphur takes place when the 'S' content of organic matter is less than.

(a) 0.60 % (b) 0. 45%

(c) 0.30% (c) 0.15%

505. VA micorrhiza promotes mainly the plant uptake of

(a) Nitrogen (b) Phosphorus

(c) Potassium (d) Molybde-num

506. Which of the following rural development project was launched before independence :

(a) Etawah Pilot Project (b) Cooperative Movement

(c) Community Development (d) National Extension Service

507. Lab to Land Programme was started by :

(a) Govt. of India (b) Govt. of U.P.

(c) Smt. Indira Gandhi (d) ICAR

508. Read the following statements and mark your answer according to the code given below :

I. Soil acidity increases with exchangeable Al^{+++}

II. Soil acidity increases with exchangeable H.

III. Soil alkalinity increases with exchangeable calcium.

IV. Soil alkalinity increases with exchangeable aluminium.

Answer Code :

(a) I & III are correct (b) I & II are correct

(b) II & III are correct (d) III & IV are correct

509. Line passing through the least cost points in the isoquants map is called :

(a) Isocline (b) Ridge line

(c) Expansion path (d) Isoquant curve

510. Read the following statements and mark your answer according to the code given below :

I. Newspaper is a popular medium of communication in rural areas.

II. Newspaper is written form of communication.

III. Film show is audio-visual form of communication .

IV. Film show is popular medium of communication in rural areas.

Answer Code :

(a) I & III are correct (b) II & III are correct

(c) III & IV are correct (d) I & IV are correct .

❑❑❑

Answers

1	d	26	a	51	a	76	b
2	a	27	c	52	b	77	a
3	c	28	b	53	b	78	b
4	b	29	c	54	c	79	b
5	a	30	c	55	d	80	a
6	c	31	b	56	d	81	b
7	c	32	d	57	b(weed)	82	c
8	a	33	d	58	b	83	d
9	c	34	b	59	a	84	d
10	b	35	a	60	c	85	c
11	b	36	c	61	b	86	b
12	d	37	c	62	a	87	b
13	a	38	d	63	d	88	d
14	c	39	c	64	a	89	b
15	d	40	a	65	a	90	d
16	d	41	c	66	a	91	d
17	c	42	c	67	b	92	a
18	c	43	d	68	a	93	a
19	b	44	b	69	c	94	d
20	a	45	b	70	d	95	a
21	a	46	d	71	c	96	d
22	a	47	d	72	c	97	c
23	d	48	d	73	c	98	a
24	d	49	a	74	d	99	b
25	c	50	c	75	d	100	d

101	a	128	d	155	c	182	a
102	b	129	a	156	a	183	a
103	d	130	c	157	b	184	a
104	a	131	b	158	a	185	a
105	b	132	c	159	c	186	b
106	b	133	b	160	c	187	a
107	a	134	a	161	d	188	b
108	d	135	c	162	a	189	a
109	c	136	d	163	a	190	a
110	b	137	-	164	d	191	a
111	d	138	b	165	c	192	d
112	d	139	c	166	c	193	b
113	b	140	c	167	b	194	b
114	d	141	d	168	b	195	b
115	d	142	a	169	b	196	a
116	c $\frac{1960}{3}$, $\frac{66}{1}$, $\frac{71}{4}$, $\frac{72}{2}$	143	a	170	c	197	a
117	c	144	b	171	b	198	a
118	d	145	a	172	a	199	d
119	c	146	b	173	b	200	b
120	b	147	d	174	b	201	a
121	d	148	b	175	b	202	c
122	b	149	d	176	c	203	b
123	a	150	a	177	c	204	c
124	c	151	c	178	c	205	c
125	b	152	c	179	a	206	a
126	d	153	a	180	b	207	d
127	b	154	c	181	c	208	b

209	c	236	c	263	c	290	b
210	b	237	d	264	d	291	b
211	c	238	c	265	c	292	c
212	a	239	b	266	a	293	b
213	c	240	d	267	b	294	b
214	b	241	c	268	b	295	a
215	b	242	b	269	c	296	c
216	a	243	d	270	d	297	c
217	c	244	b	271	a	298	b
218	c	245	a	272	c	299	c
219	c	246	d	273	c	300	d
220	d	247	b	274	d	301	d
221	d	248	a	275	c	302	a
222	d	249	d	276	c	303	b
223	a	250	b	277	d	304	b
224	c	251	c	278	b	305	a
225	c	252	a	279	b	306	b
226	b	253	c	280	b	307	b
227	c	254	c	281	b	308	d
228	b	255	d	282	d	309	b
229	d	256	a	283	d	310	b
230	b	257	d	284	d	311	a
231	a	258	c	285	d	312	c
232	d	259	c	286	d	313	c
233	a	260	a	287	a	314	d
234	c	261	b	288	a	315	a
235	c	262	b	289	a	316	b

317	d	344	a	371	c	398	d
318	c	345	a	372	b	399	d
319	d	346	d	373	c	400	d
320	d	347	c	374	c	401	b
321	b	348	b	375	c	402	d
322	d	349	d	376	b	403	b
323	b	350	a	377	b	404	c
324	a	351	c	378	d	405	b
325	c	352	a	379	a	406	c
326	a	353	b	380	a	407	a
327	a	354	b	381	d	408	c
328	b	355	d	382	b	409	a
329	d	356	b	383	d	410	d
330	c	357	a	384	c	411	c
331	c	358	d	385	a	412	a
332	b	359	b	386	b	413	b
333	a	360	a	387	c	414	c
334	a	361	b	388	d	415	a
335	c	362	a	389	a	416	a
336	d	363	b	390	d	417	c
337	d	364	b	391	d	418	a
338	c	365	c	392	c	419	d
339	c	366	c	393	b	420	d
340	a	367	c	394	c	421	a
341	c	368	d	395	b	422	b
342	b	369	a	396	d	423	b
343	d	370	c	397	c	424	a